AF885348

Foreword

This manual is designed to fill a doctrinal gap. It has been 20 years since the Army published a field manual devoted exclusively to counterinsurgency operations. For the Marine Corps it has been 25 years. With our Soldiers and Marines fighting insurgents in Afghanistan and Iraq, it is essential that we give them a manual that provides principles and guidelines for counterinsurgency operations. Such guidance must be grounded in historical studies. However, it also must be informed by contemporary experiences.

This manual takes a general approach to counterinsurgency operations. The Army and Marine Corps recognize that every insurgency is contextual and presents its own set of challenges. You cannot fight former Saddamists and Islamic extremists the same way you would have fought the Viet Cong, Moros, or Tupamaros; the application of principles and fundamentals to deal with each varies considerably. Nonetheless, all insurgencies, even today's highly adaptable strains, remain wars amongst the people. They use variations of standard themes and adhere to elements of a recognizable revolutionary campaign plan. This manual therefore addresses the common characteristics of insurgencies. It strives to provide those conducting counterinsurgency campaigns with a solid foundation for understanding and addressing specific insurgencies.

A counterinsurgency campaign is, as described in this manual, a mix of offensive, defensive, and stability operations conducted along multiple lines of operations. It requires Soldiers and Marines to employ a mix of familiar combat tasks and skills more often associated with nonmilitary agencies. The balance between them depends on the local situation. Achieving this balance is not easy. It requires leaders at all levels to adjust their approach constantly. They must ensure that their Soldiers and Marines are ready to be greeted with either a handshake or a hand grenade while taking on missions only infrequently practiced until recently at our combat training centers. Soldiers and Marines are expected to be nation builders as well as warriors. They must be prepared to help reestablish institutions and local security forces and assist in rebuilding infrastructure and basic services. They must be able to facilitate establishing local governance and the rule of law. The list of such tasks is long; performing them involves extensive coordination and cooperation with many intergovernmental, host-nation, and international agencies. Indeed, the responsibilities of leaders in a counterinsurgency campaign are daunting; however, the discussions in this manual alert leaders to the challenges of such campaigns and suggest general approaches for grappling with those challenges.

Conducting a successful counterinsurgency campaign requires a flexible, adaptive force led by agile, well-informed, culturally astute leaders. It is our hope that this manual provides the guidelines needed to succeed in operations that are exceedingly difficult and complex. Our Soldiers and Marines deserve nothing less.

DAVID H. PETRAEUS
Lieutenant General, U.S. Army
Commander
U.S. Army Combined Arms Center

JAMES F. AMOS
Lieutenant General, U.S. Marine Corps
Deputy Commandant
Combat Development and Integration

FM 3-24
*MCWP 3-33.5

Field Manual
No. 3-24

Headquarters
Department of the Army
Washington, DC

Marine Corps Warfighting Publication
No. 3-33.5

Headquarters
Marine Corps Combat Development Command
Department of the Navy
Headquarters
United States Marine Corps
Washington, DC

15 December 2006

COUNTERINSURGENCY

Contents

Distribution Restriction: Approved for public release; distribution is unlimited.

***This publication supersedes FMI 3-07.22, 1 October 2004, and MCWP 3-33.5, 29 January 1980.**

Marine Corps PCN: 143 000124 00

Figures

Tables

Vignettes

This publication is available at
Army Knowledge Online (www.us.army.mil) and
General Dennis J. Reimer Training and Doctrine
Digital Library at (www.train.army.mil).

Preface

This field manual/Marine Corps warfighting publication establishes doctrine (fundamental principles) for military operations in a counterinsurgency (COIN) environment. It is based on lessons learned from previous counterinsurgencies and contemporary operations. It is also based on existing interim doctrine and doctrine recently developed.

Counterinsurgency operations generally have been neglected in broader American military doctrine and national security policies since the end of the Vietnam War over 30 years ago. This manual is designed to reverse that trend. It is also designed to merge traditional approaches to COIN with the realities of a new international arena shaped by technological advances, globalization, and the spread of extremist ideologies—some of them claiming the authority of a religious faith.

The manual begins with a description of insurgencies and counterinsurgencies. The first chapter includes a set of principles and imperatives necessary for successful COIN operations. Chapter 2 discusses nonmilitary organizations commonly involved in COIN operations and principles for integrating military and civilian activities. Chapter 3 addresses aspects of intelligence specific to COIN operations. The next two chapters discuss the design and execution of those operations. Developing host-nation security forces, an essential aspect of successful COIN operations, is the subject of chapter 6. Leadership and ethical concerns are addressed in chapter 7. Chapter 8, which concerns sustainment of COIN operations, concludes the basic manual. The appendixes contain useful supplemental information. Appendix A discusses factors to consider during the planning, preparation, execution, and assessment of a COIN operation. Appendixes B and C contain supplemental intelligence information. Appendix D addresses legal concerns. Appendix E describes the role of airpower.

Doctrine by definition is broad in scope and involves principles, tactics, techniques, and procedures applicable worldwide. Thus, this publication is not focused on any region or country and is not intended to be a standalone reference. Users should assess information from other sources to help them decide how to apply the doctrine in this publication to the specific circumstances facing them.

The primary audience for this manual is leaders and planners at the battalion level and above. This manual applies to the United States Marine Corps, the Active Army, the Army National Guard/Army National Guard of the United States, and the United States Army Reserve unless otherwise stated.

This publication contains copyrighted material. Copyrighted material is identified with footnotes. Other sources are identified in the source notes.

Terms that have joint, Army, or Marine Corps definitions are identified in both the glossary and the text. FM 3-24 is not the proponent field manual (the authority) for any Army term. For definitions in the text, the term is italicized and the number of the proponent manual follows the definition.

Headquarters, U.S. Army Training and Doctrine Command is the proponent for this publication. The preparing agency is the Combined Arms Doctrine Directorate, U.S. Army Combined Arms Center. Send written comments and recommendations on DA Form 2028 (Recommended Changes to Publications and Blank Forms) directly to Commander, U.S. Army Combined Arms Center and Fort Leavenworth, ATTN: ATZL-CD (FM 3-24), 201 Reynolds Avenue (Building 285), Fort Leavenworth, Kansas 66027-1352. Send comments and recommendations by e-mail to web-cadd@leavenworth.army.mil. Follow the DA Form 2028 format or submit an electronic DA Form 2028.

ACKNOWLEDGEMENTS

Counterinsurgency Warfare: Theory and Practice, David Galula. Copyright © 1964 by Frederick A. Praeger, Inc. Reproduced with permission of Greenwood Publishing Group, Inc., Westport, CT.

"Battle Lessons, What the Generals Don't Know," Dan Baum. *The New Yorker*, Jan 17, 2005. Reproduced with permission of Dan Baum.

Defeating Communist Insurgency: The Lessons of Malaya and Vietnam, Sir Robert Thompson. Copyright © 1966 by Robert Thompson. Reproduced with permission of Hailer Publishing.

Introduction

This is a game of wits and will. You've got to be learning and adapting constantly to survive.

General Peter J. Schoomaker, USA, 2004

The United States possesses overwhelming conventional military superiority. This capability has pushed its enemies to fight U.S. forces unconventionally, mixing modern technology with ancient techniques of insurgency and terrorism. Most enemies either do not try to defeat the United States with conventional operations or do not limit themselves to purely military means. They know that they cannot compete with U.S. forces on those terms. Instead, they try to exhaust U.S. national will, aiming to win by undermining and outlasting public support. Defeating such enemies presents a huge challenge to the Army and Marine Corps. Meeting it requires creative efforts by every Soldier and Marine.

Throughout its history, the U.S. military has had to relearn the principles of counterinsurgency (COIN) while conducting operations against adaptive insurgent enemies. It is time to institutionalize Army and Marine Corps knowledge of this longstanding form of conflict. This publication's purpose is to help prepare Army and Marine Corps leaders to conduct COIN operations anywhere in the world. It provides a foundation for study before deployment and the basis for operations in theater. Perhaps more importantly, it provides techniques for generating and incorporating lessons learned during those operations—an essential requirement for success against today's adaptive foes. Using these techniques and processes can keep U.S. forces more agile and adaptive than their irregular enemies. Knowledge of the history and principles of insurgency and COIN provides a solid foundation that informed leaders can use to assess insurgencies. This knowledge can also help them make appropriate decisions on employing all instruments of national power against these threats.

All insurgencies are different; however, broad historical trends underlie the factors motivating insurgents. Most insurgencies follow a similar course of development. The tactics used to successfully defeat them are likewise similar in most cases. Similarly, history shows that some tactics that are usually successful against conventional foes may fail against insurgents.

One common feature of insurgencies is that the government that is being targeted generally takes awhile to recognize that an insurgency is occurring. Insurgents take advantage of that time to build strength and gather support. Thus, counterinsurgents often have to "come from behind" when fighting an insurgency. Another common feature is that forces conducting COIN operations usually begin poorly. Western militaries too often neglect the study of insurgency. They falsely believe that armies trained to win large conventional wars are automatically prepared to win small, unconventional ones. In fact, some capabilities required for conventional success—for example, the ability to execute operational maneuver and employ massive firepower—may be of limited utility or even counterproductive in COIN operations. Nonetheless, conventional forces beginning COIN operations often try to use these capabilities to defeat insurgents; they almost always fail.

The military forces that successfully defeat insurgencies are usually those able to overcome their institutional inclination to wage conventional war against insurgents. They learn how to practice COIN and apply that knowledge. This publication can help to compress the learning curve. It is a tool for planners, trainers, and field commanders. Using it can help leaders begin the learning process sooner and build it on a larger knowledge base. Learning done before deployment results in fewer lives lost and less national treasure spent relearning past lessons in combat.

In COIN, the side that learns faster and adapts more rapidly—the better learning organization—usually wins. Counterinsurgencies have been called learning competitions. Thus, this publication identifies "Learn and Adapt" as a modern COIN imperative for U.S. forces. However, Soldiers and Marines cannot wait until they are alerted to deploy to prepare for a COIN mission. Learning to conduct complex COIN operations begins with study beforehand. This publication is a good place to start. The annotated bibliography lists a number of other sources; however, these are only a sample of the vast amount of available information on this subject. Adapting occurs as Soldiers and Marines apply what they have learned through study and experience, assess the results of their actions, and continue to learn during operations.

As learning organizations, the Army and Marine Corps encourage Soldiers and Marines to pay attention to the rapidly changing situations that characterize COIN operations. Current tactics, techniques, and procedures sometimes do not achieve the desired results. When that happens, successful leaders engage in a directed search for better ways to defeat the enemy. To win, the Army and Marine Corps must rapidly develop an institutional consensus on new doctrine, publish it, and carefully observe its impact on mission accomplishment. This learning cycle should repeat continuously as U.S. counterinsurgents seek to learn faster than the insurgent enemy. The side that learns faster and adapts more rapidly wins.

Just as there are historical principles underlying success in COIN, there are organizational traits shared by most successful learning organizations. Forces that learn COIN effectively have generally—

- Developed COIN doctrine and practices locally.
- Established local training centers during COIN operations.
- Regularly challenged their assumptions, both formally and informally.
- Learned about the broader world outside the military and requested outside assistance in understanding foreign political, cultural, social and other situations beyond their experience.
- Promoted suggestions from the field.
- Fostered open communication between senior officers and their subordinates.
- Established rapid avenues of disseminating lessons learned.
- Coordinated closely with governmental and nongovernmental partners at all command levels.
- Proved open to soliciting and evaluating advice from the local people in the conflict zone.

These are not always easy practices for an organization to establish. Adopting them is particularly challenging for a military engaged in a conflict. However, these traits are essential for any military confronting an enemy who does not fight using conventional tactics and who adapts while waging irregular warfare. Learning organizations defeat insurgencies; bureaucratic hierarchies do not.

Promoting learning is a key responsibility of commanders at all levels. The U.S. military has developed first-class lessons-learned systems that allow for collecting and rapidly disseminating information from the field. But these systems only work when commanders promote their use and create a command climate that encourages bottom-up learning. Junior leaders in the field often informally disseminate lessons based on their experiences. However, incorporating this information into institutional lessons learned, and then into doctrine, requires commanders to encourage subordinates to use institutional lessons-learned processes.

Ironically, the nature of counterinsurgency presents challenges to traditional lessons-learned systems; many nonmilitary aspects of COIN do not lend themselves to rapid tactical learning. As this publication explains, performing the many nonmilitary tasks in COIN requires knowledge of many diverse, complex subjects. These include governance, economic development, public administration, and the rule of law. Commanders with a deep-rooted knowledge of these subjects can help subordinates understand challenging, unfamiliar environments and adapt more rapidly to changing situations. Reading this publication is a first stop to developing this knowledge.

COIN campaigns are often long and difficult. Progress can be hard to measure, and the enemy may appear to have many advantages. Effective insurgents rapidly adapt to changing circumstances. They cleverly use the tools of the global information revolution to magnify the effects of their actions. The often carry out barbaric acts and do not observe accepted norms of behavior. However, by focusing on efforts to secure the safety and support of the local populace, and through a concerted effort to truly function as learning organizations, the Army and Marine Corps can defeat their insurgent enemies.

Chapter 1

Insurgency and Counterinsurgency

Counterinsurgency is not just thinking man's warfare—it is the graduate level of war.

Special Forces Officer in Iraq, 2005

This chapter provides background information on insurgency and counterinsurgency. The first half describes insurgency, while the second half examines the more complex challenge of countering it. The chapter concludes with a set of principles and imperatives that contribute to success in counterinsurgency.

OVERVIEW

1-1. Insurgency and counterinsurgency (COIN) are complex subsets of warfare. Globalization, technological advancement, urbanization, and extremists who conduct suicide attacks for their cause have certainly influenced contemporary conflict; however, warfare in the 21st century retains many of the characteristics it has exhibited since ancient times. Warfare remains a violent clash of interests between organized groups characterized by the use of force. Achieving victory still depends on a group's ability to mobilize support for its political interests (often religiously or ethnically based) and to generate enough violence to achieve political consequences. Means to achieve these goals are not limited to conventional forces employed by nation-states.

1-2. Insurgency and its tactics are as old as warfare itself. Joint doctrine defines an *insurgency* as an organized movement aimed at the overthrow of a constituted government through the use of subversion and armed conflict (JP 1-02). Stated another way, an insurgency is an organized, protracted politico-military struggle designed to weaken the control and legitimacy of an established government, occupying power, or other political authority while increasing insurgent control. *Counterinsurgency* is military, paramilitary, political, economic, psychological, and civic actions taken by a government to defeat insurgency (JP 1-02). These definitions are a good starting point, but they do not properly highlight a key paradox: though insurgency and COIN are two sides of a phenomenon that has been called revolutionary war or internal war, they are distinctly different types of operations. In addition, insurgency and COIN are included within a broad category of conflict known as irregular warfare.

1-3. Political power is the central issue in insurgencies and counterinsurgencies; each side aims to get the people to accept its governance or authority as legitimate. Insurgents use all available tools—political (including diplomatic), informational (including appeals to religious, ethnic, or ideological beliefs), military, and economic—to overthrow the existing authority. This authority may be an established government or an interim governing body. Counterinsurgents, in turn, use all instruments of national power to sustain the established or emerging government and reduce the likelihood of another crisis emerging.

1-4. Long-term success in COIN depends on the people taking charge of their own affairs and consenting to the government's rule. Achieving this condition requires the government to eliminate as many causes of the insurgency as feasible. This can include eliminating those extremists whose beliefs prevent them from ever reconciling with the government. Over time, counterinsurgents aim to enable a country or regime to provide the security and rule of law that allow establishment of social services and growth of economic activity. COIN thus involves the application of national power in the political, military, economic, social, information, and infrastructure fields and disciplines. Political and military leaders and planners should never underestimate its scale and complexity; moreover, they should recognize that the Armed Forces cannot succeed in COIN alone.

ASPECTS OF INSURGENCY

1-5. Governments can be overthrown in a number of ways. An unplanned, spontaneous explosion of popular will, for example, might result in a revolution like that in France in 1789. At another extreme is the coup d'etat, where a small group of plotters replace state leaders with little support from the people at large. Insurgencies generally fall between these two extremes. They normally seek to achieve one of two goals: to overthrow the existing social order and reallocate power within a single state, or to break away from state control and form an autonomous entity or ungoverned space that they can control. Insurgency is typically a form of internal war, one that occurs primarily within a state, not between states, and one that contains at least some elements of civil war.

1-6. The exception to this pattern of internal war involves resistance movements, where indigenous elements seek to expel or overthrow what they perceive to be a foreign or occupation government. Such a resistance movement could be mounted by a legitimate government in exile as well as by factions competing for that role.

1-7. Even in internal war, the involvement of outside actors is expected. During the Cold War, the Soviet Union and the United States participated in many such conflicts. Today, outside actors are often transnational organizations motivated by ideologies based on extremist religious or ethnic beliefs. These organizations exploit the unstable internal conditions plaguing failed and failing states. Such outside involvement, however, does not change one fact: the long-term objective for all sides remains acceptance of the legitimacy of one side's claim to political power by the people of the state or region.

1-8. The terrorist and guerrilla tactics common to insurgency have been among the most common approaches to warfare throughout history. Any combatant prefers a quick, cheap, overwhelming victory over a long, bloody, protracted struggle. But to succeed against superior resources and technology, weaker actors have had to adapt. The recent success of U.S. military forces in major combat operations undoubtedly will lead many future opponents to pursue asymmetric approaches. Because the United States retains significant advantages in fires and technical surveillance, a thinking enemy is unlikely to choose to fight U.S. forces in open battle. Some opponents have attempted to do so, such as in Panama in 1989 and Iraq in 1991 and 2003. They were defeated in conflicts measured in hours or days. Conversely, other opponents have offset America's fires and surveillance advantages by operating close to civilians, as Somali clans did in 1993 and insurgents in Iraq have done since mid-2003; these enemies have been more successful in achieving their aims. This situation does not mean that counterinsurgents do not face open warfare. Although insurgents frequently use nonviolent means like political mobilization and work stoppages (strikes), they do resort to conventional military operations when conditions seem right.

1-9. The contest of internal war is not "fair"; many of the "rules" favor insurgents. That is why insurgency has been a common approach used by the weak against the strong. At the beginning of a conflict, insurgents typically hold the strategic initiative. Though they may resort to violence because of regime changes or government actions, insurgents generally initiate the conflict. Clever insurgents strive to disguise their intentions. When these insurgents are successful at such deception, potential counterinsurgents are at a disadvantage. A coordinated reaction requires political and military leaders to recognize that an insurgency exists and to determine its makeup and characteristics. While the government prepares to respond, the insurgents gain strength and foster increasing disruption throughout the state or region. The government normally has an initial advantage in resources; however, that edge is counterbalanced by the requirement to maintain order and protect the population and critical resources. Insurgents succeed by sowing chaos and disorder anywhere; the government fails unless it maintains a degree of order everywhere.

1-10. For the reasons just mentioned, maintaining security in an unstable environment requires vast resources, whether host nation, U.S., or multinational. In contrast, a small number of highly motivated insurgents with simple weapons, good operations security, and even limited mobility can undermine security over a large area. Thus, successful COIN operations often require a high ratio of security forces to the protected population. (See paragraph1-67.) For that reason, protracted COIN operations are hard to sustain. The effort requires a firm political will and substantial patience by the government, its people, and the countries providing support.

1-11. Revolutionary situations may result from regime changes, external interventions, or grievances carefully nurtured and manipulated by unscrupulous leaders. Sometimes societies are most prone to unrest not when conditions are the worst, but when the situation begins to improve and people's expectations rise. For example, when major combat operations conclude, people may have unrealistic expectations of the United States' capability to improve their lives. The resulting discontent can fuel unrest and insurgency. At such times, the influences of globalization and the international media may create a sense of relative deprivation, contributing to increased discontent as well.

1-12. The information environment is a critical dimension of such internal wars, and insurgents attempt to shape it to their advantage. One way they do this is by carrying out activities, such as suicide attacks, that may have little military value but create fear and uncertainty within the populace and government institutions. These actions are executed to attract high-profile media coverage or local publicity and inflate perceptions of insurgent capabilities. Resulting stories often include insurgent fabrications designed to undermine the government's legitimacy.

1-13. Insurgents have an additional advantage in shaping the information environment. Counterinsurgents seeking to preserve legitimacy must stick to the truth and make sure that words are backed up by deeds; insurgents, on the other hand, can make exorbitant promises and point out government shortcomings, many caused or aggravated by the insurgency. Ironically, as insurgents achieve more success and begin to control larger portions of the populace, many of these asymmetries diminish. That may produce new vulnerabilities that adaptive counterinsurgents can exploit.

1-14. Before most COIN operations begin, insurgents have seized and exploited the initiative, to some degree at the least. Therefore, counterinsurgents undertake offensive and defensive operations to regain the initiative and create a secure environment. However, killing insurgents—while necessary, especially with respect to extremists—by itself cannot defeat an insurgency. Gaining and retaining the initiative requires counterinsurgents to address the insurgency's causes through stability operations as well. This initially involves securing and controlling the local populace and providing for essential services. As security improves, military resources contribute to supporting government reforms and reconstruction projects. As counterinsurgents gain the initiative, offensive operations focus on eliminating the insurgent cadre, while defensive operations focus on protecting the populace and infrastructure from direct attacks. As counterinsurgents establish military ascendancy, stability operations expand across the area of operations (AO) and eventually predominate. Victory is achieved when the populace consents to the government's legitimacy and stops actively and passively supporting the insurgency.

The Evolution of Insurgency

1-15. Insurgency has taken many forms over time. Past insurgencies include struggles for independence against colonial powers, the rising up of ethnic or religious groups against their rivals, and resistance to foreign invaders. Students and practitioners of COIN must begin by understanding the specific circumstances of their particular situation. The history of this form of warfare shows how varied and adaptive it can be, and why students must understand that they cannot focus on countering just one insurgent approach. This is particularly true when addressing a continually complex, changing situation like that of Iraq in 2006.

1-16. Insurgencies and counterinsurgencies have been common throughout history, but especially since the beginning of the 20th century. The United States began that century by defeating the Philippine Insurrection. The turmoil of World War I and its aftermath produced numerous internal wars. Trotsky and Lenin seized power in Russia and then defended the new regime against counterrevolutionaries. T.E. Lawrence and Arab forces used guerrilla tactics to overcome the Ottoman Turks during the Arab Revolt.

1-17. Before World War I, insurgencies were mostly conservative; insurgents were usually concerned with defending hearth, home, monarchies, and traditional religion. Governments were seldom able to completely defeat these insurgencies; violence would recur when conditions favored a rebellion. For example, the history of the British Isles includes many recurring insurgencies by subjugated peoples based on ethnic identities. Another example of a conservative insurgency is the early 19th century Spanish uprising against Napoleon that sapped French strength and contributed significantly to Napoleon's defeat.

1-18. Since World War I, insurgencies have generally had more revolutionary purposes. The Bolshevik takeover of Russia demonstrated a conspiratorial approach to overthrowing a government; it spawned a communist movement that supported further "wars of national liberation." Lawrence's experiences in the Arab Revolt made him a hero and also provide some insights for today.

1-19. The modern era of insurgencies and internal wars began after World War II. Many of the resistance movements against German and Japanese occupation continued after the Axis defeat in 1945. As nationalism rose, the imperial powers declined. Motivated by nationalism and communism, people began forming governments viewed as more responsive to their needs. The development of increasingly lethal and portable killing technologies dramatically increased the firepower available to insurgent groups. As important was the increase in the news media's ability to get close to conflicts and transmit imagery locally and globally. In 1920, T.E. Lawrence noted, "The printing press is the greatest weapon in the armory of the modern commander." Today, he might have added, "and the modern insurgent," though certainly the Internet and compact storage media like cassettes, compact disks, and digital versatile disks (DVDs) have become more important in recent years.

1-20. Thus, 20th century events transformed the purpose and character of most insurgencies. Most 19th century insurgencies were local movements to sustain the status quo. By the mid-20th century they had become national and transnational revolutionary movements. Clausewitz thought that wars by an armed populace could only serve as a strategic defense; however, theorists after World War II realized that insurgency could be a decisive form of warfare. This era spawned the Maoist, Che Guevara-type focoist, and urban approaches to insurgency.

1-21. While some Cold War insurgencies persisted after the Soviet Union's collapse, many new ones appeared. These new insurgencies typically emerged from civil wars or the collapse of states no longer propped up by Cold War rivalries. Power vacuums breed insurgencies. Similar conditions exist when regimes are changed by force or circumstances. Recently, ideologies based on extremist forms of religious or ethnic identities have replaced ideologies based on secular revolutionary ideals. These new forms of old, strongly held beliefs define the identities of the most dangerous combatants in these new internal wars. These conflicts resemble the wars of religion in Europe before and after the Reformation of the 16th century. People have replaced nonfunctioning national identities with traditional sources of unity and identity. When countering an insurgency during the Cold War, the United States normally focused on increasing a threatened but friendly government's ability to defend itself and on encouraging political and economic reforms to undercut support for the insurgency. Today, when countering an insurgency growing from state collapse or failure, counterinsurgents often face a more daunting task: helping friendly forces reestablish political order and legitimacy where these conditions may no longer exist.

1-22. Interconnectedness and information technology are new aspects of this contemporary wave of insurgencies. Using the Internet, insurgents can now link virtually with allied groups throughout a state, a region, and even the entire world. Insurgents often join loose organizations with common objectives but different motivations and no central controlling body, which makes identifying leaders difficult.

1-23. Today's operational environment also includes a new kind of insurgency, one that seeks to impose revolutionary change worldwide. Al Qaeda is a well-known example of such an insurgency. This movement seeks to transform the Islamic world and reorder its relationships with other regions and cultures. It is notable for its members' willingness to execute suicide attacks to achieve their ends. Such groups often feed on local grievances. Al Qaeda-type revolutionaries are willing to support causes they view as compatible with their own goals through the provision of funds, volunteers, and sympathetic and targeted propaganda. While the communications and technology used for this effort are often new and modern, the grievances and methods sustaining it are not. As in other insurgencies, terrorism, subversion, propaganda, and open warfare are the tools of such movements. Today, these time-tested tools have been augmented by the precision munition of extremists—suicide attacks. Defeating such enemies requires a global, strategic response—one that addresses the array of linked resources and conflicts that sustain these movements while tactically addressing the local grievances that feed them.

Insurgents and Their Motives

1-24. Each insurgency is unique, although there are often similarities among them. In all cases, insurgents aim to force political change; any military action is secondary and subordinate, a means to an end. Few insurgencies fit neatly into any rigid classification. In fact, counterinsurgent commanders may face a confusing and shifting coalition of many kinds of opponents, some of whom may be at odds with one another. Examining the specific type of insurgency they face enables commanders and staffs to build a more accurate picture of the insurgents and the thinking behind their overall approach. Such an examination identifies the following:

- Root cause or causes of the insurgency.
- Extent to which the insurgency enjoys internal and external support.
- Basis (including the ideology and narrative) on which insurgents appeal to the target population.
- Insurgents' motivation and depth of commitment.
- Likely insurgent weapons and tactics.
- Operational environment in which insurgents seek to initiate and develop their campaign and strategy.

Insurgent Approaches

1-25. Counterinsurgents have to determine not only their opponents' motivation but also the approach being used to advance the insurgency. This information is essential to developing effective programs that attack the insurgency's root causes. Analysis of the insurgents' approach shapes counterinsurgent military options. Insurgent approaches include, but are not limited to, the following:

- Conspiratorial.
- Military-focused.
- Urban.
- Protracted popular war.
- Identity-focused.
- Composite and coalition.

Conspiratorial

1-26. A conspiratorial approach involves a few leaders and a militant cadre or activist party seizing control of government structures or exploiting a revolutionary situation. In 1917, Lenin used this approach in carrying out the Bolshevik Revolution. Such insurgents remain secretive as long as possible. They emerge only when success can be achieved quickly. This approach usually involves creating a small, secretive, "vanguard" party or force. Insurgents who use this approach successfully may have to create security forces and generate mass support to maintain power, as the Bolsheviks did.

Military-Focused

1-27. Users of military-focused approaches aim to create revolutionary possibilities or seize power primarily by applying military force. For example, the focoist approach, popularized by figures like Che Guevera, asserts that an insurrection itself can create the conditions needed to overthrow a government. Focoists believe that a small group of guerrillas operating in a rural environment where grievances exist can eventually gather enough support to achieve their aims. In contrast, some secessionist insurgencies have relied on major conventional forces to try to secure their independence. Military-focused insurgencies conducted by Islamic extremist groups or insurgents in Africa or Latin America have little or no political structure; they spread their control through movement of combat forces rather than political subversion.

Urban

1-28. Organizations like the Irish Republican Army, certain Latin American groups, and some Islamic extremist groups in Iraq have pursued an urban approach. This approach uses terrorist tactics in urban areas to accomplish the following:

- Sow disorder.
- Incite sectarian violence.
- Weaken the government.
- Intimidate the population.
- Kill government and opposition leaders.
- Fix and intimidate police and military forces, limiting their ability to respond to attacks.
- Create government repression.

1-29. Protracted urban terrorism waged by small, independent cells requires little or no popular support. It is difficult to counter. Historically, such activities have not generated much success without wider rural support. However, as societies have become more urbanized and insurgent networks more sophisticated, this approach has become more effective. When facing adequately run internal security forces, urban insurgencies typically assume a conspiratorial cellular structure recruited along lines of close association—family, religious affiliation, political party, or social group.

Protracted Popular War

1-30. Protracted conflicts favor insurgents, and no approach makes better use of that asymmetry than the protracted popular war. The Chinese Communists used this approach to conquer China after World War II. The North Vietnamese and Algerians adapted it to fit their respective situations. And some Al Qaeda leaders suggest it in their writings today. This approach is complex; few contemporary insurgent movements apply its full program, although many apply parts of it. It is, therefore, of more than just historical interest. Knowledge of it can be a powerful aid to understanding some insurgent movements.

Mao Zedong's Theory of Protracted War

1-31. Mao's Theory of Protracted War outlines a three-phased, politico-military approach:

- **Strategic defensive**, when the government has a stronger correlation of forces and insurgents must concentrate on survival and building support.
- **Strategic stalemate**, when force correlations approach equilibrium and guerrilla warfare becomes the most important activity.
- **Strategic counteroffensive**, when insurgents have superior strength and military forces move to conventional operations to destroy the government's military capability.

1-32. Phase I, strategic defensive, is a period of latent insurgency that allows time to wear down superior enemy strength while the insurgency gains support and establishes bases. During this phase, insurgent leaders develop the movement into an effective clandestine organization. Insurgents use a variety of subversive techniques to psychologically prepare the populace to resist the government or occupying power. These techniques may include propaganda, demonstrations, boycotts, and sabotage. In addition, movement leaders organize or develop cooperative relationships with legitimate political action groups, youth groups, trade unions, and other front organizations. Doing this develops popular support for later political and military activities. Throughout this phase, the movement leadership—

- Recruits, organizes, and trains cadre members.
- Infiltrates key government organizations and civilian groups.
- Establishes cellular intelligence, operations, and support networks.
- Solicits and obtains funds.
- Develops sources for external support.

Subversive activities are frequently executed in an organized pattern, but major combat is avoided. The primary military activity is terrorist strikes. These are executed to gain popular support, influence recalcitrant individuals, and sap enemy strength. In the advanced stages of this phase, the insurgent organization may establish a counterstate that parallels the established authority. (A *counterstate* [or shadow government] is a competing structure that a movement sets up to replace the government. It includes the administrative and bureaucratic trappings of political power and performs the normal functions of a government.)

1-33. Phase II, strategic stalemate, begins with overt guerrilla warfare as the correlation of forces approaches equilibrium. In a rural-based insurgency, guerrillas normally operate from a relatively secure base area in insurgent-controlled territory. In an urban-based insurgency, guerrillas operate clandestinely, using a cellular organization. In the political arena, the movement concentrates on undermining the people's support of the government and further expanding areas of control. Subversive activities can take the form of clandestine radio broadcasts, newspapers, and pamphlets that openly challenge the control and legitimacy of the established authority. As the populace loses faith in the established authority the people may decide to actively resist it. During this phase, a counterstate may begin to emerge to fill gaps in governance that the host-nation (HN) government is unwilling or unable to address. Two recent examples are Moqtada al Sadr's organization in Iraq and Hezbollah in Lebanon. Sadr's Mahdi Army provides security and some services in parts of southern Iraq and Baghdad under Sadr's control. (In fact, the Mahdi Army created gaps by undermining security and services; then it moved to solve the problem it created.) Hezbollah provides essential services and reconstruction assistance for its constituents as well as security. Each is an expression of Shiite identity against governments that are pluralist and relatively weak.

1-34. Phase III, strategic counteroffensive, occurs as the insurgent organization becomes stronger than the established authority. Insurgent forces transition from guerrilla warfare to conventional warfare. Military forces aim to destroy the enemy's military capability. Political actions aim to completely displace all government authorities. If successful, this phase causes the government's collapse or the occupying power's withdrawal. Without direct foreign intervention, a strategic offensive takes on the characteristics of a full-scale civil war. As it gains control of portions of the country, the insurgent movement becomes responsible for the population, resources, and territory under its control. To consolidate and preserve its gains, an effective insurgent movement continues the phase I activities listed in paragraph 1-32. In addition it—

- Establishes an effective civil administration.
- Establishes an effective military organization.
- Provides balanced social and economic development.
- Mobilizes the populace to support the insurgent organization.
- Protects the populace from hostile actions.

1-35. Effectively applying Maoist strategy does not require a sequential or complete application of all three stages. The aim is seizing political power; if the government's will and capability collapse early in the process, so much the better. If unsuccessful in a later phase, the insurgency might revert to an earlier one. Later insurgents added new twists to this strategy, to include rejecting the need to eventually switch to large-scale conventional operations. For example, the Algerian insurgents did not achieve much military success of any kind; instead they garnered decisive popular support through superior organizational skills and astute propaganda that exploited French mistakes. These and other factors, including the loss of will in France, compelled the French to withdraw.

The North Vietnamese Dau Trahn

1-36. The Vietnamese conflict offers another example of the application of Mao's strategy. The North Vietnamese developed a detailed variant of it known as dau tranh ("the struggle") that is most easily described in terms of logical lines of operations (LLOs). In this context, a *line of operations* is a logical line that connects actions on nodes and/or decisive points related in time and purpose with an objective (JP 1-02). LLOs can also be described as an operational framework/planning construct used to define the concept of multiple, and often disparate, actions arranged in a framework unified by purpose. (Chapters 4 and 5 discuss LLOs typically used in COIN operations.) Besides modifying Mao's three phases, dau tranh delineated LLOs for achieving political objectives among the enemy population, enemy soldiers, and friendly forces. The "general offensive–general uprising" envisioned in this approach did not occur during the Vietnam War; however, the approach was designed to achieve victory by whatever means were effective.

It did not attack a single enemy center of gravity; instead it put pressure on several, asserting that, over time, victory would result in one of two ways: from activities along one LLO or the combined effects of efforts along several. North Vietnamese actions after their military failure in the 1968 Tet offensive demonstrate this approach's flexibility. At that time, the North Vietnamese shifted their focus from defeating U.S. forces in Vietnam to weakening U.S. will at home. These actions expedited U.S. withdrawal and laid the groundwork for the North Vietnamese victory in 1975.

Complexity and the Shifting Mosaic

1-37. Protracted popular war approaches are conducted along multiple politico-military LLOs and are locally configured. Insurgents may use guerrilla tactics in one province while executing terrorist attacks and an urban approach in another. There may be differences in political activities between villages in the same province. The result is more than just a "three-block war": it is a shifting "mosaic war" that is difficult for counterinsurgents to envision as a coherent whole. In such situations, an effective COIN strategy must be multifaceted and flexible.

Identity-Focused

1-38. The identity-focused approach mobilizes support based on the common identity of religious affiliation, clan, tribe, or ethnic group. Some movements may be based on an appeal to a religious identity, either separately from or as part of other identities. This approach is common among contemporary insurgencies and is sometimes combined with the military-focused approach. The insurgent organization may not have the dual military/political hierarchy evident in a protracted popular war approach. Rather, communities often join the insurgent movement as a whole, bringing with them their existing social/military hierarchy. Additionally, insurgent leaders often try to mobilize the leadership of other clans and tribes to increase the movement's strength.

Composite Approaches and Coalitions

1-39. As occurred in Iraq, contemporary insurgents may use different approaches at different times, applying tactics that take best advantage of circumstances. Insurgents may also apply a composite approach that includes tactics drawn from any or all of the other approaches. In addition—and as in Iraq at present—different insurgent forces using different approaches may form loose coalitions when it serves their interests; however, these same movements may fight among themselves, even while engaging counterinsurgents. Within a single AO, there may be multiple competing entities, each seeking to maximize its survivability and influence—and this situation may be duplicated several times across a joint operations area. This reality further complicates both the mosaic that counterinsurgents must understand and the operations necessary for victory.

Mobilization Means and Causes

1-40. The primary struggle in an internal war is to mobilize people in a struggle for political control and legitimacy. Insurgents and counterinsurgents seek to mobilize popular support for their cause. Both try to sustain that struggle while discouraging support for their adversaries. Two aspects of this effort are mobilization means and causes.

Mobilization Means

1-41. There are five means to mobilize popular support:

- Persuasion.
- Coercion.
- Reaction to abuses.
- Foreign support.
- Apolitical motivations.

A mixture of them may motivate any one individual.

Persuasion

1-42. In times of turmoil, political, social, security, and economic benefits can often entice people to support one side or the other. Ideology and religion are means of persuasion, especially for the elites and leadership. In this case, legitimacy derives from the consent of the governed, though leaders and led can have very different motivations. In Iraq, for example, an issue that motivated fighters in some Baghdad neighborhoods in 2004 was lack of adequate sewer, water, electricity, and trash services. Their concerns were totally disconnected from the overall Ba'athist goal of expelling U.S. forces and retaining Sunni Arab power.

Coercion

1-43. The struggle in Iraq has produced many examples of how insurgent coercion can block government success. In the eyes of some, a government that cannot protect its people forfeits the right to rule. Legitimacy is accorded to the element that can provide security, as citizens seek to ally with groups that can guarantee their safety. In some areas of Iraq and Afghanistan, for instance, militias established themselves as extragovernmental arbiters of the populace's physical security—in some case, after first undermining that security.

1-44. Insurgents may use coercive force to provide security for people or to intimidate them and the legitimate security forces into active or passive support. Kidnapping or killing local leaders or their families is a common insurgent tactic to discourage working with the government. Militias sometimes use the promise of security, or the threat to remove it, to maintain control of cities and towns. Such militias may be sectarian or based on political parties. The HN government must recognize and remove the threat to sovereignty and legitimacy posed by extragovernmental organizations of this type. (The dangers of militias are further described in paragraphs 3-112 and 3-113.)

Reaction to Abuses

1-45. Though firmness by security forces is often necessary to establish a secure environment, a government that exceeds accepted local norms and abuses its people or is tyrannical generates resistance to its rule. People who have been maltreated or have had close friends or relatives killed by the government, particularly by its security forces, may strike back at their attackers. Security force abuses and the social upheaval caused by collateral damage from combat can be major escalating factors for insurgencies.

Foreign Support

1-46. Foreign governments can provide the expertise, international legitimacy, and money needed to start or intensify a conflict. For example, although there was little popular support for the renewal of fighting in Chechnya in 1999, the conflict resumed anyway because foreign supporters and warlords had enough money to hire a guerrilla army. Also of note, nongovernmental organizations (NGOs), even those whose stated aims are impartial and humanitarian, may wittingly or unwittingly support insurgents. For example, funds raised overseas for professed charitable purposes can be redirected to insurgent groups.

Apolitical Motivations

1-47. Insurgencies attract criminals and mercenaries. Individuals inspired by the romanticized image of the revolutionary or holy warrior and others who imagine themselves as fighters for a cause might also join. It is important to note that political solutions might not satisfy some of them enough to end their participation. Fighters who have joined for money will probably become bandits once the fighting ends unless there are jobs for them. This category also includes opportunists who exploit the absence of security to engage in economically lucrative criminal activity, such as kidnapping and theft. True extremists are unlikely to be reconciled to any other outcome than the one they seek; therefore, they must be killed or captured.

Causes

1-48. A cause is a principle or movement militantly defended or supported. Insurgent leaders often seek to adopt attractive and persuasive causes to mobilize support. These causes often stem from the unresolved contradictions existing within any society or culture. Frequently, contradictions are based on real problems. However, insurgents may create artificial contradictions using propaganda and misinformation. Insurgents can gain more support by not limiting themselves to a single cause. By selecting an assortment of causes and tailoring them for various groups within the society, insurgents increase their base of sympathetic and complicit support.

1-49. Insurgents employ deep-seated, strategic causes as well as temporary, local ones, adding or deleting them as circumstances demand. Leaders often use a bait-and-switch approach. They attract supporters by appealing to local grievances; then they lure followers into the broader movement. Without an attractive cause, an insurgency might not be able to sustain itself. But a carefully chosen cause is a formidable asset; it can provide a fledgling movement with a long-term, concrete base of support. The ideal cause attracts the most people while alienating the fewest and is one that counterinsurgents cannot co-opt.

1-50. Potential insurgents can capitalize on a number of potential causes. Any country ruled by a small group without broad, popular participation provides a political cause for insurgents. Exploited or repressed social groups—be they entire classes, ethnic or religious groups, or small elites—may support larger causes in reaction to their own narrower grievances. Economic inequities can nurture revolutionary unrest. So can real or perceived racial or ethnic persecution. For example, Islamic extremists use perceived threats to their religion by outsiders to mobilize support for their insurgency and justify terrorist tactics. As previously noted, effective insurgent propaganda can also turn an artificial problem into a real one.

1-51. Skillful counterinsurgents can deal a significant blow to an insurgency by appropriating its cause. Insurgents often exploit multiple causes, however, making counterinsurgents' challenges more difficult. In the end, any successful COIN operation must address the legitimate grievances insurgents use to generate popular support. These may be different in each local area, in which case a complex set of solutions will be needed.

Mobilizing Resources

1-52. Insurgents resort to such tactics as guerrilla warfare and terrorism for any number of reasons. These may include disadvantages in manpower or organization, relatively limited resources compared to the government, and, in some cases, a cultural predisposition to an indirect approach to conflict. To strengthen and sustain their effort once manpower is mobilized, insurgents require money, supplies, and weapons.

1-53. Weapons are especially important. In some parts of the world, lack of access to weapons may forestall insurgencies. Unfortunately, there is widespread availability of weapons in many areas, with especially large surpluses in the most violent regions of the world. Explosive hazards, such as mines and improvised explosive devices, are likely to be common weapons in insurgencies. (See FMI 3-34.119/MCIP 3-17.01 for more information on improvised explosive devices.) Insurgents can obtain weapons through legal or illegal purchases or from foreign sources. A common tactic is to capture them from government forces. Skillful counterinsurgents cut off the flow of arms into the AO and eliminate their sources.

1-54. Income is essential not only for insurgents to purchase weapons but also to pay recruits and bribe corrupt officials. Money and supplies can be obtained through many sources. Foreign support has already been mentioned. Local supporters or international front organizations may provide donations. Sometimes legitimate businesses are established to furnish funding. In areas controlled by insurgents, confiscation or taxation might be utilized. Another common source of funding is criminal activity.

INSURGENCY AND CRIME

1-55. Funding greatly influences an insurgency's character and vulnerabilities. The insurgents' approach determines the movement's requirements. Protracted popular war approaches that emphasize mobilization of the masses require the considerable resources needed to build and maintain a counterstate. In comparison, the military-focused approach, which emphasizes armed action, needs only the resources necessary to sustain a military campaign. A conspiratorial or urban approach requires even less support.

1-56. Sustainment requirements often drive insurgents into relationships with organized crime or into criminal activity themselves. Reaping windfall profits and avoiding the costs and difficulties involved in securing external support makes illegal activity attractive to insurgents. Taxing a mass base usually yields low returns. In contrast, kidnapping, extortion, bank robbery, and drug trafficking—four favorite insurgent activities—are very lucrative. The activities of the Fuerzas Armadas Revolucionarias de Colombia (FARC) illustrate this point: profits from single kidnappings often total millions of U.S. dollars. For the Maoist Communist Party of Nepal, directly taxing the mass base proved inferior to other criminal forms of "revolutionary taxation," such as extortion and kidnapping. Drugs retain the highest potential for obtaining large profits from relatively small investments. In the 1990s, insurgents in Suriname, South America, were asked why they were selling gold at half the market price; they responded that the quick profits provided seed money to invest in the drug trade, from which they "could make real money." Similarly, failed and failing states with rich natural resources like oil or poppies (which provide the basis for heroin) are particularly lucrative areas for criminal activity. State failure precipitated by violent regime change further encourages criminal activity because of the collapse of law enforcement, the courts, and penal systems.

1-57. Devoting exceptional amounts of time and effort to fund-raising requires an insurgent movement to shortchange ideological or armed action. Indeed, the method of raising funds is often at the heart of debates on characterizing movements as diverse as the Provisional Irish Republican Army in Ulster and the FARC in Colombia. The first has been involved in all sorts of criminal activity for many years; however, it remains committed to its ideological aims. The second, through its involvement in the drug trade, has become the richest self-sustaining insurgent group in history; yet it continues to claim to pursue "Bolivarian" and "socialist" or "Marxist-Leninist" ends. FARC activities, though, have increasingly been labeled "narcoterrorist" or simply criminal by a variety of critics.

1-58. Throughout history, many insurgencies have degenerated into criminality. This occurred as the primary movements disintegrated and the remaining elements were cast adrift. Such disintegration is desirable; it replaces a dangerous, ideologically inspired body of disaffiliated individuals with a less dangerous but more diverse body, normally of very uneven character. The first is a security threat, the second a law-and-order concern. This should not be interpreted, of course, as denigrating the armed capacity of a law-and-order threat. Successful counterinsurgents are prepared to address this disintegration. They also recognize that the ideal approach eliminates both the insurgency and any criminal threats its elimination produces.

Elements of Insurgency

1-59. Though insurgencies take many forms, most share some common attributes. An insurgent organization normally consists of five elements:

- Movement leaders.
- Combatants (main, regional, and local forces [including militias]).
- Political cadre (also called militants or the party).
- Auxiliaries (active followers who provide important support services).
- Mass base (the bulk of the membership).

1-60. The proportion of each element relative to the larger movement depends on the strategic approach the insurgency adopts. A conspiratorial approach does not pay much attention to combatants or a mass base. Military-focused insurgencies downplay the importance of a political cadre and emphasize military action to generate popular support. The people's war approach is the most complex: if the state presence has been eliminated, the elements exist openly; if the state remains a continuous or occasional presence, the elements maintain a clandestine existence.

Movement Leaders

1-61. Movement leaders provide strategic direction to the insurgency. They are the "idea people" and the planners. They usually exercise leadership through force of personality, the power of revolutionary ideas, and personal charisma. In some insurgencies, they may hold their position through religious, clan, or tribal authority.

Combatants

1-62. Combatants (sometimes called "foot soldiers") do the actual fighting and provide security. They are often mistaken for the movement itself; however, they exist only to support the insurgency's broader political agenda and to maintain local control. Combatants protect and expand the counterstate, if the insurgency sets up such an institution. They also protect training camps and networks that facilitate the flow of money, instructions, and foreign and local fighters.

Political Cadre

1-63. The cadre forms the political core of the insurgency. They are actively engaged in the struggle to accomplish insurgent goals. They may also be designated as a formal party to signify their political importance. The cadre implement guidance and procedures provided by the movement leaders. Modern noncommunist insurgencies rarely, if ever, use the term "cadre"; however these movements usually include a group that performs similar functions. Additionally, movements based on religious extremism usually include religious and spiritual advisors among their cadre.

1-64. The cadre assesses grievances in local areas and carries out activities to satisfy them. They then attribute the solutions they have provided to the insurgency. As the insurgency matures, deeds become more important to make insurgent slogans meaningful to the population. Larger societal issues, such as foreign presence, facilitate such political activism because insurgents can blame these issues for life's smaller problems. Destroying the state bureaucracy and preventing national reconstruction after a conflict (to sow disorder and sever legitimate links with the people) are also common insurgent tactics. In time, the cadre may seek to replace that bureaucracy and assume its functions in a counterstate.

Auxiliaries

1-65. Auxiliaries are active sympathizers who provide important support services. They do not participate in combat operations. Auxiliaries may do the following:

- Run safe houses.
- Store weapons and supplies.
- Act as couriers.
- Provide passive intelligence collection.
- Give early warning of counterinsurgent movements.
- Provide funding from lawful and unlawful sources.
- Provide forged or stolen documents and access or introductions to potential supporters.

Mass Base

1-66. The mass base consists of the followers of the insurgent movement—the supporting populace. Mass base members are often recruited and indoctrinated by the cadre. However, in many politically charged situations or identity-focused insurgencies, such active pursuit is not necessary. Mass base members may continue in their normal positions in society. Many, however, lead clandestine lives for the insurgent movement. They may even pursue full-time positions within the insurgency. For example, combatants normally begin as members of the mass base. In tribal- or clan-based insurgencies, such roles are particularly hard to define. There is no clear cadre in those movements, and people drift between combatant, auxiliary, and follower status as needed.

Employing the Elements

1-67. The movement leaders provide the organizational and managerial skills needed to transform mobilized individuals and communities into an effective force for armed political action. The result is a contest of resource mobilization and force deployment. No force level guarantees victory for either side. During previous conflicts, planners assumed that combatants required a 10 or 15 to 1 advantage over insurgents to win. However, no predetermined, fixed ratio of friendly troops to enemy combatants ensures success in COIN. The conditions of the operational environment and the approaches insurgents use vary too widely. A better force requirement gauge is troop density, the ratio of security forces (including the host nation's military and police forces as well as foreign counterinsurgents) to inhabitants. Most density recommendations fall within a range of 20 to 25 counterinsurgents for every 1000 residents in an AO. Twenty counterinsurgents per 1000 residents is often considered the minimum troop density required for effective COIN operations; however as with any fixed ratio, such calculations remain very dependent upon the situation.

1-68. As in any conflict, the size of the force needed to defeat an insurgency depends on the situation. However, COIN is manpower intensive because counterinsurgents must maintain widespread order and security. Moreover, counterinsurgents typically have to adopt different approaches to address each element of the insurgency. For example, auxiliaries might be co-opted by economic or political reforms, while fanatic combatants will most likely have to be killed or captured.

DYNAMICS OF AN INSURGENCY

1-69. Insurgencies are also shaped by several common dynamics:

- Leadership.
- Objectives.
- Ideology and narrative.
- Environment and geography.
- External support and sanctuaries.
- Phasing and timing.

These make up a framework that can be used to assess the insurgency's strengths and weaknesses. Although these dynamics can be examined separately, studying their interaction is necessary to fully understand an insurgency.

1-70. The interplay of these dynamics influences an insurgency's approach and organization. Effective counterinsurgents identify the organizational pattern these dynamics form and determine if it changes. For example, insurgents operating in an urban environment usually form small, cohesive, secretive organizations. In contrast, insurgents following a military-focused strategy often operate in a rural environment and exploit international support to a greater extent. A change in location or the amount of external support might lead insurgents to adjust their approach and organization.

Leadership

1-71. Leadership is critical to any insurgency. An insurgency is not simply random violence; it is directed and focused violence aimed at achieving a political objective. It requires leadership to provide vision, direction, guidance, coordination, and organizational coherence. Successful insurgent leaders make their cause known to the people and gain popular support. Their key tasks are to break the ties between the people and the government and to establish credibility for their movement. Their education, background, family and social connections, and experiences contribute to their ability to organize and inspire the people who form the insurgency.

1-72. Some insurgent movements have their roots in a clash of cultures over power and preeminence. Others begin as the tangible manifestation of some form of political estrangement. In either case, alienated elite members advance alternatives to existing conditions. As their movement grows, leaders decide which approach to adopt. The level of decentralization of responsibility and authority drives the insurgency's structure and operational procedures. Extreme decentralization results in a movement that rarely functions as a coherent body. It is, however, capable of inflicting substantial casualties and damage. Loose networks find it difficult to create a viable counterstate; they therefore have great difficulty seizing political power. However, they are also very hard to destroy and can continue to sow disorder, even when degraded. It takes very little coordination to disrupt most states.

1-73. Many contemporary insurgencies are identity-based. These insurgencies are often led by traditional authority figures, such as tribal sheikhs, local warlords, or religious leaders. As the Indonesian Dar 'ul Islam rebellions of 1948 and 1961 demonstrate, traditional authority figures often wield enough power to single-handedly drive an insurgency. This is especially true in rural areas. Identity-focused insurgencies can be defeated in some cases by co-opting the responsible traditional authority figure; in others, the authority figures have to be discredited or eliminated. Accurately determining whether a leader can be co-opted is crucial. Failed attempts to co-opt traditional leaders can backfire if those leaders choose to oppose the counterinsurgency. Their refusal to be co-opted can strengthen their standing as they gain power and influence among insurgents.

Objectives

1-74. Effective analysis of an insurgency requires identifying its strategic, operational, and tactical objectives. The strategic objective is the insurgents' desired end state. Operational objectives are those that insurgents pursue to destroy government legitimacy and progressively establish their desired end state. Tactical objectives are the immediate aims of insurgent acts. Objectives can be psychological or physical. One example of a psychological objective is discouraging support for the government by assassinating local officials. An example of a physical objective is the disruption of government services by damaging or seizing a key facility. These tactical acts are often linked to higher purposes; in fact, tactical actions by both insurgents and counterinsurgents frequently have strategic effects.

Ideology and Narrative

1-75. Ideas are a motivating factor in insurgent activities. Insurgencies can gather recruits and amass popular support through ideological appeal (including religious or other cultural identifiers). Promising potential recruits often include individuals receptive to the message that the West is dominating their region through puppet governments and local surrogates. The insurgent group channels anti-Western anger and provides members with identity, purpose, and community, in addition to physical, economic, and psychological security. The movement's ideology explains its followers' difficulties and provides a means to remedy those ills. The most powerful ideologies tap latent, emotional concerns of the populace. Examples of these concerns include religiously based objectives, a desire for justice, ethnic aspirations, and a goal of liberation from foreign occupation. Ideology provides a prism, including a vocabulary and analytical categories, through which followers perceive their situation.

1-76. The central mechanism through which ideologies are expressed and absorbed is the narrative. A narrative is an organizational scheme expressed in story form. Narratives are central to representing identity, particularly the collective identity of religious sects, ethnic groupings, and tribal elements. Stories about a community's history provide models of how actions and consequences are linked. Stories are often the basis for strategies and actions, as well as for interpreting others' intentions. Insurgent organizations like Al Qaeda use narratives very effectively in developing legitimating ideologies. In the Al Qaeda narrative, for example, Osama bin Laden depicts himself as a man purified in the mountains of Afghanistan who is gathering and inspiring followers and punishing infidels. In the collective imagination of Bin Laden and his followers, they are agents of Islamic history who will reverse the decline of the umma [Muslim community] and bring about its inevitable triumph over Western imperialism. For them, Islam can be renewed both politically and theologically only through jihad [holy war] as they define it.

1-77. Though most insurgencies have been limited to nation-states, there have been numerous transnational insurgencies. Likewise, external powers have tried to tap into or create general upheaval by coordinating national insurgencies to give them a transnational character. Al Qaeda's ongoing activities also attempt to leverage religious identity to create and support a transnational array of insurgencies. Operational-level commanders address elements of the transnational movement within their joint operations areas. Other government agencies and higher level officials deal with the national-strategic response to such threats.

1-78. As noted earlier, insurgent groups often employ religious concepts to portray their movement favorably and mobilize followers in pursuit of their political goals. For example, the Provisional Irish Republican Army frequently used Roman Catholic iconography in its publications and proclamations, although many of its members were not devout Catholics. In other cases, a religious ideology may be the source of an insurgent group's political goals. This is the case in Al Qaeda's apparent quest to "reestablish the Caliphate." For many Moslems, the Caliphate produces a positive image of the golden age of Islamic civilization. This image mobilizes support for Al Qaeda among some of the most traditional Muslims while concealing the details of the movement's goal. In fact, Al Qaeda's leaders envision the "restored Caliphate" as a totalitarian state similar to the pre-2002 Taliban regime in Afghanistan.

1-79. Religious extremist insurgents, like many secular radicals and some Marxists, frequently hold an all-encompassing worldview; they are ideologically rigid and uncompromising, seeking to control their members' private thought, expression, and behavior. Seeking power and believing themselves to be ideologically pure, violent religious extremists often brand those they consider insufficiently orthodox as enemies. For example, extreme, violent groups like Al Qaeda routinely attack Islamic sects that profess beliefs inconsistent with their religious dogma. Belief in an extremist ideology fortifies the will of believers. It confirms the idea, common among hard-core transnational terrorists, that using unlimited means is appropriate to achieve their often unlimited goals. Some ideologies, such as the one underlying the culture of martyrdom, maintain that using such means will be rewarded.

1-80. Cultural knowledge is essential to waging a successful counterinsurgency. American ideas of what is "normal" or "rational" are not universal. To the contrary, members of other societies often have different notions of rationality, appropriate behavior, level of religious devotion, and norms concerning gender. Thus, what may appear abnormal or strange to an external observer may appear as self-evidently normal to a group member. For this reason, counterinsurgents—especially commanders, planners, and small-unit leaders—should strive to avoid imposing their ideals of normalcy on a foreign cultural problem.

1-81. Many religious extremists believe that the conversion, subjugation, or destruction of their ideological opponents is inevitable. Violent extremists and terrorists are often willing to use whatever means necessary, even violence against their own followers, to meet their political goals. Nevertheless, they often pursue their ends in highly pragmatic ways based on realistic assumptions. Not all Islamic insurgents or terrorists are fighting for a global revolution. Some are pursing regional goals, such as a establishing a Sunni Arab-dominated Iraq or replacing Israel with an Arab Palestinian state. And militant groups with nationalist as well as religious agendas seek cease fires and participate in elections when such actions support their interests.

1-82. In that light, commanders must consider the presence of religious extremism in the insurgents' ideology when evaluating possible friendly and enemy courses of action. Enemy courses of action that may appear immoral or irrational to Westerners may be acceptable to extremists. Moreover, violent extremists resist changing their worldview; for them, coexistence or compromise is often unacceptable, especially when the movement is purist (like Al Qaeda), in an early stage, or small. However, some extremists are willing to overlook their worldview to achieve short-term goals. Terrorist groups, regardless of their ideology, have cooperated with seemingly incompatible groups. For example, the Palestinian group Black September used German terrorists to perform reconnaissance of the Olympic Village before its 1972 attack on Israeli athletes. Currently, the Taliban is engaged in the drug trade in South Asia. Al Qaeda cooperates with a variety of diverse groups to improve its global access as well.

1-83. The rigid worldview of such extremist groups means that friendly actions intended to create good will among the populace are unlikely to affect them. Similarly, if a group's ideology is so strong that it dominates all other issues, dialog and negotiation will probably prove unproductive. The challenge for counterinsurgents in such cases is to identify the various insurgent groups and determine their motivations. Commanders can then determine the best course of action for each group. This includes identifying the groups with goals flexible enough to allow productive negotiations and determining how to eliminate the extremists without alienating the populace.

Environment and Geography

1-84. Environment and geography, including cultural and demographic factors, affect all participants in a conflict. The manner in which insurgents and counterinsurgents adapt to these realities creates advantages and disadvantages for each. The effects of these factors are immediately visible at the tactical level. There they are perhaps the predominant influence on decisions regarding force structure and doctrine (including tactics, techniques, and procedures). Insurgencies in urban environments present different planning considerations from insurgencies in rural environments. Border areas contiguous to states that may wittingly or unwittingly provide external support and sanctuary to insurgents create a distinct vulnerability for counterinsurgents.

External Support and Sanctuaries

1-85. Access to external resources and sanctuaries has always influenced the effectiveness of insurgencies. External support can provide political, psychological, and material resources that might otherwise be limited or unavailable. Such assistance does not need to come just from neighboring states; countries from outside the region seeking political or economic influence can also support insurgencies. Insurgencies may turn to transnational criminal elements for funding or use the Internet to create a support network among NGOs. Ethnic or religious communities in other states may also provide a form of external support and sanctuary, particularly for transnational insurgencies.

1-86. The meaning of the term sanctuary is evolving. Sanctuaries traditionally were physical safe havens, such as base areas, and this form of safe haven still exists. But insurgents today can also draw on "virtual" sanctuaries in the Internet, global financial systems, and the international media. These virtual sanctuaries can be used to try to make insurgent actions seem acceptable or laudable to internal and external audiences.

1-87. Historically, sanctuaries in neighboring countries have provided insurgents places to rebuild and reorganize without fear of counterinsurgent interference. Modern target acquisition and intelligence-gathering technology make insurgents in isolation, even in neighboring states, more vulnerable than those hidden among the population. Thus, contemporary insurgencies often develop in urban environments, leveraging formal and informal networks for action. Understanding these networks is vital to defeating such insurgencies.

1-88. Insurgencies can also open up sanctuaries within a state over which the host nation's forces cannot extend control or significant influence. In these sanctuaries, nonstate actors with intentions hostile to the host nation or United States can develop unimpaired. When it is to their advantage, such elements provide support for insurgencies. The issue of sanctuaries thus cannot be ignored during planning. Effective COIN operations work to eliminate all sanctuaries.

1-89. Changes in the security environment since the end of the Cold War and the terrorist attacks of 11 September 2001 have increased concerns about the role of nonstate actors in insurgencies. Nonstate actors, such as transnational terrorist organizations, often represent a security threat beyond the areas they inhabit. Some pose a direct concern for the United States and its partners. These nonstate actors often team with insurgents and, in this sense, profit from the conflict.

1-90. A feature of today's operational environment deserving mention is the effort by Islamic extremists, including those that advocate violence, to spread their influence through the funding and use of entities that share their views or facilitate them to varying degrees. These entities may or may not be threats themselves; however, they can provide passive or active support to local or distant insurgencies. Examples include the following:

- Religious schools and mosques.
- NGOs.
- Political parties.
- Business and financial institutions.
- Militia organizations.
- Terrorist training camps and organizations.

Phasing and Timing

1-91. Insurgencies often pass through common phases of development, such as those listed in paragraph 1-31. However, not all insurgencies experience such phased development, and progression through all phases is not required for success. Moreover, a single insurgent movement may be in different phases in different parts of a country. Insurgencies under pressure can also revert to an earlier phase. They then resume development when favorable conditions return. Indeed, this flexibility is the key strength of a phased approach, which provides fallback positions for insurgents when threatened. The protracted popular war phases may not provide a complete template for understanding contemporary insurgencies; however, they do explain the shifting mosaic of activities usually present in some form.

1-92. Versions of protracted popular war have been used by movements as diverse as communist and Islamist insurgencies because the approach is sound and based on mass mobilization—which is a common requirement. Strategic movement from one phase to another does not end the operational and tactical activities typical of earlier phases; it incorporates them. The North Vietnamese explicitly recognized this fact in their doctrine, as was discussed in paragraph 1-36. Their approach emphasized that all forms of warfare occur simultaneously, even as a particular form is paramount. Debates about Vietnam that focus on whether U.S. forces should have concentrated on guerrilla or conventional operations ignore this complexity. In fact, forces that win a mosaic war are those able to respond to both types of operations, often simultaneously.

1-93. The phases of protracted popular war do not necessarily apply to the conspiratorial or military-focused approach. These approaches emphasize quick or armed action and minimize political organization. In many ways, these approaches are less difficult to counter. However, long-term political objectives, as evidenced in the protracted popular war approach, are major parts of any insurgent approach. Effective counterinsurgents understand their overall importance and address them appropriately.

INSURGENT NETWORKS

1-94. A network is a series of direct and indirect ties from one actor to a collection of others. Insurgents use technological, economic, and social means to recruit partners into their networks. Networking is a tool available to territorially rooted insurgencies, such as the FARC in Colombia. It extends the range and variety of both their military and political actions. Other groups have little physical presence in their target countries and exist almost entirely as networks. Networked organizations are difficult to destroy. In addition, they tend to heal, adapt, and learn rapidly. However, such organizations have a limited ability to attain strategic success because they cannot easily muster and focus power. The best outcome they can expect is to create a security vacuum leading to a collapse of the targeted regime's will and then to gain in the competition for the spoils. However, their enhanced abilities to sow disorder and survive present particularly difficult problems for counterinsurgents.

INSURGENT VULNERABILITIES

1-95. While this chapter so far has stressed the difficulties insurgencies present, they do have vulnerabilities that skilled counterinsurgents can exploit. Chapters 4 and 5 discuss how to do this. However, some potential vulnerabilities are worth highlighting here:

- Insurgents' need for secrecy.
- Inconsistencies in the mobilization message.
- Need to establish a base of operations.
- Reliance on external support.
- Need to obtain financial resources.
- Internal divisions.
- Need to maintain momentum.
- Informants within the insurgency.

Secrecy

1-96. Any group beginning from a position of weakness that intends to use violence to pursue its political aims must initially adopt a covert approach for its planning and activities. This practice can become counterproductive once an active insurgency begins. Excessive secrecy can limit insurgent freedom of action, reduce or distort information about insurgent goals and ideals, and restrict communication within the insurgency. Some insurgent groups try to avoid the effects of too much secrecy by splitting into political and military wings. This allows the movement to address the public (political) requirements of an insurgency while still conducting clandestine (military) actions. An example is the insurgency in Northern Ireland, comprised of Sinn Fein (its political wing) and the Irish Republican Army (its military wing). Hamas and Hezbollah also use this technique.

Mobilization and Message

1-97. In the early stages of an insurgency, a movement may be tempted to go to almost any extremes to attract followers. To mobilize their base of support, insurgent groups use a combination of propaganda and intimidation, and they may overreach in both. Effective counterinsurgents use information operations (IO) to exploit inconsistencies in the insurgents' message as well as their excessive use of force or intimidation. The insurgent cause itself may also be a vulnerability. Counterinsurgents may be able to "capture" an insurgency's cause and exploit it. For example, an insurgent ideology based on an extremist interpretation of a holy text can be countered by appealing to a moderate interpretation of the same text. When a credible religious or other respected leader passes this kind of message, the counteraction is even more effective.

Base of Operations

1-98. Insurgents can experience serious difficulties finding a viable base of operations. A base too far from the major centers of activity may be secure but risks being out of touch with the populace. It may also be vulnerable to isolation. A base too near centers of government activity risks opening the insurgency to observation and perhaps infiltration. Bases close to national borders can be attractive when they are beyond the reach of counterinsurgents yet safe enough to avoid suspicions of the neighboring authority or population. Timely, resolute counterinsurgent actions to exploit poor enemy base locations and eliminate or disrupt good ones can significantly weaken an insurgency.

External Support

1-99. Insurgent movements do not control the geographic borders of a country. In fact, insurgencies often rely heavily on freedom of movement across porous borders. Insurgencies usually cannot sustain themselves without substantial external support. An important feature of many transnational terrorist groups is the international nature of their basing. Terrorists may train in one country and fight or conduct other types of operations in another country. The movement of fighters and their support is vulnerable to intervention or attack.

Financial Weakness

1-100. All insurgencies require funding to some extent. Criminal organizations are possible funding sources; however, these groups may be unreliable. Such cooperation may attract undue attention from HN authorities and create vulnerabilities to counterinsurgent intelligence operations. In addition, cooperating with criminals may not be ideologically consistent with the movement's core beliefs, although it often does not prevent such cooperation. Funding from outside donors may come with a political price that affects the overall aim of an insurgency and weakens its popular appeal.

1-101. Counterinsurgents can exploit insurgent financial weaknesses. Controls and regulations that limit the movement and exchange of materiel and funds may compound insurgent financial vulnerabilities. These counters are especially effective when an insurgency receives funding from outside the state.

Internal Divisions

1-102. Counterinsurgents remain alert for signs of divisions within an insurgent movement. A series of successes by counterinsurgents or errors by insurgent leaders can cause some insurgents to question their cause or challenge their leaders. In addition, relations within an insurgency do not remain harmonious when factions form to vie for power. Rifts between insurgent leaders, if identified, can be exploited. Offering amnesty or a seemingly generous compromise can also cause divisions within an insurgency and present opportunities to split or weaken it.

Maintaining Momentum

1-103. Controlling the pace and timing of operations is vital to the success of any insurgency. Insurgents control when the conflict begins and have some measure of control over subsequent activity. However, many insurgencies have failed to capitalize on their initial opportunities. Others have allowed counterinsurgents to dictate the pace of events and scope of activities. If insurgents lose momentum, counterinsurgents can regain the strategic initiative.

Informants

1-104. Nothing is more demoralizing to insurgents than realizing that people inside their movement or trusted supporters among the public are deserting or providing information to government authorities. Counterinsurgents may attract deserters or informants by arousing fear of prosecution or by offering rewards. However, informers must be confident that the government can protect them and their families against retribution.

ASPECTS OF COUNTERINSURGENCY

1-105. The purpose of America's ground forces is to fight and win the Nation's wars. Throughout history, however, the Army and Marine Corps have been called on to perform many tasks beyond pure combat; this has been particularly true during the conduct of COIN operations. COIN requires Soldiers and Marines to be ready both to fight and to build—depending on the security situation and a variety of other factors. The full spectrum operations doctrine (described in FM 3-0) captures this reality.

1-106. All full spectrum operations executed overseas—including COIN operations—include offensive, defensive, and stability operations that commanders combine to achieve the desired end state. The exact mix varies depending on the situation and the mission. Commanders weight each operation based on their assessment of the campaign's phase and the situation in their AO. They shift the weight among these operations as necessary to address situations in different parts of the AO while continuing to pursue their overall objectives. (See figure 1-1.)

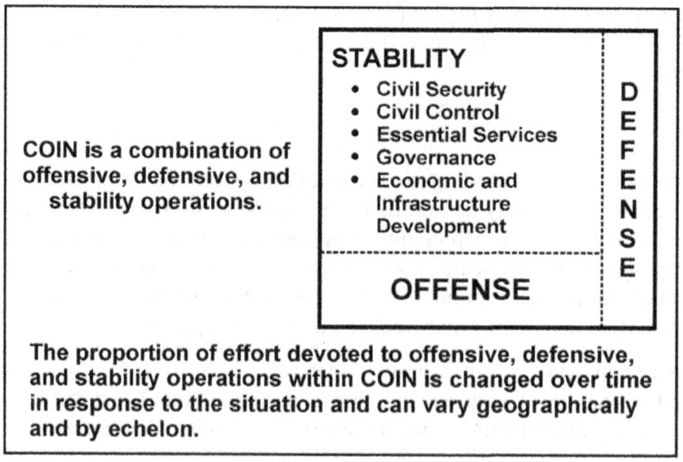

Figure 1-1. Aspects of counterinsurgency operations

1-107. Offensive and defensive operations are integral to COIN. COIN differs from peacekeeping operations in this regard; indeed, this is a key point. In peacekeeping operations, combat is not expected and the goal is an absence of violence. In COIN, such an absence may actually mask insurgent preparations for combat. This was the case, for example, in the Sadr City area of Baghdad in 2003.

1-108. In almost every case, counterinsurgents face a populace containing an active minority supporting the government and an equally small militant faction opposing it. Success requires the government to be accepted as legitimate by most of that uncommitted middle, which also includes passive supporters of both sides. (See figure 1-2.) Because of the ease of sowing disorder, it is usually not enough for counterinsurgents to get 51 percent of popular support; a solid majority is often essential. However, a passive populace may be all that is necessary for a well-supported insurgency to seize political power.

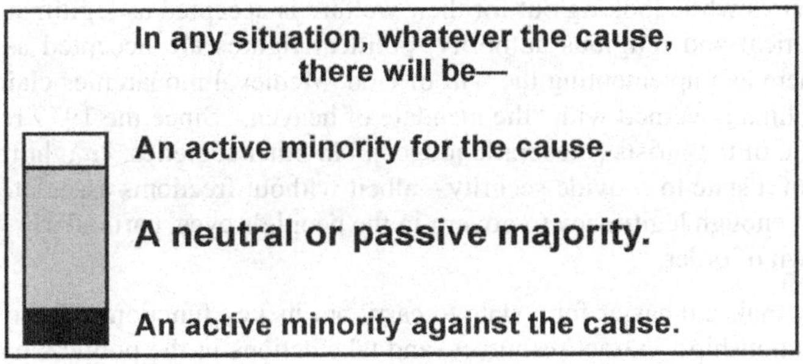

Figure 1-2. Support for an insurgency

1-109. Counterinsurgents must be prepared to identify their opponents and their approach to insurgency. Counterinsurgents must also understand the broader context within which they are operating. A mission to assist a functioning government offers different options from situations where no such viable entity exists or where a regime has been changed by conflict. The last two situations add complex sovereignty and national reconstruction issues to an already complex mission. The state of the infrastructure determines the resources required for reconstruction. The level of violence is a factor in determining how agencies outside the Department of Defense support COIN operations. An extremely violent environment may hamper their freedom of movement. These agencies may elect to operate only from secure areas within or outside of the country.

1-110. The rest of this publication describes how to conduct COIN operations. The following discussion addresses some general themes that shape the following chapters:

- Historical principles for COIN.
- Contemporary imperatives for COIN.
- Paradoxes of COIN operations.
- Successful and unsuccessful COIN practices.

1-111. The historical principles and contemporary imperatives derived from the historical record and detailed below provide some guideposts for forces engaged in COIN operations. However, COIN operations are complicated, and even following the principles and imperatives does not guarantee success. This paradox is present in all forms of warfare but is most obvious in COIN. The following principles and imperatives are presented in the belief that understanding them helps illuminate the challenges inherent in defeating an insurgency.

HISTORICAL PRINCIPLES FOR COUNTERINSURGENCY

1-112. The following principles are derived from past insurgencies.

Legitimacy Is the Main Objective

1-113. The primary objective of any COIN operation is to foster development of effective governance by a legitimate government. Counterinsurgents achieve this objective by the balanced application of both military and nonmilitary means. All governments rule through a combination of consent and coercion. Governments described as "legitimate" rule primarily with the consent of the governed; those described as "illegitimate" tend to rely mainly or entirely on coercion. Citizens of the latter obey the state for fear of the consequences of doing otherwise, rather than because they voluntarily accept its rule. A government that derives its powers from the governed tends to be accepted by its citizens as legitimate. It still uses coercion—for example, against criminals—but most of its citizens voluntarily accept its governance.

1-114. In Western liberal tradition, a government that derives its just powers from the people and responds to their desires while looking out for their welfare is accepted as legitimate. In contrast, theocratic societies fuse political and religious authority; political figures are accepted as legitimate because the populace views them as implementing the will of God. Medieval monarchies claimed "the divine right of kings." Imperial China governed with "the mandate of heaven." Since the 1979 revolution, Iran has operated under the "rule of the jurists [theocratic judges]." In other societies, "might makes right." And sometimes, the ability of a state to provide security—albeit without freedoms associated with Western democracies—can give it enough legitimacy to govern in the people's eyes, particularly if they have experienced a serious breakdown of order.

1-115. Legitimacy makes it easier for a state to carry out its key functions. These include the authority to regulate social relationships, extract resources, and take actions in the public's name. Legitimate governments can develop these capabilities more easily; this situation usually allows them to competently manage, coordinate, and sustain collective security as well as political, economic, and social development. Conversely, illegitimate states (sometimes called "police states") typically cannot regulate society or can do so only by applying overwhelming coercion. Legitimate governance is inherently stable; the societal support it engenders allows it to adequately manage the internal problems, change, and conflict that affect individual and collective well-being. Conversely, governance that is not legitimate is inherently unstable; as soon as the state's coercive power is disrupted, the populace ceases to obey it. Thus legitimate governments tend to be resilient and exercise better governance; illegitimate ones tend to be fragile and poorly administered.

1-116. Six possible indicators of legitimacy that can be used to analyze threats to stability include the following:

- The ability to provide security for the populace (including protection from internal and external threats).
- Selection of leaders at a frequency and in a manner considered just and fair by a substantial majority of the populace.
- A high level of popular participation in or support for political processes.
- A culturally acceptable level of corruption.
- A culturally acceptable level and rate of political, economic, and social development.
- A high level of regime acceptance by major social institutions.

1-117. Governments scoring high in these categories probably have the support of a sufficient majority of the population. Different cultures, however, may see acceptable levels of development, corruption, and participation differently. And for some societies, providing security and some basic services may be enough for citizens to grant a government legitimacy; indeed, the importance of security in situations where violence has escalated cannot be overemphasized. In such cases, establishing security can win the people's confidence and enable a government to develop legitimacy in other areas.

1-118. In working to understand the problem, commanders and staffs determine what the HN population defines as effective and legitimate governance. This understanding continues to evolve as information is developed. Commanders and staffs must continually diagnose what they understand legitimacy to mean to the HN population. The population's expectations will influence all ensuing operations. Additionally, planners may also consider perceptions of legitimacy held by outside supporters of the HN government and the insurgents. Differences between U.S., local, and international visions of legitimacy can further complicate operations. But the most important attitude remains that of the HN population. In the end, its members determine the ultimate victor.

1-119. The presence of the rule of law is a major factor in assuring voluntary acceptance of a government's authority and therefore its legitimacy. A government's respect for preexisting and impersonal legal rules can provide the key to gaining it widespread, enduring societal support. Such government respect for rules—ideally ones recorded in a constitution and in laws adopted through a credible, democratic process—is the essence of the rule of law. As such, it is a powerful potential tool for counterinsurgents.

1-120. Military action can address the symptoms of a loss of legitimacy. In some cases, it can eliminate substantial numbers of insurgents. However, success in the form of a durable peace requires restoring legitimacy, which, in turn, requires the use of all instruments of national power. A COIN effort cannot achieve lasting success without the HN government achieving legitimacy.

Unity of Effort Is Essential

1-121. Unity of effort must be present at every echelon of a COIN operation. Otherwise, well-intentioned but uncoordinated actions can cancel each other or provide vulnerabilities for insurgents to exploit. Ideally, a single counterinsurgent leader has authority over all government agencies involved in COIN operations. Usually, however, military commanders work to achieve unity of effort through liaison with leaders of a wide variety of nonmilitary agencies. The U.S. Ambassador and country team, along with senior HN representatives, must be key players in higher level planning; similar connections are needed throughout the chain of command.

1-122. NGOs often play an important role at the local level. Many such agencies resist being overtly involved with military forces; however, efforts to establish some kind of liaison are needed. The most important connections are those with joint, interagency, multinational, and HN organizations. The goal of these connections is to ensure that, as much as possible, objectives are shared and actions and messages synchronized. Achieving this synergy is essential.

Political Factors Are Primary

1-123. General Chang Ting-chen of Mao Zedong's central committee once stated that revolutionary war was 80 percent political action and only 20 percent military. Such an assertion is arguable and certainly depends on the insurgency's stage of development; it does, however, capture the fact that political factors have primacy in COIN. At the beginning of a COIN operation, military actions may appear predominant as security forces conduct operations to secure the populace and kill or capture insurgents; however, political objectives must guide the military's approach. Commanders must, for example, consider how operations contribute to strengthening the HN government's legitimacy and achieving U.S. political goals. This means that political and diplomatic leaders must actively participate throughout the conduct (planning, preparation, execution, and assessment) of COIN operations. The political and military aspects of insurgencies are so bound together as to be inseparable. Most insurgent approaches recognize that fact. Military actions executed without properly assessing their political effects at best result in reduced effectiveness and at worst are counterproductive. Resolving most insurgencies requires a political solution; it is thus imperative that counterinsurgent actions do not hinder achieving that political solution.

Counterinsurgents Must Understand the Environment

1-124. Successful conduct of COIN operations depends on thoroughly understanding the society and culture within which they are being conducted. Soldiers and Marines must understand the following about the population in the AO:

- Organization of key groups in the society.
- Relationships and tensions among groups.
- Ideologies and narratives that resonate with groups.
- Values of groups (including tribes), interests, and motivations.
- Means by which groups (including tribes) communicate.

- The society's leadership system.

1-125. In most COIN operations in which U.S. forces participate, insurgents hold a distinct advantage in their level of local knowledge. They speak the language, move easily within the society, and are more likely to understand the population's interests. Thus, effective COIN operations require a greater emphasis on certain skills, such as language and cultural understanding, than does conventional warfare. The interconnected, politico-military nature of insurgency and COIN requires immersion in the people and their lives to achieve victory. Specifically, successful COIN operations require Soldiers and Marines at every echelon to possess the following within the AO's cultural context:

- A clear appreciation of the essential nature and nuances of the conflict.
- An understanding of the motivation, strengths, and weaknesses of the insurgents.
- Knowledge of the roles of other actors in the AO.

Without this understanding of the environment, intelligence cannot be understood and properly applied.

Intelligence Drives Operations

1-126. Without good intelligence, counterinsurgents are like blind boxers wasting energy flailing at unseen opponents and perhaps causing unintended harm. With good intelligence, counterinsurgents are like surgeons cutting out cancerous tissue while keeping other vital organs intact. Effective operations are shaped by timely, specific, and reliable intelligence, gathered and analyzed at the lowest possible level and disseminated throughout the force.

1-127. Because of the dispersed nature of COIN operations, counterinsurgents' own actions are a key generator of intelligence. A cycle develops where operations produce intelligence that drives subsequent operations. Reporting by units, members of the country team, and associated civilian agencies is often of greater importance than reporting by specialized intelligence assets. These factors, along with the need to generate a favorable tempo (rate of military operations), drive the requirement to produce and disseminate intelligence at the lowest practical level. (Chapter 3 addresses intelligence in COIN.)

Insurgents Must be Isolated from Their Cause and Support

1-128. It is easier to separate an insurgency from its resources and let it die than to kill every insurgent. Clearly, killing or capturing insurgents will be necessary, especially when an insurgency is based in religious or ideological extremism. However, killing every insurgent is normally impossible. Attempting to do so can also be counterproductive in some cases; it risks generating popular resentment, creating martyrs that motivate new recruits, and producing cycles of revenge.

1-129. Dynamic insurgencies can replace losses quickly. Skillful counterinsurgents must thus cut off the sources of that recuperative power. Some sources can be reduced by redressing the social, political, and economic grievances that fuel the insurgency. Physical support can be cut off by population control or border security. International or local legal action might be required to limit financial support. Urban insurgents, however, are especially difficult to isolate from their cause and sources of support. They may operate in small, compartmentalized cells that are usually independent or semi-independent. These cells often have their own support mechanisms and few, if any, ties to the population that counterinsurgents can track.

1-130. As the HN government increases its legitimacy, the populace begins to assist it more actively. Eventually, the people marginalize and stigmatize insurgents to the point that the insurgency's claim to legitimacy is destroyed. However, victory is gained not when this isolation is achieved, but when the victory is permanently maintained by and with the people's active support and when insurgent forces have been defeated.

Security Under the Rule of Law is Essential

1-131. The cornerstone of any COIN effort is establishing security for the civilian populace. Without a secure environment, no permanent reforms can be implemented and disorder spreads. To establish legitimacy, commanders transition security activities from combat operations to law enforcement as quickly as feasible. When insurgents are seen as criminals, they lose public support. Using a legal system established in line with local culture and practices to deal with such criminals enhances the HN government's legitimacy. Soldiers and Marines help establish HN institutions that sustain that legal regime, including police forces, court systems, and penal facilities. It is important to remember that the violence level must be reduced enough for police forces to maintain order prior to any transition; otherwise, COIN forces will be unable to secure the populace and may lose the legitimacy gained by the transition.

1-132. Illegitimate actions are those involving the use of power without authority—whether committed by government officials, security forces, or counterinsurgents. Such actions include unjustified or excessive use of force, unlawful detention, torture, and punishment without trial. Efforts to build a legitimate government though illegitimate actions are self-defeating, even against insurgents who conceal themselves amid noncombatants and flout the law. Moreover, participation in COIN operations by U.S. forces must follow United States law, including domestic laws, treaties to which the United States is party, and certain HN laws. (See appendix D.) Any human rights abuses or legal violations committed by U.S. forces quickly become known throughout the local populace and eventually around the world. Illegitimate actions undermine both long- and short-term COIN efforts.

1-133. Every action by counterinsurgents leaves a "forensic trace" that may be required sometime later in a court of law. Counterinsurgents document all their activities to preserve, wherever possible, a chain of evidence. Accurate documentation can also be an important means to counter insurgent propaganda.

Counterinsurgents Should Prepare for a Long-Term Commitment

1-134. Insurgencies are protracted by nature. Thus, COIN operations always demand considerable expenditures of time and resources. The populace may prefer the HN government to the insurgents; however, people do not actively support a government unless they are convinced that the counterinsurgents have the means, ability, stamina, and will to win. The insurgents' primary battle is against the HN government, not the United States; however, U.S. support can be crucial to building public faith in that government's viability. The populace must have confidence in the staying power of both the counterinsurgents and the HN government. Insurgents and local populations often believe that a few casualties or a few years will cause the United States to abandon a COIN effort. Constant reaffirmations of commitment, backed by deeds, can overcome that perception and bolster faith in the steadfastness of U.S. support. But even the strongest U.S. commitment will not succeed if the populace does not perceive the HN government as having similar will and stamina. U.S. forces must help create that capacity and sustain that impression.

1-135. Preparing for a protracted COIN effort requires establishing headquarters and support structures designed for long-term operations. Planning and commitments should be based on sustainable operating tempo and personnel tempo limits for the various components of the force. (Operating tempo and personnel tempo are defined in the glossary.) Even in situations where the U.S. goal is reducing its military force levels as quickly as possible, some support for HN institutions usually remains for a long time.

1-136. At the strategic level, gaining and maintaining U.S. public support for a protracted deployment is critical. Only the most senior military officers are involved in this process at all. It is properly a political activity. However, military leaders typically take care to ensure that their actions and statements are forthright. They also ensure that the conduct of operations neither makes it harder for elected leaders to maintain public support nor undermines public confidence.

Contemporary Imperatives of Counterinsurgency

1-137. Recent COIN experiences have identified an important set of additional imperatives to keep in mind for success.

Manage Information and Expectations

1-138. Information and expectations are related; skillful counterinsurgents manage both. To limit discontent and build support, the HN government and any counterinsurgents assisting it create and maintain a realistic set of expectations among the populace, friendly military forces, and the international community. IO (including psychological operations and the related activities of public affairs and civil-military operations) are key tools to accomplish this. Achieving steady progress toward a set of reasonable expectations can increase the populace's tolerance for the inevitable inconveniences entailed by ongoing COIN operations. Where a large U.S. force is present to help establish a regime, such progress can extend the period before an army of liberation becomes perceived as an army of occupation.

1-139. U.S. forces start with a built-in challenge because of their reputation for accomplishment, what some call the "man on the moon syndrome." This refers to the expressed disbelief that a nation able to put a man on the moon cannot quickly restore basic services. U.S. agencies trying to fan enthusiasm for their efforts should avoid making unrealistic promises. In some cultures, failure to deliver promised results is automatically interpreted as deliberate deception, rather than good intentions gone awry. In other cultures, exorbitant promises are normal and people do not expect them to be kept. Effective counterinsurgents understand local norms; they use locally tailored approaches to control expectations. Managing expectations also involves demonstrating economic and political progress to show the populace how life is improving. Increasing the number of people who feel they have a stake in the success of the state and its government is a key to successful COIN operations. In the end, victory comes, in large measure, by convincing the populace that their life will be better under the HN government than under an insurgent regime.

1-140. Both counterinsurgents and the HN government ensure that their deeds match their words. They also understand that any action has an information reaction. Counterinsurgents and the HN government carefully consider that impact on the many audiences involved in the conflict and on the sidelines. They work actively to shape responses that further their ends. In particular, messages to different audiences must be consistent. In the global information environment, people in the AO can access the Internet and satellite television to determine the messages counterinsurgents are sending to the international community and the U.S. public. Any perceived inconsistency reduces credibility and undermines COIN efforts.

Use the Appropriate Level of Force

1-141. Any use of force generates a series of reactions. There may be times when an overwhelming effort is necessary to destroy or intimidate an opponent and reassure the populace. Extremist insurgent combatants often have to be killed. In any case, however, counterinsurgents should calculate carefully the type and amount of force to be applied and who wields it for any operation. An operation that kills five insurgents is counterproductive if collateral damage leads to the recruitment of fifty more insurgents.

1-142. In a COIN environment, it is vital for commanders to adopt appropriate and measured levels of force and apply that force precisely so that it accomplishes the mission without causing unnecessary loss of life or suffering. Normally, counterinsurgents can use escalation of force/force continuum procedures to minimize potential loss of life. These procedures are especially appropriate during convoy operations and at checkpoints and roadblocks. *Escalation of force* (Army)/*force continuum* (Marine Corps) refers to using lesser means of force when such use is likely to achieve the desired effects and Soldiers and Marines can do so without endangering themselves, others, or mission accomplishment. Escalation of force/force continuum procedures do not limit the right of self-defense, including the use of deadly force when such force is necessary to defend against a hostile act or demonstrated hostile intent. Commanders ensure that their Soldiers and Marines are properly trained in such procedures and, more importantly, in methods of shaping situations so that small-unit leaders have to make fewer split-second, life-or-death decisions.

1-143. Who wields force is also important. If the HN police have a reasonable reputation for competence and impartiality, it is better for them to execute urban raids; the populace is likely to view that application of force as more legitimate. This is true even if the police are not as well armed or as capable as military units. However, local circumstances affect this decision. If the police are seen as part of an ethnic or sectarian group oppressing the general population, their use may be counterproductive. Effective counterinsurgents thus understand the character of the local police and popular perceptions of both police and military units. This understanding helps ensure that the application of force is appropriate and reinforces the rule of law.

Learn and Adapt

1-144. An effective counterinsurgent force is a learning organization. Insurgents constantly shift between military and political phases and tactics. In addition, networked insurgents constantly exchange information about their enemy's vulnerabilities—even with insurgents in distant theaters. However, skillful counterinsurgents can adapt at least as fast as insurgents. Every unit needs to be able to make observations, draw and apply lessons, and assess results. Commanders must develop an effective system to circulate best practices throughout their command. Combatant commanders might also need to seek new laws or policies that authorize or resource necessary changes. Insurgents shift their AOs looking for weak links, so widespread competence is required throughout the counterinsurgent force.

Empower the Lowest Levels

1-145. *Mission command* is the conduct of military operations through decentralized execution based upon mission orders for effective mission accomplishment. Successful mission command results from subordinate leaders at all echelons exercising disciplined initiative within the commander's intent to accomplish missions. It requires an environment of trust and mutual understanding (FM 6-0). It is the Army's and Marine Corps' preferred method for commanding and controlling forces during all types of operations. Under mission command, commanders provide subordinates with a mission, their commander's intent, a concept of operations, and resources adequate to accomplish the mission. Higher commanders empower subordinates to make decisions within the commander's intent. They leave details of execution to their subordinates and expect them to use initiative and judgment to accomplish the mission.

1-146. Mission command is ideally suited to the mosaic nature of COIN operations. Local commanders have the best grasp of their situations. Under mission command, they are given access to or control of the resources needed to produce timely intelligence, conduct effective tactical operations, and manage IO and civil-military operations. Thus, effective COIN operations are decentralized, and higher commanders owe it to their subordinates to push as many capabilities as possible down to their level. Mission command encourages the initiative of subordinates and facilitates the learning that must occur at every level. It is a major characteristic of a COIN force that can adapt and react at least as quickly as the insurgents.

Support the Host Nation

1-147. U.S. forces committed to a COIN effort are there to assist a HN government. The long-term goal is to leave a government able to stand by itself. In the end, the host nation has to win on its own. Achieving this requires development of viable local leaders and institutions. U.S. forces and agencies can help, but HN elements must accept responsibilities to achieve real victory. While it may be easier for U.S. military units to conduct operations themselves, it is better to work to strengthen local forces and institutions and then assist them. HN governments have the final responsibility to solve their own problems. Eventually all foreign armies are seen as interlopers or occupiers; the sooner the main effort can transition to HN institutions, without unacceptable degradation, the better.

Paradoxes of Counterinsurgency Operations

1-148. The principles and imperatives discussed above reveal that COIN presents a complex and often unfamiliar set of missions and considerations. In many ways, the conduct of COIN is counterintuitive to the traditional U.S. view of war—although COIN operations have actually formed a substantial part of the U.S. military experience. Some representative paradoxes of COIN are presented here as examples of the different mindset required. These paradoxes are offered to stimulate thinking, not to limit it. The applicability of the thoughts behind the paradoxes depends on a sense of the local situation and, in particular, the state of the insurgency. For example, the admonition "Sometimes, the More Force Used, the Less Effective It Is" does not apply when the enemy is "coming over the barricades"; however, that thought is applicable when increased security is achieved in an area. In short, these paradoxes should not be reduced to a checklist; rather, they should be used with considerable thought.

Sometimes, the More You Protect Your Force, the Less Secure You May Be

1-149. Ultimate success in COIN is gained by protecting the populace, not the COIN force. If military forces remain in their compounds, they lose touch with the people, appear to be running scared, and cede the initiative to the insurgents. Aggressive saturation patrolling, ambushes, and listening post operations must be conducted, risk shared with the populace, and contact maintained. The effectiveness of establishing patrol bases and operational support bases should be weighed against the effectiveness of using larger unit bases. (FM 90-8 discusses saturation patrolling and operational support bases.) These practices ensure access to the intelligence needed to drive operations. Following them reinforces the connections with the populace that help establish real legitimacy.

Sometimes, the More Force Is Used, the Less Effective It Is

1-150. Any use of force produces many effects, not all of which can be foreseen. The more force applied, the greater the chance of collateral damage and mistakes. Using substantial force also increases the opportunity for insurgent propaganda to portray lethal military activities as brutal. In contrast, using force precisely and discriminately strengthens the rule of law that needs to be established. As noted above, the key for counterinsurgents is knowing when more force is needed—and when it might be counterproductive. This judgment involves constant assessment of the security situation and a sense of timing regarding insurgents' actions.

The More Successful the Counterinsurgency Is, the Less Force Can Be Used and the More Risk Must Be Accepted

1-151. This paradox is really a corollary to the previous one. As the level of insurgent violence drops, the requirements of international law and the expectations of the populace lead to a reduction in direct military actions by counterinsurgents. More reliance is placed on police work, rules of engagement may be tightened, and troops may have to exercise increased restraint. Soldiers and Marines may also have to accept more risk to maintain involvement with the people.

Sometimes Doing Nothing Is the Best Reaction

1-152. Often insurgents carry out a terrorist act or guerrilla raid with the primary purpose of enticing counterinsurgents to overreact, or at least to react in a way that insurgents can exploit—for example, opening fire on a crowd or executing a clearing operation that creates more enemies than it takes off the streets. If an assessment of the effects of a course of action determines that more negative than positive effects may result, an alternative should be considered—potentially including not acting.

Some of the Best Weapons for Counterinsurgents Do Not Shoot

1-153. Counterinsurgents often achieve the most meaningful success in garnering public support and legitimacy for the HN government with activities that do not involve killing insurgents (though, again, killing clearly will often be necessary). Arguably, the decisive battle is for the people's minds; hence synchronizing IO with efforts along the other LLOs is critical. Every action, including uses of force, must be "wrapped in a bodyguard of information." While security is essential to setting the stage for overall progress, lasting victory comes from a vibrant economy, political participation, and restored hope. Particularly after security has been achieved, dollars and ballots will have more important effects than bombs and bullets. This is a time when "money is ammunition." Depending on the state of the insurgency, therefore, Soldiers and Marines should prepare to execute many nonmilitary missions to support COIN efforts. Everyone has a role in nation building, not just Department of State and civil affairs personnel.

The Host Nation Doing Something Tolerably Is Normally Better than Us Doing It Well

1-154. It is just as important to consider who performs an operation as to assess how well it is done. Where the United States is supporting a host nation, long-term success requires establishing viable HN leaders and institutions that can carry on without significant U.S. support. The longer that process takes, the more U.S. public support will wane and the more the local populace will question the legitimacy of their own forces and government. General Creighton Abrams, the U.S. commander in Vietnam in 1971, recognized this fact when he said, "There's very clear evidence,...in some things, that we helped too much. And we *retarded* the Vietnamese by doing it.... *We* can't run this thing.... *They've* got to run it. The nearer we get to that the better off *they* are and the better off *we* are." T.E. Lawrence made a similar observation while leading the Arab Revolt against the Ottoman Empire in 1917: "Do not try to do too much with your own hands. Better the Arabs do it tolerably than that you do it perfectly. It is their war, and you are to help them, not to win it for them." However, a key word in Lawrence's advice is "tolerably." If the host nation cannot perform tolerably, counterinsurgents supporting it may have to act. Experience, knowledge of the AO, and cultural sensitivity are essential to deciding when such action is necessary.

If a Tactic Works this Week, It Might Not Work Next Week; If It Works in this Province, It Might Not Work in the Next

1-155. Competent insurgents are adaptive. They are often part of a widespread network that communicates constantly and instantly. Insurgents quickly adjust to successful COIN practices and rapidly disseminate information throughout the insurgency. Indeed, the more effective a COIN tactic is, the faster it may become out of date because insurgents have a greater need to counter it. Effective leaders at all levels avoid complacency and are at least as adaptive as their enemies. There is no "silver bullet" set of COIN procedures. Constantly developing new practices is essential.

Tactical Success Guarantees Nothing

1-156. As important as they are in achieving security, military actions by themselves cannot achieve success in COIN. Insurgents that never defeat counterinsurgents in combat still may achieve their strategic objectives. Tactical actions thus must be linked not only to strategic and operational military objectives but also to the host nation's essential political goals. Without those connections, lives and resources may be wasted for no real gain.

Many Important Decisions Are Not Made by Generals

1-157. Successful COIN operations require competence and judgment by Soldiers and Marines at all levels. Indeed, young leaders—so-called "strategic corporals"—often make decisions at the tactical level that have strategic consequences. Senior leaders set the proper direction and climate with thorough training and clear guidance; then they trust their subordinates to do the right thing. Preparation for tactical-level leaders requires more than just mastering Service doctrine; they must also be trained and educated to adapt to their local situations, understand the legal and ethical implications of their actions, and exercise initiative and sound judgment in accordance with their senior commanders' intent.

SUCCESSFUL AND UNSUCCESSFUL COUNTERINSURGENCY PRACTICES

1-158. Table 1-1 lists some practices that have contributed significantly to success or failure in past counterinsurgencies.

SUMMARY

1-159. COIN is an extremely complex form of warfare. At its core, COIN is a struggle for the population's support. The protection, welfare, and support of the people are vital to success. Gaining and maintaining that support is a formidable challenge. Achieving these aims requires synchronizing the efforts of many nonmilitary and HN agencies in a comprehensive approach.

1-160. Designing operations that achieve the desired end state requires counterinsurgents to understand the culture and the problems they face. Both insurgents and counterinsurgents are fighting for the support of the populace. However, insurgents are constrained by neither the law of war nor the bounds of human decency as Western nations understand them. In fact, some insurgents are willing to commit suicide and kill innocent civilians in carrying out their operations—and deem this a legitimate option. They also will do anything to preserve their greatest advantage, the ability to hide among the people. These amoral and often barbaric enemies survive by their wits, constantly adapting to the situation. Defeating them requires counterinsurgents to develop the ability to learn and adapt rapidly and continuously. This manual emphasizes this "Learn and Adapt" imperative as it discusses ways to gain and maintain the support of the people.

1-161. Popular support allows counterinsurgents to develop the intelligence necessary to identify and defeat insurgents. Designing and executing a comprehensive campaign to secure the populace and then gain its support requires carefully coordinating actions along several LLOs over time to produce success. One of these LLOs is developing HN security forces that can assume primary responsibility for combating the insurgency. COIN operations also place distinct burdens on leaders and logisticians. All of these aspects of COIN are described and analyzed in the chapters that follow.

Table 1-1. Successful and unsuccessful counterinsurgency operational practices

Successful practices	*Unsuccessful practices*
• Emphasize intelligence. • Focus on the population, its needs, and its security. • Establish and expand secure areas. • Isolate insurgents from the populace (population control). • Conduct effective, pervasive, and continuous information operations. • Provide amnesty and rehabilitation for those willing to support the new government. • Place host-nation police in the lead with military support as soon as the security situation permits. • Expand and diversify the host-nation police force. • Train military forces to conduct counterinsurgency operations. • Embed quality advisors and special forces with host-nation forces. • Deny sanctuary to insurgents. • Encourage strong political and military cooperation and information sharing. • Secure host-nation borders. • Protect key infrastructure.	• Overemphasize killing and capturing the enemy rather than securing and engaging the populace. • Conduct large-scale operations as the norm. • Concentrate military forces in large bases for protection. • Focus special forces primarily on raiding. • Place low priority on assigning quality advisors to host-nation forces. • Build and train host-nation security forces in the U.S. military's image. • Ignore peacetime government processes, including legal procedures. • Allow open borders, airspace, and coastlines.

This page intentionally left blank.

Chapter 2

Unity of Effort: Integrating Civilian and Military Activities

Essential though it is, the military action is secondary to the political one, its primary purpose being to afford the political power enough freedom to work safely with the population.

David Galula, *Counterinsurgency Warfare*, 1964[1]

This chapter begins with the principles involved in integrating the activities of military and civilian organizations during counterinsurgency operations. It then describes the categories of organizations usually involved. After that, it discusses assignment of responsibilities and mechanisms used to integrate civilian and military activities. It concludes by listing information commanders need to know about civilian agencies operating in their area of operations.

INTEGRATION

2-1. Military efforts are necessary and important to counterinsurgency (COIN) efforts, but they are only effective when integrated into a comprehensive strategy employing all instruments of national power. A successful COIN operation meets the contested population's needs to the extent needed to win popular support while protecting the population from the insurgents. Effective COIN operations ultimately eliminate insurgents or render them irrelevant. Success requires military forces engaged in COIN operations to—

- Know the roles and capabilities of U.S., intergovernmental, and host-nation (HN) partners.
- Include other participants, including HN partners, in planning at every level.
- Support civilian efforts, including those of nongovernmental organizations (NGOs) and intergovernmental organizations (IGOs).
- As necessary, conduct or participate in political, social, informational, and economic programs.

2-2. The integration of civilian and military efforts is crucial to successful COIN operations. All efforts focus on supporting the local populace and HN government. Political, social, and economic programs are usually more valuable than conventional military operations in addressing the root causes of conflict and undermining an insurgency. COIN participants come from many backgrounds. They may include military personnel, diplomats, police, politicians, humanitarian aid workers, contractors, and local leaders. All must make decisions and solve problems in a complex and extremely challenging environment.

2-3. Controlling the level of violence is a key aspect of the struggle. A high level of violence often benefits insurgents. The societal insecurity that violence brings discourages or precludes nonmilitary organizations, particularly external agencies, from helping the local populace. A more benign security environment allows civilian agencies greater opportunity to provide their resources and expertise. It thereby relieves military forces of this burden.

2-4. An essential COIN task for military forces is fighting insurgents; however, these forces can and should use their capabilities to meet the local populace's fundamental needs as well. Regaining the populace's active and continued support for the HN government is essential to deprive an insurgency of its power and appeal. The military forces' primary function in COIN is protecting that populace. However,

[1] © 1964 by Frederick A. Praeger, Inc. Reproduced with permission of Greenwood Publishing Group, Inc., Westport, CT.

employing military force is not the only way to provide civil security or defeat insurgents. Indeed, excessive use of military force can frequently undermine policy objectives at the expense of achieving the overarching political goals that define success. This dilemma places tremendous importance on the measured application of force.

2-5. Durable policy success requires balancing the measured use of force with an emphasis on nonmilitary programs. Political, social, and economic programs are most commonly and appropriately associated with civilian organizations and expertise; however, effective implementation of these programs is more important than who performs the tasks. If adequate civilian capacity is not available, military forces fill the gap. COIN programs for political, social, and economic well-being are essential to developing the local capacity that commands popular support when accurately perceived. COIN is also a battle of ideas. Insurgents seek to further their cause by creating misperceptions of COIN efforts. Comprehensive information programs are necessary to amplify the messages of positive deeds and to counter insurgent propaganda.

2-6. COIN is fought among the populace. Counterinsurgents take upon themselves responsibility for the people's well-being in all its manifestations. These include the following:

- Security from insurgent intimidation and coercion, as well as from nonpolitical violence and crime.
- Provision for basic economic needs.
- Provision of essential services, such as water, electricity, sanitation, and medical care.
- Sustainment of key social and cultural institutions.
- Other aspects that contribute to a society's basic quality of life.

Effective COIN programs address all aspects of the local populace's concerns in a unified fashion. Insurgents succeed by maintaining turbulence and highlighting local grievances the COIN effort fails to address. COIN forces succeed by eliminating turbulence and helping the host nation meet the populace's basic needs.

2-7. When the United States commits to helping a host nation defeat an insurgency, success requires applying the instruments of national power along multiple lines of operations. (Normally, these are logical lines of operations [LLOs], as described in chapter 5.) Since efforts along one LLO often affect progress in others, uncoordinated actions are frequently counterproductive.

2-8. LLOs in COIN focus primarily on the populace. Each line depends on the others. The interdependence of the lines is total: if one fails, the mission fails. Many LLOs require applying capabilities usually resident in civilian organizations, such as—

- U.S. government agencies other than the Department of Defense (DOD).
- Other nations' defense and nondefense agencies and ministries.
- IGOs, such as the United Nations and its subordinate organizations.
- NGOs.
- Private corporations.
- Other organizations that wield diplomatic, informational, and economic power.

These civilian organizations bring expertise that complements that of military forces. At the same time, civilian capabilities cannot be employed effectively without the security that military forces provide. Effective COIN leaders understand the interdependent relationship of all participants, military and civilian. COIN leaders orchestrate their efforts to achieve unity of effort and coherent results.

Unity of Command

2-9. Unity of command is the preferred doctrinal method for achieving unity of effort by military forces. Where possible, COIN leaders achieve unity of command by establishing and maintaining the formal command or support relationships discussed in FM 3-0. Unity of command should extend to all military forces supporting a host nation. The ultimate objective of these arrangements is for military forces, police, and other security forces to establish effective control while attaining a monopoly on the legitimate use of violence within the society. Command and control of all U.S. Government organizations engaged in a COIN mission should be exercised by a single leader through a formal command and control system. (FM 6-0, chapter 5, discusses a commander's command and control system.)

2-10. The relationships and authorities between military and nonmilitary U.S. Government agencies are usually given in the document directing an agency to support the operation. (The document is usually a memorandum of agreement or understanding.) Commanders exercise only the authority those documents allow; however, the terms in those documents may form the basis for establishing some form of relationship between commanders and agency chiefs.

2-11. As important as unity of command is to military operations, it is one of the most sensitive and difficult-to-resolve issues in COIN. The participation of U.S. and multinational military forces in COIN missions is inherently problematic, as it influences perceptions of the capacity and legitimacy of local security forces. Although unity of command of military forces may be desirable, it may be impractical due to political considerations. Political sensitivities about the perceived subordination of national forces to those of other states or IGOs often preclude strong command relationships; however, the agreements that establish a multinational force provide a source for determining possible authorities and command, support, or other relationships.

2-12. The differing goals and fundamental independence of NGOs and local organizations usually prevent formal relationships governed by command authority. In the absence of such relationships, military leaders seek to persuade and influence other participants to contribute to achieving COIN objectives. Informal or less authoritative relationships include coordination and liaison. In some cases, direct interaction among various organizations may be impractical or undesirable. Basic awareness and general information sharing may be the most that can be accomplished. When unity of command with part or all of the force, including nonmilitary elements, is not possible, commanders work to achieve unity of effort through cooperation and coordination among all elements of the force—even those not part of the same command structure.

UNITY OF EFFORT

2-13. Achieving unity of effort is the goal of command and support relationships. All organizations contributing to a COIN operation should strive, or be persuaded to strive, for maximum unity of effort. Informed, strong leadership forms the foundation for achieving it. Leadership in this area focuses on the central problems affecting the local populace. A clear understanding of the desired end state should infuse all efforts, regardless of the agencies or individuals charged with their execution. Given the primacy of political considerations, military forces often support civilian efforts. However, the mosaic nature of COIN operations means that lead responsibility shifts among military, civilian, and HN authorities. Regardless, military leaders should prepare to assume local leadership for COIN efforts. The organizing imperative is focusing on what needs to be done, not on who does it.

"Hand Shake Con" in Operation Provide Comfort

Operation Provide Comfort provided relief to the Kurdish refugees from northern Iraq and protection for humanitarian relief efforts. It began on 6 April 1991 and ended 24 July 1991. General Anthony C. Zinni, USMC, the multinational force commander, relates the following conversation regarding his command and control arrangements:

"[Regarding command and control relationships with other multinational contingents, t]he Chairman of the Joint Chiefs of Staff asked me… 'The lines in your command chart, the command relationships, what are they? OpCon [operational control]? TaCon [tactical control]? Command?' 'Sir, we don't ask, because no one can sign up to any of that stuff.' 'Well, how do you do business?' 'Hand Shake Con. That's it.' No memoranda of agreement. No memoranda of understanding.... [T]he relationships are worked out on the scene, and they aren't pretty. And you don't really want to try to capture them,...distill them, and say as you go off in the future, you're going to have this sort of command relationship.... [I]t is Hand Shake Con and that's the way it works. It is consultative. It is behind-the-scene."

2-14. Countering an insurgency begins with understanding the complex environment and the numerous competing forces within it. Gaining an understanding of the environment—including the insurgents, affected populace, and different counterinsurgent organizations—is essential to an integrated COIN operation. The complexity of identifying the insurgency's causes and integrating actions addressing them across multiple, interrelated LLOs requires understanding the civilian and military capabilities, activities, and end state. Various agencies acting to reestablish stability may differ in goals and approaches, based on their experience and institutional culture. When their actions are allowed to adversely affect each other, the populace suffers and insurgents identify grievances to exploit. Integrated actions are essential to defeat the ideologies professed by insurgents. A shared understanding of the operation's purpose provides a unifying theme for COIN efforts. Through a common understanding of that purpose, the COIN leadership can design an operation that promotes effective collaboration and coordination among all agencies and the affected populace.

COORDINATION AND LIAISON

2-15. Many organizations can contribute to successful COIN operations. An insurgency's complex diplomatic, informational, military, and economic context precludes military leaders from commanding all contributing organizations—and they should not try to do so. Interagency partners, NGOs, and private organizations have many interests and agendas that military forces cannot control. Additionally, local legitimacy is frequently affected by the degree to which local institutions are perceived as independent and capable without external support. Nevertheless, military leaders should make every effort to ensure that COIN actions are as well integrated as possible. Active leadership by military leaders is imperative to effect coordination, establish liaison (formal and informal), and share information. Influencing and persuading groups outside a commander's authority requires skill and often subtlety. As actively as commanders pursue unity of effort, they should also be mindful of their prominence and recognize the wisdom of acting indirectly and in ways that allow credit for success to go to others—particularly local individuals and organizations.

2-16. Many groups often play critical roles in influencing the outcome of a COIN effort but are beyond the control of military forces or civilian governing institutions. These groups include the following:

- Local leaders.
- Informal associations.
- Religious groups.
- Families.
- Tribes.
- Some private enterprises.
- Some humanitarian groups.
- The media.

Commanders remain aware of the influence of such groups and are prepared to work with, through, or around them.

KEY COUNTERINSURGENCY PARTICIPANTS AND THEIR LIKELY ROLES

2-17. Likely participants in COIN operations include the following:

- U.S. military forces.
- Multinational (including HN) forces.
- U.S. Government agencies.
- Other governments' agencies.
- NGOs.
- IGOs.
- Multinational corporations and contractors.
- HN civil and military authorities (including local leaders).

U.S. MILITARY FORCES

2-18. Military forces play an extensive role in COIN efforts. Demanding and complex, COIN draws heavily on a broad range of the joint force's capabilities and requires a different mix of offensive, defensive, and stability operations from that expected in major combat operations. Air, land, and maritime components all contribute to successful operations and to the vital effort to separate insurgents from the people they seek to control. The Army and Marine Corps usually furnish the principal U.S. military contributions to COIN forces. Special operations forces (SOF) are particularly valuable due to their specialized capabilities:

- Civil affairs.
- Psychological operations.
- Intelligence.
- Language skills.
- Region-specific knowledge.

SOF can provide light, agile, high-capability teams able to operate discreetly in local communities. SOF can also conduct complex counterterrorist operations.

2-19. U.S. military forces are vastly capable. Designed primarily for conventional warfare, they nonetheless have the capabilities essential to successfully conduct COIN operations. The most important military assets in COIN are disciplined Soldiers and Marines with adaptive, self-aware, and intelligent leaders. Military forces also have capabilities particularly relevant to common COIN requirements. These capabilities include the following:

- Dismounted infantry.
- Human intelligence.
- Language specialists.
- Military police.
- Civil affairs.
- Engineers.
- Medical units.
- Logistic support.
- Legal affairs.
- Contracting elements.

All are found in the Army; most are found in the Marine Corps. To a limited degree, they are also found in the Air Force and Navy.

2-20. U.S. forces help HN military, paramilitary, and police forces conduct COIN operations, including area security and local security operations. U.S. forces provide advice and help find, disperse, capture, and defeat insurgent forces. Concurrently, they emphasize training HN forces to perform essential defense functions. These are the central tasks of foreign internal defense, a core SOF task. The current and more extensive national security demands for such efforts require conventional forces of all Services be prepared to contribute to establishing and training local security forces.

2-21. Land combat forces, supported by air and maritime forces, conduct full spectrum operations to disrupt or destroy insurgent military capabilities. Land forces use offensive operations to disrupt insurgent efforts to establish base areas and consolidate their forces. They conduct defensive operations to provide area and local security. They conduct stability operations to thwart insurgent efforts to control or disrupt people's lives and routine activities. In all applications of combat power, commanders first ensure that likely costs do not outweigh or undermine other more important COIN efforts.

2-22. Most valuable to long-term success in winning the support of the populace are the contributions land forces make by conducting stability operations. *Stability operations* is an overarching term encompassing various military missions, tasks, and activities conducted outside the United States in coordination with other instruments of national power to maintain or reestablish a safe and secure environment, provide essential governmental services, emergency infrastructure reconstruction, and humanitarian relief (JP 1-02).

Forces engaged in stability operations establish, safeguard, or restore basic civil services. They act directly but also support government agencies. Success in stability operations enables the local populace and HN government agencies to resume or develop the capabilities needed to conduct COIN operations and create conditions that permit U.S. military forces to disengage.

2-23. Military forces also use their capabilities to enable the efforts of nonmilitary participants. Logistics, transportation, equipment, personnel, and other assets can support interagency partners and other civilian organizations.

MULTINATIONAL (INCLUDING HOST-NATION) MILITARY FORCES

2-24. The U.S. Government prefers that U.S. military forces operate with other nations' forces and not alone. Thus, Soldiers and Marines normally function as part of a multinational force. In COIN operations, U.S. forces usually operate with the security forces of the local populace or host nation. Each multinational participant provides capabilities and strengths that U.S forces may not have. Many other countries' military forces bring cultural backgrounds, historical experiences, and other capabilities that can be particularly valuable to COIN efforts.

2-25. However, nations join coalitions for various reasons. Although the missions of multinational partners may appear similar to those of the United States, rules of engagement, home-country policies, and sensitivities may differ among partners. U.S. military leaders require a strong cultural and political awareness of HN and other multinational military partners.

NONMILITARY COUNTERINSURGENCY PARTICIPANTS

2-26. Many nonmilitary organizations may support a host nation as it confronts an insurgency. Some of these organizations are discussed below. (JP 3-08, volume II, discusses nonmilitary organizations often associated with joint operations.)

U.S. Government Organizations

2-27. Commanders' situational awareness includes being familiar with other U.S. Government organizations participating in the COIN effort and aware of their capabilities. Planning includes determining which organizations are supporting the force or operating in their area of operations (AO). Commanders and leaders of other U.S. Government organizations should collaboratively plan and coordinate actions to avoid conflict or duplication of effort. Within the U.S. Government, key organizations include—

- Department of State.
- U.S. Agency for International Development (USAID).
- Central Intelligence Agency.
- Department of Justice.
- Drug Enforcement Administration (under Department of Justice).
- Department of the Treasury.
- Department of Homeland Security.
- Department of Energy.
- Department of Agriculture.
- Department of Commerce.
- Department of Transportation.
- U.S. Coast Guard (under Department of Homeland Security).
- Federal Bureau of Investigation (under Department of Justice).
- Immigration Customs Enforcement (under Department of Homeland Security).

Other Governments' Agencies

2-28. Agencies of other national governments (such as ministries of defense, foreign affairs, development, and justice) are likely to actively participate in COIN operations. The list of possible participants from other countries is too long to list. Leaders of U.S. contingents must work closely with their multinational counterparts to become familiar with agencies that may operate in their AO. To the degree possible, military leaders should use U.S. civilian representatives to establish appropriate relationships and awareness of their multinational counterparts.

Nongovernmental Organizations

2-29. Joint doctrine defines a *nongovernmental organization* as a private, self-governing, not-for-profit organization dedicated to alleviating human suffering; and/or promoting education, health care, economic development, environmental protection, human rights, and conflict resolution; and/or encouraging the establishment of democratic institutions and civil society (JP 1-02). There are several thousand NGOs of many different types. Their activities are governed by their organizing charters and their members' motivations. Some NGOs receive at least part of their funding from national governments or IGOs. Some may become implementing partners in accordance with grants or contracts. (For example, USAID provides some NGO funding.) In these cases, the funding organization often gains oversight and authority over how the funds are used.

2-30. Some NGOs maintain strict independence from governments and belligerents and do not want to be seen directly associating with military forces. Gaining the support of and coordinating operations with these NGOs can be difficult. Establishing basic awareness of these groups and their activities may be the most commanders can achieve. NGOs play important roles in resolving insurgencies, however. Many NGOs arrive before military forces and remain afterwards. They can support lasting stability. To the greatest extent possible, commanders try to complement and not override their capabilities. Building a complementary, trust-based relationship is vital.

2-31. Examples of NGOs include—

- International Committee of the Red Cross.
- World Vision.
- Médecins sans Frontières (Doctors Without Borders).
- Cooperative for Assistance and Relief Everywhere (CARE).
- Oxford Committee for Famine Relief (OXFAM).
- Save the Children.
- Mercy Corps.
- Academy for Educational Development.

Intergovernmental Organizations

2-32. Joint doctrine defines an *intergovernmental organization* as an organization created by a formal agreement (for example, a treaty) between two or more governments. It may be established on a global, regional, or functional basis for wide-ranging or narrowly defined purposes. IGOs are formed to protect and promote national interests shared by member states (JP 1-02). The most notable IGO is the United Nations (UN). Regional organizations like the Organization of American States and European Union may be involved in some COIN operations. The UN in particular has many subordinate and affiliated agencies active worldwide. Depending on the situation and HN needs, any number of UN organizations may be present, such as the following:

- Office of the Chief of Humanitarian Affairs.
- Department of Peacekeeping Operations.
- World Food Program.
- UN Refugee Agency (known as UNHCR, the acronym for its director, the UN High Commissioner for Refugees).
- UN High Commissioner for Human Rights.
- UN Development Program.

Multinational Corporations and Contractors

2-33. Multinational corporations often engage in reconstruction, economic development, and governance activities. At a minimum, commanders should know which companies are present in their AO and where those companies are conducting business. Such information can prevent fratricide and destruction of private property.

2-34. Recently, private contractors from firms providing military-related services have become more prominent in theaters of operations. This category includes armed contractors providing many different security services to the U.S. Government, NGOs, and private businesses. Many businesses market expertise in areas related to supporting governance, economics, education, and other aspects of civil society as well. Providing capabilities similar to some NGOs, these firms often obtain contracts through government agencies.

2-35. When contractors or other businesses are being paid to support U.S. military or other government agencies, the principle of unity of command should apply. Commanders should be able to influence contractors' performance through U.S. Government contract supervisors. When under contract to the United States, contractors should behave as an extension of the organizations or agencies for which they work. Commanders should identify contractors operating in their AO and determine the nature of their contract, existing accountability mechanisms, and appropriate coordination relationships.

Host-Nation Civil Authorities

2-36. Sovereignty issues are among the most difficult for commanders conducting COIN operations, both in regard to forces contributed by other nations and by the host nation. Often, commanders are required to lead through coordination, communication, and consensus, in addition to traditional command practices. Political sensitivities must be acknowledged. Commanders and subordinates often act as diplomats as well as warriors. Within military units, legal officers and their staffs are particularly valuable for clarifying legal arrangements with the host nation. To avoid adverse effects on operations, commanders should address all sovereignty issues through the chain of command to the U.S. Ambassador. As much as possible, sovereignty issues should be addressed before executing operations. Examples of key sovereignty issues include the following:

- Aerial ports of debarkation.
- Basing.
- Border crossings.
- Collecting and sharing information.
- Protection (tasks related to preserving the force).
- Jurisdiction over members of the U.S. and multinational forces.
- Location and access.
- Operations in the territorial waters, both sea and internal.
- Overflight rights.
- Police operations, including arrest, detention, penal, and justice authority and procedures.
- Railheads.
- Seaports of debarkation.

2-37. Commanders create coordinating mechanisms, such as committees or liaison elements, to facilitate cooperation and build trust with HN authorities. HN military or nonmilitary representatives should have leading roles in such mechanisms. These organizations facilitate operations by reducing sensitivities and misunderstandings while removing impediments. Sovereignty issues can be formally resolved with the host nation by developing appropriate technical agreements to augment existing or recently developed status of forces agreements. In many cases, security assistance organizations, NGOs, and IGOs have detailed local knowledge and reservoirs of good will that can help establish a positive, constructive relationship with the host nation.

2-38. Coordination and support should exist down to local levels (such as villages and neighborhoods). Soldiers and Marines should be aware of the political and societal structures in their AOs. Political structures usually have designated leaders responsible to the government and people. However, the societal structure may include informal leaders who operate outside the political structure. These leaders may be—

- Economic (such as businessmen).
- Theological (such as clerics and lay leaders).
- Informational (such as newspaper publishers or journalists).
- Family based (such as elders or patriarchs).

Some societal leaders may emerge due to charisma or other intangible influences. Commanders should identify the key leaders and the manner in which they are likely to influence COIN efforts.

KEY RESPONSIBILITIES IN COUNTERINSURGENCY

2-39. Participants best qualified and able to accomplish nonmilitary tasks are not always available. The realistic division of labor does not match the preferred division of labor. In those cases, military forces perform those tasks. Sometimes forces have the skills required; other times they learn them during execution.

PREFERRED DIVISION OF LABOR

2-40. In COIN it is always preferred for civilians to perform civilian tasks. Whenever possible, civilian agencies or individuals with the greatest applicable expertise should perform a task. Legitimate local authorities should receive special preference. There are many U.S. agencies and civilian IGOs with more expertise in meeting the fundamental needs of a population under assault than military forces have; however, the ability of such agencies to deploy to foreign countries in sustainable numbers and with ready access to necessary resources is usually limited. The violence level in the AO also affects civilian agencies' ability to operate. The more violent the environment, the more difficult it is for civilians to operate effectively. Hence, the preferred or ideal division of labor is frequently unattainable. The more violent the insurgency, the more unrealistic is this preferred division of labor.

REALISTIC DIVISION OF LABOR

2-41. By default, U.S. and multinational military forces often possess the only readily available capability to meet many of the local populace's fundamental needs. Human decency and the law of war require land forces to assist the populace in their AOs. Leaders at all levels prepare to address civilian needs. Commanders identify people in their units with regional and interagency expertise, civil-military competence, and other critical skills needed to support a local populace and HN government. Useful skill sets may include the following:

- Knowledge, cultural understanding, and appreciation of the host nation and region.
- Functional skills needed for interagency and HN coordination (for example, liaison, negotiation, and appropriate social or political relationships).
- Language skills needed for coordination with the host nation, NGOs, and multinational partners.
- Knowledge of basic civic functions such as governance, infrastructure, public works, economics, and emergency services.

2-42. U.S. Government agencies and IGOs rarely have the resources and capabilities needed to address all COIN tasks. Success requires adaptable leaders who prepare to perform required tasks with available resources. These leaders understand that long-term security cannot be imposed by military force alone; it requires an integrated, balanced application of effort by all participants with the goal of supporting the local populace and achieving legitimacy for the HN government. David Galula wisely notes, "To confine soldiers to purely military functions while urgent and vital tasks have to be done, and nobody else is available to undertake them, would be senseless. The soldier must then be prepared to become…a social worker, a civil engineer, a schoolteacher, a nurse, a boy scout. But only for as long as he cannot be replaced, for it is better to entrust civilian tasks to civilians." Galula's last sentence is important. Military forces can perform civilian tasks but often not as well as the civilian agencies with people trained in those skills. Further, military forces performing civilian tasks are not performing military tasks. Diverting them from those tasks should be a temporary measure, one taken to address urgent circumstances.

TRANSITIONS

2-43. Regardless of the division of labor, an important recurring feature of COIN is transitioning responsibility and participation in key LLOs. As consistently and conscientiously as possible, military leaders ensure continuity in meeting the needs of the HN government and local populace. The same general guidelines governing battle handovers apply to COIN transitions. Whether the transition is between military units or from a military unit to a civilian agency, all involved must clearly understand the tasks and responsibilities being passed. Maintaining unity of effort is particularly important during transitions, especially between organizations of different capabilities and capacities. Relationships tend to break down during transitions. A transition is not a single event where all activity happens at once. It is a rolling process of little handoffs between different actors along several streams of activities. There are usually multiple transitions for any one stream of activity over time. Using the coordination mechanisms discussed below can help create and sustain the links that support effective transitions without compromising unity of effort.

CIVILIAN AND MILITARY INTEGRATION MECHANISMS

2-44. Applying the principle of unity of effort is possible in many organizational forms. The first choice should be to identify existing coordination mechanisms and incorporate them into comprehensive COIN efforts. This includes existing U.S. Government, multinational, and HN mechanisms. Context is extremely important. Although many of these structures exist and are often employed in other types of missions (such as peacekeeping or humanitarian relief), there is an acute and fundamental difference in an insurgency environment. The nature of the conflict and its focus on the populace make civilian and military unity a critical enabling aspect of a COIN operation. The following discussion highlights some of the well-established, general mechanisms for civilian and military integration. Many civil-military organizations and mechanisms have been created for specific missions. Although the names and acronyms differ, in their general outlines they usually reflect the concepts discussed below.

2-45. The U.S. Government influences events worldwide by effectively employing the instruments of national power: diplomatic, informational, military, and economic. These instruments are coordinated by the appropriate executive branch officials, often with assistance from the National Security Council (NSC) staff.

2-46. The NSC is the President's principal forum for considering national security and foreign policy matters. It serves as the President's principal means for coordinating policy among various interagency organizations. At the strategic level, the NSC directs the creation of the interagency political-military plan for COIN. The NSC staff, guided by the deputies and principals, assists in integrating interagency processes to develop the plan for NSC approval. (See JP 1.)

JOINT INTERAGENCY COORDINATION GROUP

2-47. *Interagency coordination,* within the context of Department of Defense involvement, is the coordination that occurs between elements of Department of Defense and engaged U.S. Government agencies for the purpose of achieving an objective (JP 1-02). Joint interagency coordination groups (JIACGs) help combatant commanders conduct COIN operations by providing interagency support of plans, operations, contingencies, and initiatives. The goal of a JIACG is to provide timely, usable information and advice from an interagency perspective to the combatant commander by information sharing, integration, synchronization, training, and exercises. JIACGs may include representatives from other federal departments and agencies and state and local authorities, as well as liaison officers from other commands and DOD components. The interagency representatives and liaison officers are the subject matter experts for their respective agencies and commands. They provide the critical bridge between the combatant commander and interagency organizations. (See JP 3-08, volume I.)

COUNTRY TEAM

2-48. At the HN level, the U.S. country team is the primary interagency coordinating structure for COIN. (See figure 2-1.) The country team is the senior in-country coordinating and supervising body, headed by the U.S. chief of mission, usually the Ambassador. It is composed of the senior member of each represented department or agency. In a foreign country, the chief of mission is the highest U.S. civil authority.

The Foreign Service Act assigns the chief of mission to a foreign country responsibility for the direction, coordination, and supervision of all government executive branch employees in that country except for service members and employees under the command of a U.S. area military commander. As the senior U.S. Government official permanently assigned in the host nation, the chief of mission is responsible to the President for policy oversight of all United States government programs. The chief of mission leads the country team and is responsible for integrating U.S. efforts in support of the host nation. As permanently established interagency organizations, country teams represent a priceless COIN resource. They often provide deep reservoirs of local knowledge and interaction with the HN government and population.

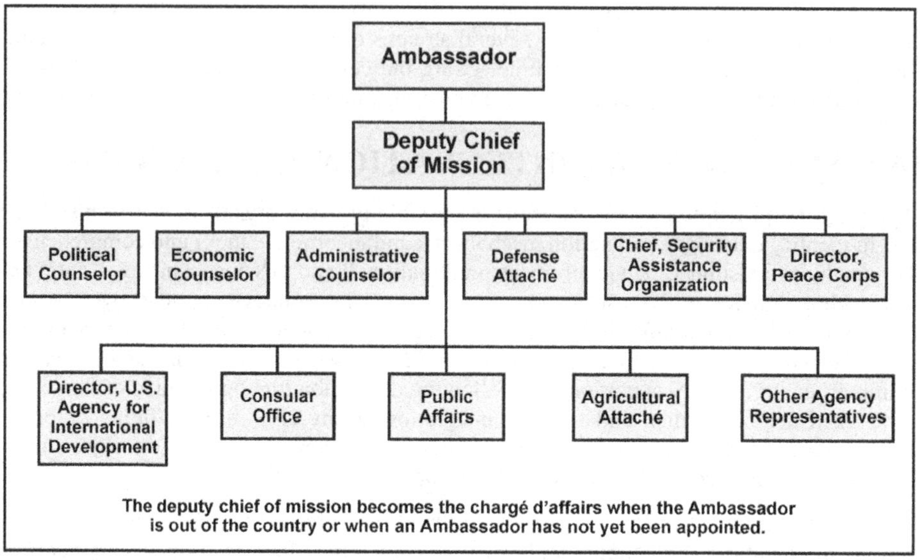

Figure 2-1. Sample country team

2-49. The more extensive the U.S. participation is in a COIN and the more dispersed U.S. forces are throughout a country, the greater the need for additional mechanisms to extend civilian oversight and assistance. However, given the limited resources of the Department of State and the other U.S. Government agencies, military forces often represent the country team in decentralized and diffuse operational environments. Operating with a clear understanding of the guiding political aims, members of the military at all levels must be prepared to exercise judgment and act without the benefit of immediate civilian oversight and control. At each subordinate political level of the HN government, military and civilian leaders should establish a coordinating structure, such as an area coordination center or civil-military operations center (CMOC), that includes representatives of the HN government and security forces, as well as U.S. and multinational forces and agencies. CMOCs facilitate the integration of military and political actions. Below the national level, additional structures where military commanders and civilian leaders can meet directly with local leaders to discuss issues may be established. Where possible, IGOs and NGOs should be encouraged to participate in coordination meetings to ensure their actions are integrated with military and HN plans. (See JP 3-07.1 for additional information about COIN planning and coordination organizations.)

2-50. In practice, the makeup of country teams varies widely, depending on the U.S. departments and agencies represented in country, the desires of the Ambassador, and the HN situation. During COIN, country team members meet regularly to coordinate U.S. Government diplomatic, informational, military, and economic activities in the host nation to ensure unity of effort. The interagency representatives usually include at least the following:

- Departments of State, Defense, Justice, and Treasury.
- USAID.
- Central Intelligence Agency.
- Drug Enforcement Administration.

Participation of other U.S. Government organizations depends on the situation.

2-51. In almost all bilateral missions, DOD is represented on the country team by the U.S. defense attaché's office or the security assistance organization. They are key military sources of information for interagency coordination in foreign countries. (Security assistance organizations are called by various names, such as the office of defense cooperation, security assistance office, or military group. The choice is largely governed by HN preference.)

Provincial Reconstruction Teams in Afghanistan

A model for civil-military cooperation is the provincial reconstruction teams (PRTs) first fielded in 2003 in Afghanistan. PRTs were conceived as a means to extend the reach and enhance the legitimacy of the central government into the provinces of Afghanistan at a time when most assistance was limited to the nation's capital. Though PRTs were staffed by a number of coalition and NATO allied countries, they generally consisted of 50 to 300 troops as well as representatives from multinational development and diplomatic agencies. Within U.S. PRTs, USAID and Department of State leaders and the PRT commander formed a senior team that coordinated the policies, strategies, and activities of each agency towards a common goal. In secure areas, PRTs maintained a low profile. In areas where coalition combat operations were underway, PRTs worked closely with maneuver units and local government entities to ensure that shaping operations achieved their desired effects. Each PRT leadership team received tremendous latitude to determine its own strategy. However, each PRT used its significant funding and diverse expertise to pursue activities that fell into one of three general logical lines of operations: pursue security sector reform, build local governance, or execute reconstruction and development.

2-52. The country team determines how the United States can effectively apply interagency capabilities to assist a HN government in creating a complementary institutional capacity to deal with an insurgency. Efforts to support local officials and build HN capacity must be integrated with information operations so HN citizens are aware of their government's efforts. In addition, interagency capabilities must be applied at the tactical level to give commanders access to options such capabilities make available. The Civil Operations and Revolutionary—later Rural—Development Support approach developed during the Vietnam War provides a positive example of integrated civilian and military structures that reached every level of the COIN effort.

CORDS and Accelerated Pacification in Vietnam

During the Vietnam War, one of the most valuable and successful elements of COIN was the Civil Operations and Revolutionary—later Rural—Development Support (CORDS) program. CORDS was created in 1967 to integrate U.S. civilian and military support of the South Vietnamese government and people. CORDS achieved considerable success in supporting and protecting the South Vietnamese population and in undermining the communist insurgents' influence and appeal, particularly after implementation of accelerated pacification in 1968.

Pacification was the process by which the government asserted its influence and control in an area beset by insurgents. It included local security efforts, programs to distribute food and medical supplies, and lasting reforms (like land redistribution). In 1965, U.S. civilian contributions to pacification consisted of several civilian agencies (among them, the Central Intelligence Agency, Agency for International Development, U.S. Information Service, and Department of State). Each developed its own programs. Coordination was uneven. The U.S. military contribution to pacification consisted of thousands of advisors. By early 1966, there were military advisory teams in all of South Vietnam's 44 provinces and most of its 243 districts. But there were two separate chains of command for military and civilian pacification efforts, making it particularly difficult for the civilian-run pacification program to function.

In 1967, President Lyndon B. Johnson established CORDS within the Military Assistance Command, Vietnam (MACV), which was commanded by General William Westmoreland, USA. The purpose of CORDS was to establish closer integration of civilian and military efforts. Robert Komer was appointed to run the program, with a three-star-equivalent rank. Civilians, including an assistant chief of staff for CORDS, were integrated into military staffs at all levels. This placed civilians in charge of military personnel and resources. Komer was energetic, strong-willed, and persistent in getting the program started. Nicknamed "Blowtorch Bob" for his aggressive style, Komer was modestly successful in leading improvements in pacification before the 1968 Tet offensive.

In mid-1968, the new MACV commander, General Creighton Abrams, USA, and his new civilian deputy, William Colby, used CORDS as the implementing mechanism for an accelerated pacification program that became the priority effort for the United States. Significant allocations of personnel helped make CORDS effective. In this, the military's involvement was key. In September 1969—the high point of the pacification effort in terms of total manpower—there were 7,601 advisors assigned to province and district pacification teams. Of these 6,464 were military.

The effectiveness of CORDS was a function of integrated civilian and military teams at every level of society in Vietnam. From district to province to national level, U.S. advisors and U.S. interagency partners worked closely with their Vietnamese counterparts. The entire effort was well established under the direction of the country team, led by Ambassador Ellsworth Bunker. General Abrams and his civilian deputy were clear in their focus on pacification as the priority and ensured that military and civilian agencies worked closely together. Keen attention was given to the ultimate objective of serving the needs of the local populace. Success in meeting basic needs of the populace led, in turn, to improved intelligence that facilitated an assault on the Viet Cong political infrastructure. By early 1970, statistics indicated that 93 percent of South Vietnamese lived in "relatively secure" villages, an increase of almost 20 percent from the middle of 1968. By 1972, pacification had largely uprooted the insurgency from among the South Vietnamese population and forced the communists to rely more heavily on infiltrating conventional forces from North Vietnam and employing them in irregular and conventional operations.

In 1972, South Vietnamese forces operating with significant support from U.S. airpower defeated large-scale North Vietnamese conventional attacks. Unfortunately, a North Vietnamese conventional assault succeeded in 1975 after the withdrawal of U.S. forces, ending of U.S. air support, and curtailment of U.S. funding to South Vietnam.

Pacification, once it was integrated under CORDS, was generally led, planned, and executed well. CORDS was a successful synthesis of military and civilian efforts. It is a useful model to consider for other COIN operations.

Civil-Military Operations Center

2-53. Another mechanism for bringing elements together for coordination is the CMOC. CMOCs can be established at all levels of command. CMOCs coordinate the interaction of U.S. and multinational military forces with a wide variety of civilian agencies. A CMOC is not designed, nor should it be used as, a command and control element. However, it is useful for transmitting the commander's guidance to other agencies, exchanging information, and facilitating complementary efforts.

2-54. Overall management of a CMOC may be assigned to a multinational force commander, shared by a U.S. and a multinational commander, or shared by a U.S. commander and a civilian agency head. A CMOC can be used to build on-site, interagency coordination to achieve unity of effort. There is no established CMOC structure; its size and composition depend on the situation. However, CMOCs are organic to Army civil affairs organizations, from civil affairs command to company. Senior civil affairs officers normally serve as the CMOC director and deputy director. Other military participants usually include civil affairs, legal, operations, logistic, engineering, and medical representatives of the supported headquarters. Civilian members of a CMOC may include representatives of the following:

- U.S. Government organizations.
- Multinational partners.
- IGOs.
- HN or other local organizations.
- NGOs.

(For more information on CMOCs, see FM 3-05.401/CRP 3-33.1A.)

TACTICAL-LEVEL INTERAGENCY CONSIDERATIONS

2-55. Tactical units may find interagency expertise pushed to their level when they are responsible for large AOs in a COIN environment. Tactical units down to company level must be prepared to integrate their efforts with civilian organizations.

2-56. To ensure integration of interagency capabilities, units should coordinate with all interagency representatives and organizations that enter their AO. Despite the best efforts to coordinate, the fog and friction inherent in COIN will often lead to civilian organizations entering an AO without prior coordination. (Table 2-1 is a suggested list for coordinating with interagency and other nonmilitary organizations.)

Table 2-1. Example interagency coordination checklist

• Identify organizational structures and leadership. • Identify key objectives, responsibilities, capabilities, and programs. • Develop common courses of action or options for inclusion in planning, movement coordination, and security briefings. • Determine how to ensure coordination and communications before and during the execution of the organization's activities in the unit's area of operations. • Develop relationships that enable the greatest possible integration. • Assign liaison officers to the most important civilian organizations. • Define problems in clear and unambiguous terms. • Determine the intended duration of operations. • Determine the location of bases of operations. • Determine the number, names, and descriptions of personnel. • Determine the type, color, number, and license numbers of civilian vehicles. • Identify other agency resources in the area of operations. • Identify local groups and the agencies with whom they are working. • Establish terms of reference or operating procedures, especially in the event of incidents that result in casualties. • Identify funding for interagency projects.

SUMMARY

2-57. President John F. Kennedy noted, "You [military professionals] must know something about strategy and tactics and…logistics, but also economics and politics and diplomacy and history. You must know everything you can know about military power, and you must also understand the limits of military power. You must understand that few of the important problems of our time have…been finally solved by military power alone." Nowhere is this insight more relevant than in COIN. Successful COIN efforts require unity of effort in bringing all instruments of national power to bear. Civilian agencies can contribute directly to military operations, particularly by providing information. That theme is developed further in the next chapter.

Chapter 3
Intelligence in Counterinsurgency

Everything good *that happens seems to come from good intelligence.*
General Creighton W. Abrams Jr., USA, 1970

Effective, accurate, and timely intelligence is essential to the conduct of any form of warfare. This maxim applies especially to counterinsurgency operations; the ultimate success or failure of the mission depends on the effectiveness of the intelligence effort. This chapter builds upon previous concepts to further describe insurgencies, requirements for intelligence preparation of the battlefield and predeployment planning, collection and analysis of intelligence in counterinsurgency, intelligence fusion, and general methodology for integrating intelligence with operations. This chapter does not supersede processes in U.S. military doctrine (see FM 2-0, FM 34-130/FMFRP 3-23-2, and FMI 2-91.4) but instead provides specific guidance for counterinsurgency.

SECTION I – INTELLIGENCE CHARACTERISTICS IN COUNTERINSURGENCY

3-1. Counterinsurgency (COIN) is an intelligence-driven endeavor. The function of intelligence in COIN is to facilitate understanding of the operational environment, with emphasis on the populace, host nation, and insurgents. Commanders require accurate intelligence about these three areas to best address the issues driving the insurgency. Both insurgents and counterinsurgents require an effective intelligence capability to be successful. Both attempt to create and maintain intelligence networks while trying to neutralize their opponent's intelligence capabilities.

3-2. Intelligence in COIN is about people. U.S. forces must understand the people of the host nation, the insurgents, and the host-nation (HN) government. Commanders and planners require insight into cultures, perceptions, values, beliefs, interests and decision-making processes of individuals and groups. These requirements are the basis for collection and analytical efforts.

3-3. Intelligence and operations feed each other. Effective intelligence drives effective operations. Effective operations produce information, which generates more intelligence. Similarly, ineffective or inaccurate intelligence produces ineffective operations, which produce the opposite results.

3-4. All operations have an intelligence component. All Soldiers and Marines collect information whenever they interact with the populace. Operations should therefore always include intelligence collection requirements.

3-5. Insurgencies are local. They vary greatly in time and space. The insurgency one battalion faces will often be different from that faced by an adjacent battalion. The mosaic nature of insurgencies, coupled with the fact that all Soldiers and Marines are potential intelligence collectors, means that all echelons both produce and consume intelligence. This situation results in a bottom-up flow of intelligence. This pattern also means that tactical units at brigade and below require a lot of support for both collection and analysis, as their organic intelligence structure is often inadequate.

3-6. COIN occurs in a joint, interagency, and multinational environment at all echelons. Commanders and staffs must coordinate intelligence collection and analysis with foreign militaries, foreign and U.S. intelligence services, and other organizations.

SECTION II – PREDEPLOYMENT PLANNING AND INTELLIGENCE PREPARATION OF THE BATTLEFIELD

3-7. *Intelligence preparation of the battlefield* is the systematic, continuous process of analyzing the threat and environment in a specific geographic area. Intelligence preparation of the battlefield (IPB) is designed to support the staff estimate and military decision-making process. Most intelligence requirements are generated as a result of the IPB process and its interrelation with the decision-making process (FM 34-130). Planning for deployment begins with a thorough mission analysis, including IPB. IPB is accomplished in four steps:

- Define the operational environment.
- Describe the effects of the operational environment.
- Evaluate the threat.
- Determine threat courses of action.

3-8. The purpose of planning and IPB before deployment is to develop an understanding of the operational environment. This understanding drives planning and predeployment training. Predeployment intelligence must be as detailed as possible. It should focus on the host nation, its people, and insurgents in the area of operations (AO). Commanders and staffs use predeployment intelligence to establish a plan for addressing the underlying causes of the insurgency and to prepare their units to interact with the populace appropriately. The goal of planning and preparation is for commanders and their subordinates not to be surprised by what they encounter in theater.

3-9. IPB in COIN operations follows the methodology described in FM 34-130/FMFRP 3-23-2. However, it places greater emphasis on civil considerations, especially people and leaders in the AO, than does IPB for conventional operations. IPB is continuous and its products are revised throughout the mission. Nonetheless, predeployment products are of particular importance for the reasons explained above. Whenever possible, planning and preparation for deployment includes a thorough and detailed IPB. IPB in COIN requires personnel to work in areas like economics, anthropology, and governance that may be outside their expertise. Therefore, integrating staffs and drawing on the knowledge of nonintelligence personnel and external experts with local and regional knowledge are critical to effective preparation.

3-10. Deployed units are the best sources of intelligence. Deploying units should make an effort to reach forward to deployed units. The Secret Internet Protocol Router Network (SIPRNET) allows deploying units to immerse themselves virtually in the situation in theater. Government agencies, such as the Department of State, U.S. Agency for International Development, and intelligence agencies, can often provide country studies and other background information as well.

3-11. *Open-source intelligence* is information of potential intelligence value that is available to the general public (JP 1-02). It is important to predeployment IPB. In many cases, background information on the populations, cultures, languages, history, and governments of states in an AO is in open sources. Open sources include books, magazines, encyclopedias, Web sites, tourist maps, and atlases. Academic sources, such as journal articles and university professors, can also be of great benefit.

DEFINE THE OPERATIONAL ENVIRONMENT

3-12. The *operational environment* is a composite of the conditions, circumstances, and influences that affect the employment of capabilities and bear on the decisions of the commander. (JP 1-02). At the tactical and operational levels, defining the operational environment involves defining a unit's AO and determining an area of interest. The *area of interest* is area of concern to the commander, including the area of influence, areas adjacent thereto, and extending into enemy territory to the objectives of current or planned operations. This area also includes areas occupied by enemy forces who could jeopardize the accomplishment of the mission (JP 1-02).

3-13. AOs may be relatively static, but people and information flow through AOs continuously. Therefore, when defining an area of interest, commanders take into account physical geography and civil considerations, particularly human factors. AOs often cut across physical lines of communications, such as roads, as well as areas that are tribally, economically, or culturally defined. For instance, tribal and family groups in Iraq and Afghanistan cross national borders into neighboring countries. The cross-border ties allow insurgents safe haven outside of their country and aid them in cross-border smuggling. The area of interest can be large relative to the AO; it must often account for various influences that affect the AO, such as—

- Family, tribal, ethnic, religious, or other links that go beyond the AO.
- Communication links to other regions.
- Economic links to other regions.
- Media influence on the local populace, U.S. public, and multinational partners.
- External financial, moral, and logistic support for the enemy.

3-14. At the combatant command level, the area of interest may be global, if, for example, enemy forces have an international financial network or are able to affect popular support within the U.S. or multinational partners. At the tactical level, commands must be aware of activities in neighboring regions and population centers that affect the population in their AO.

3-15. As explained in chapter 2, another consideration for predeployment planning and defining the operational environment is understanding the many military and nonmilitary organizations involved in the COIN effort. Intelligence planners determine the non-Department of Defense (DOD) agencies, multinational forces, nongovernmental organizations, and HN organizations in the AO. Knowledge of these organizations is needed to establish working relationships and procedures for sharing information. These relationships and procedures are critical to developing a comprehensive common operational picture and enabling unity of effort.

DESCRIBE THE EFFECTS OF THE OPERATIONAL ENVIRONMENT

3-16. This IPB step involves developing an understanding of the operational environment and is critical to the success of operations. It includes—

- Civil considerations, with emphasis on the people, history, and HN government in the AO.
- Terrain analysis (physical geography), with emphasis on the following:
  - Complex terrain.
  - Suburban and urban terrain.
  - Key infrastructure.
  - Lines of communications.
- Weather analysis, focusing on how weather affects the populace's activities, such as agriculture, smuggling, or insurgent actions.

3-17. Including all staff members in this step improves the knowledge base used to develop an understanding of the AO. For instance, civil affairs personnel receive training in analysis of populations, cultures, and economic development. These Soldiers and Marines can contribute greatly to understanding civil considerations. As another example, foreign area officers have linguistic, historical, and cultural knowledge about particular regions and have often lived there for extended periods.

3-18. The products that result from describing the effects of the operational environment influence operations at all echelons. The description informs political activities and economic policies of combatant commanders. It drives information operations (IO) and civil-military operations planning. The knowledge gained affects the way Soldiers and Marines interact with the populace.

3-19. *Civil considerations* are how the manmade infrastructure, civilian institutions, and attitudes and activities of the civilian leaders, populations, and organizations within an area of operations influence the conduct of military operations (FM 6-0). Civil considerations form one of the six categories into which relevant information is grouped for military operations. (The glossary lists the other five categories under the entry for METT-TC.) Civil considerations comprise six characteristics, expressed in the memory aid ASCOPE: areas, structures, capabilities, organizations, people, and events. (Paragraphs B-10 through B-27 discuss these characteristics.) While all characteristics of civil considerations are important, understanding the people is particularly important in COIN. In order to evaluate the people, the following six socio-cultural factors should be analyzed:

- Society.
- Social structure.
- Culture.
- Language.
- Power and authority.
- Interests.

This analysis may also identify information related to areas, structures, organizations, and events.

SOCIETY

3-20. A *society* can be defined as a population whose members are subject to the same political authority, occupy a common territory, have a common culture, and share a sense of identity. A society is not easily created or destroyed, but it is possible to do so through genocide or war.

3-21. No society is homogeneous. A society usually has a dominant culture, but can also have a vast number of secondary cultures. Different societies may share similar cultures, such as Canada and the United States do. Societies are not static, but change over time.

3-22. Understanding the societies in the AO allows counterinsurgents to achieve objectives and gain support. Commanders also consider societies outside the AO whose actions, opinions, or political influence can affect the mission.

SOCIAL STRUCTURE

3-23. Each society is composed of both social structure and culture. Social structure refers to the relations among groups of persons within a system of groups. Social structure is persistent over time. It is regular and continuous despite disturbances. The relationships among the parts hold steady, even as groups expand or contract. In a military organization, for example, the structure consists of the arrangement into groups like divisions, battalions, and companies. In a society, the social structure includes groups, institutions, organizations, and networks. Social structure involves the following:

- Arrangement of the parts that constitute society.
- Organization of social positions.
- Distribution of people within those positions.

Groups

3-24. A *group* is two or more people regularly interacting on the basis of shared expectations of others' behavior and who have interrelated statuses and roles. A social structure includes a variety of groups. These groups may be racial, ethnic, religious, or tribal. There may also be other kinship-based groups.

3-25. A *race* is a human group that defines itself or is defined by other groups as different by virtue of innate physical characteristics. Biologically, there is no such thing as race among human beings; race is a social category.

3-26. An *ethnic group* is a human community whose learned cultural practices, language, history, ancestry, or religion distinguish them from others. Members of ethnic groups see themselves as different from other groups in a society and are recognized as such by others. Religious groups may be subsets of larger ethnic groups. An ethnic group may contain members of different religions. For example, some Kurds are Muslim, while others are Christian. Other ethnic groups may be associated with a particular religion—such as Sri Lankan Sinhalese, who are almost exclusively Buddhist. Other religious groups have members of many different ethnicities. A prominent example of such a group is the Roman Catholic faith.

3-27. *Tribes* are generally defined as autonomous, genealogically structured groups in which the rights of individuals are largely determined by their ancestry and membership in a particular lineage. Tribes are essentially adaptive social networks organized by extended kinship and descent with common needs for physical and economic security.

3-28. Understanding the composition of groups in the AO is vital for effective COIN operations. This is especially true when insurgents organize around racial, ethnic, religious, or tribal identities. Furthermore, tensions or hostilities between groups may destabilize a society and provide opportunities for insurgents. Commanders should thus identify powerful groups both inside and outside their AO and obtain the following information about them:

- Formal relationships (such as treaties or alliances) between groups.
- Informal relationships (such as tolerance or friction) between groups.
- Divisions and cleavages between groups.
- Cross-cutting ties (for example, religious alignments that cut across ethnic differences) between groups.

In some cases, insurgent leaders and their followers may belong to separate groups. In others, the bulk of the population may differ from the insurgents. These characteristics may suggest courses of action aimed at reinforcing or widening seams among insurgents or between insurgents and the population.

Networks

3-29. *Networks* may be an important aspect of a social structure as well as within the insurgent organization (See paragraphs 1-94 and B-29 through B-56). Common types of networks include elite networks, prison networks, worldwide ethnic and religious communities, and neighborhood networks. Networks can have many purposes: economic, criminal, and emotional. Effective social network analysis, discussed below, considers the structure of a network and the nature of interactions between network actors.

Institutions

3-30. Groups engaged in patterned activity to complete a common task are called *institutions*. Educational institutions bring together groups and individuals whose statuses and roles concern teaching and learning. Military institutions bring together groups and individuals whose statuses and roles concern defense and security. Institutions, the basic building blocks of societies, are continuous through many generations. They continue to exist, even when the individuals who compose them are replaced.

Organizations

3-31. *Organizations*, both formal and informal, are institutions with the following characteristics:

- Bounded membership.
- Defined goals.
- Established operations.
- Fixed facilities or meeting places.
- Means of financial or logistic support.

3-32. Planners can generally group organizations into the following categories:

- **Communicating** organizations have the power to influence a population's perceptions.
- **Religious** organizations regulate norms, restrain or empower activities, reaffirm worldviews, and provide social support. A religious organization differs from a religious group. A religious group is a general category, such as Christian; a religious organization is a specific community, such as the Episcopal Church.
- **Economic** organizations provide employment, help regulate and stabilize monetary flow, assist in development, and create social networks.
- **Social** organizations provide support to the population, create social networks, and can influence ideologies. Examples include schools, civil society groups, and sports teams.

Organizations may belong to more than one category. For instance, an influential religious organization may also be a communicating organization.

3-33. Organizations may control, direct, restrain, or regulate the local populace. Thus, commanders should identify influential organizations both inside and outside of their AO. Commanders need to know which members of what groups belong to each organization and how their activities may affect military operations. The next step is to determine how these organizations affect the local populace, whose interests they fulfill, and what role they play in influencing local perceptions.

Roles and Statuses

3-34. Understanding a society requires identifying the most common roles, statuses, and institutions within the society. Individuals in a society interact as members with social positions. These positions are referred to as statuses. Most societies associate particular statuses with particular social groups, such as family, lineage, ethnicity, or religion. Statuses may be achieved by meeting certain criteria or may be ascribed by birth. Statuses are often reciprocal, such as that of husband and wife or teacher and student. Every status carries a cluster of expected behaviors known as a *role*, which includes how a person of that status is expected to think, feel, and act. A status may also include expectations about how others should treat a person of that status. Thus, in American society parents (status) have the obligation to care for their children (role) and the right to discipline them (role).

Social Norms

3-35. Violation of a role prescribed by a given status, such as failing to feed one's children, results in social disapproval. The standard of conduct for social roles is known as a *social norm*. A social norm is what people are expected to do or should do, rather than what people actually do. Norms may be either moral (incest prohibition, homicide prohibition) or customary (prayer before a meal, removing shoes before entering a house). When a person's behavior does not conform to social norms, the person may be sanctioned. Understanding the roles, statuses, and social norms of groups within an AO can clarify expected behavior and provide guidelines on how to act. Some norms that may impact military operations include the following:

- The requirement for revenge if honor is lost.
- Appropriate treatment of women and children.
- Common courtesies, such as gift giving.
- Local business practices, such as bribes and haggling.

CULTURE

3-36. Once the social structure has been thoroughly mapped out, staffs should identify and analyze the culture of the society as a whole and of each major group within the society. Social structure comprises the relationships among groups, institutions, and individuals within a society; in contrast, culture (ideas, norms, rituals, codes of behavior) provides meaning to individuals within the society. For example, families are a core institutional building block of social structure found everywhere. However, marital monogamy, expectations of a certain number of children, and willingness to live with in-laws are highly variable in different societies. They are matters of culture. Social structure can be thought of as a skeleton, with culture being the muscle on the bones. The two are mutually dependent and reinforcing. A change in one results in a change in the other.

3-37. Culture is "web of meaning" shared by members of a particular society or group within a society. (See FM 3-05.301/MCRP 3-40.6A.) Culture is—

- A system of shared beliefs, values, customs, behaviors, and artifacts that members of a society use to cope with their world and with one another.
- Learned, though a process called enculturation.
- Shared by members of a society; there is no "culture of one."
- Patterned, meaning that people in a society live and think in ways forming definite, repeating patterns.
- Changeable, through social interactions between people and groups.

- Arbitrary, meaning that Soldiers and Marines should make no assumptions regarding what a society considers right and wrong, good and bad.
- Internalized, in the sense that it is habitual, taken for granted, and perceived as "natural" by people within the society.

3-38. Culture might also be described as an "operational code" that is valid for an entire group of people. Culture conditions the individual's range of action and ideas, including what to do and not do, how to do or not do it, and whom to do it with or not to do it with. Culture also includes under what circumstances the "rules" shift and change. Culture influences how people make judgments about what is right and wrong, assess what is important and unimportant, categorize things, and deal with things that do not fit into existing categories. Cultural rules are flexible in practice. For example, the kinship system of a certain Amazonian Indian tribe requires that individuals marry a cousin. However, the definition of cousin is often changed to make people eligible for marriage.

Identity

3-39. Each individual belongs to multiple groups, through birth, assimilation, or achievement. Each group to which individuals belong influences their beliefs, values, attitudes, and perceptions. Individuals consciously or unconsciously rank their identities into primary and secondary identities. Primary identities are frequently national, racial, and religious. In contrast, secondary identities may include such things as hunter, blogger, or coffee drinker. Frequently, individuals' identities are in conflict; counterinsurgents can use these conflicts to influence key leaders' decisions.

Beliefs

3-40. Beliefs are concepts and ideas accepted as true. Beliefs can be core, intermediate, or peripheral.

3-41. *Core* beliefs are those views that are part of a person's deep identity. They are not easily changed. Examples include belief in the existence of God, the value of democratic government, the importance of individual and collective honor, and the role of the family. Core beliefs are unstated, taken for granted, resistant to change, and not consciously considered. Attempts to change the central beliefs of a culture may result in significant unintended second- and third- order consequences. Decisions to do so are made at the national-strategic level.

3-42. *Intermediate* beliefs are predicated on reference to authority figures or authoritative texts. Thus, intermediate beliefs can sometimes be influenced by co-opting opinion leaders.

3-43. From intermediate beliefs flow *peripheral* beliefs. These beliefs are open to debate, consciously considered, and easiest to change. For example, a belief about birth control may derive from an individual's beliefs about the Roman Catholic Church. Beliefs about the theory of sexual repression may come from a person's opinion of Sigmund Freud.

Values

3-44. A *value* is an enduring belief that a specific mode of conduct or end state of existence is preferable to an opposite or converse mode of conduct or end state of existence. Values include beliefs concerning such topics as toleration, stability, prosperity, social change, and self-determination. Each group to which a person belongs inculcates that person with its values and their ranking of importance. Individuals do not unquestioningly absorb all the values of the groups to which they belong; they accept some and reject others. Most individuals belong to more than one social group. The values of each group are often in conflict: religious values may conflict with generational values or gender values with organizational practices. Commanders should evaluate the values of each group in the AO. They should determine whether the values promoted by the insurgency correspond to the values of other social groups in the AO or to those of the HN government. Based on that assessment, commanders can determine whether counterinsurgents can exploit these differences in values.

Attitudes and Perceptions

3-45. *Attitudes* are affinities for and aversions to groups, persons, and objects. Attitudes affect *perception*, which is the process by which an individual selects, evaluates, and organizes information from the external environment. Commanders should consider groups' attitudes regarding the following:

- Other groups.
- Outsiders.
- HN government.
- United States.
- U.S. military.
- Globalization.

Belief Systems

3-46. The totality of the identities, beliefs, values, attitudes, and perceptions that an individual holds—and the ranking of their importance—is that person's belief system. Religions, ideologies, and all types of "isms" fall into this category. As a belief system, a religion may include such things as a concept of God, a view of the afterlife, ideas about the sacred and the profane, funeral practices, rules of conduct, and modes of worship.

3-47. A belief system acts as a filter for new information: it is the lens through which people perceive the world. What members of a particular group believe to be rational, normal, or true may appear to outsiders to be strange, irrational or illogical. Understanding the belief systems of various groups in an AO allows counterinsurgents to more effectively influence the population.

3-48. Commanders should give the belief systems of insurgents and other groups in the AO careful attention. An insurgency may frame its objectives in terms of a belief system or may use a belief system to mobilize and recruit followers. Differences between the insurgents' and civilian groups' belief systems provide opportunities for counterinsurgents to separate the insurgents from the population. If local individuals are members of more than one group, there maybe contradictions in their belief systems that can be exploited.

Cultural Forms

3-49. Cultural forms are the concrete expression of the belief systems shared by members of a particular culture. Cultural forms include rituals, symbols, ceremonies, myths, and narratives. Cultural forms are the medium for communicating ideologies, values, and norms that influence thought and behavior. Each culture constructs or invents its own cultural forms through which cultural meanings are transmitted and reproduced. A culture's belief systems can be decoded by observing and analyzing its cultural forms. Insurgent groups frequently use local cultural forms to mobilize the population. Counterinsurgents can use cultural forms to shift perceptions, gain support, or reduce support for insurgents.

3-50. The most important cultural form for counterinsurgents to understand is the narrative. A cultural narrative is a story recounted in the form of a causally linked set of events that explains an event in a group's history and expresses the values, character, or self-identity of the group. Narratives are the means through which ideologies are expressed and absorbed by members of a society. For example, at the Boston Tea Party in 1773, Samuel Adams and the Sons of Liberty dumped five tons of tea into the Boston Harbor to protest what they considered unfair British taxation. This narrative explains in part why the Revolutionary War began. However, it also tells Americans something about themselves each time they hear the story: that fairness, independence, and justice are worth fighting for. As this example indicates, narratives may not conform to historical facts or they may drastically simplify facts to more clearly express basic cultural values. (For example, Americans in 1773 were taxed less than their British counterparts and most British attempts to raise revenues from the colonies were designed to help reduce the crushing national debt incurred in their defense.) By listening to narratives, counterinsurgents can identify a society's core values. Commanders should pay particular attention to cultural narratives of the HN population pertaining to outlaws, revolutionary heroes, and historical resistance figures. Insurgents may use these narratives to mobilize the population.

3-51. Other cultural forms include ritual and symbols. A *ritual* is a stereotyped sequence of activities involving gestures, words, and objects performed to influence supernatural entities or forces on behalf of the actors' goals and interests. Rituals can be either sacred or secular. A *symbol* is the smallest unit of cultural meaning. Symbols are filled with a vast amount of information that can be decoded by a knowledgeable observer. Symbols can be objects, activities, words, relationships, events, or gestures. Institutions and organizations often use cultural symbols to amass political power or generate resistance against external groups. Commanders should pay careful attention to the meaning of common symbols in the AO and how various groups use them.

LANGUAGE

3-52. *Language* is a system of symbols that people use to communicate with one another. It is a learned element of culture. Successful communication requires more than just grammatical knowledge; it also requires understanding the social setting, appropriate behaviors towards people of different statuses, and nonverbal cues, among other things. An understanding of the social environment can facilitate effective communication, even if counterinsurgents do not speak the local language and must work through translators or interpreters.

3-53. The languages used in an AO have a major impact on operations. Languages must be identified to facilitate language training, communication aids such as phrase cards, and requisitioning of translators. Translators are critical for collecting intelligence, interacting with local citizens and community leaders, and developing IO products. (Appendix C addresses linguist support.)

3-54. The transliteration of names not normally written using the English alphabet affects all intelligence operations, especially collection, analysis, and fusion. Unfamiliar and similar place names can make it hard find places on a map and cause targeting errors. In addition, detained insurgents may be released if their name is misidentified. In countries that do not use the English alphabet, a combatant-command-wide standard for spelling names agreed upon by non-DOD agencies should be set.

POWER AND AUTHORITY

3-55. Once they have mapped the social structure and understand the culture, staffs must determine how power is apportioned and used within a society. *Power* is the probability that one actor within a social relationship will be in a position to carry out his or her own will despite resistance. Understanding power is the key to manipulating the interests of groups within a society.

3-56. There may be many formal and informal power holders in a society. The formal political system includes the following organizations:

- Central governments.
- Local governments.
- Political interest groups.
- Political parties.
- Unions.
- Government agencies.
- Regional and international political bodies.

Understanding the formal political system is necessary but not sufficient for COIN operations. Informal power holders are often more important. They may include ethno-religious groups, social elites, and religious figures.

3-57. For each group in an AO, counterinsurgents should identify the type of power the group has, what it uses the power for (such as amassing resources and protecting followers), and how it acquires and maintains power. Commanders should also determine the same information about leaders within particular groups. There are four major forms of power in a society:

- Coercive force.
- Social capital.
- Economic resources.
- Authority.

Coercive Force

3-58. *Coercion* is the ability to compel a person to act through threat of harm or by use of physical force. Coercive force can be positive or negative. An example of coercion used positively is a group providing security for its members (as in policing and defending of territory). An example of coercion used negatively is a group intimidating or threatening group members or outsiders.

3-59. One essential role of government is providing physical security for its citizens by monopolizing the use of coercive force within its territory. When a government fails to provide security to its citizens or becomes a threat to them, citizens may seek alternative security guarantees. Ethnic, political, religious, or tribal groups in the AO may provide such guarantees.

3-60. Insurgents and other nongovernmental groups may possess considerable means of coercive force. Such groups may use coercion to gain power over the population. Examples of organizations providing such force are paramilitary units, tribal militias, gangs, and organizational security personnel. Groups may use their coercive means for a variety of purposes unrelated to the insurgency. Protecting their community members, carrying out vendettas, and engaging in criminal activities are examples. What may appear to be insurgent violence against innocent civilians could in fact be related to a tribal blood feud rather than the insurgency.

Social Capital

3-61. *Social capital* refers to the power of individuals and groups to use social networks of reciprocity and exchange to accomplish their goals. In many non-Western societies, patron-client relationships are an important form of social capital. In a system based on patron-client relationships, an individual in a powerful position provides goods, services, security, or other resources to followers in exchange for political support or loyalty, thereby amassing power. Counterinsurgents must identify, where possible, groups and individuals with social capital and how they attract and maintain followers.

Economic Power

3-62. *Economic power* is the power of groups or individuals to use economic incentives and disincentives to change people's behavior. Economic systems can be formal, informal, or a mixture of both. In weak or failed states, the formal economy may not function well. In such cases, the informal economy plays a central role in people's daily lives. The informal economy refers to such activities as smuggling, black market activities, barter, and exchange. For example, in many societies, monies and other economic goods are distributed though the tribal or clan networks and are connected to indigenous patronage systems. Those groups able to provide their members with economic resources through an informal economy gain followers and may amass considerable political power. Therefore, counterinsurgents must monitor the local informal economy and evaluate the role played by various groups and individuals within it. Insurgent organizations may also attract followers through criminal activities that provide income.

Authority

3-63. *Authority* is legitimate power associated with social positions. It is justified by the beliefs of the obedient. There are three primary types of authority:

- **Rational-legal authority**, which is grounded in law and contract, codified in impersonal rules, and most commonly found in developed, Western societies.
- **Charismatic authority**, which is exercised by leaders who develop allegiance among their followers because of their unique, individual charismatic appeal, whether ideological, religious, political, or social.

- **Traditional authority**, which is usually invested in a hereditary line or particular office by a higher power.

3-64. Traditional authority relies on the precedent of history. It is a common type of authority in non-Western societies. In particular, tribal and religious forms of organization rely heavily on traditional authority. Traditional authority figures often wield enough power, especially in rural areas, to single-handedly drive an insurgency. Understanding the types of authority at work in the formal and informal political systems of the AO helps counterinsurgents identify agents of influence who can help or hinder achieving objectives.

INTERESTS

3-65. Commanders and staffs analyze the culture of the society as a whole and that of each group within the society. They identify who holds formal and informal power and why. Then they consider ways to reduce support for insurgents and gain support for the HN government.

3-66. Accomplishing these tasks requires commanders and staffs to understand the population's interests. *Interests* refer to the core motivations that drive behavior. These include physical security, basic necessities, economic well-being, political participation, and social identity. During times of instability, when the government cannot function, the groups and organizations to which people belong satisfy some or all of their interests. Understanding a group's interests allows commanders to identify opportunities to meet or frustrate those interests. A group's interests may become grievances if the HN government does not satisfy them.

Physical Security

3-67. During any period of instability, people's primary interest is physical security for themselves and their families. When HN forces fail to provide security or threaten the security of civilians, the population is likely to seek security guarantees from insurgents, militias, or other armed groups. This situation can feed support for an insurgency. However, when HN forces provide physical security, people are more likely to support the government. Commanders therefore identify the following:

- Whether the population is safe from harm.
- Whether there is a functioning police and judiciary system.
- Whether the police and courts are fair and nondiscriminatory.
- Who provides security for each group when no effective, fair government security apparatus exists.

The provision of security by the HN government must occur in conjunction with political and economic reform.

Essential Services

3-68. *Essential services* provide those things needed to sustain life. Examples of these essential needs are food, water, clothing, shelter, and medical treatment. Stabilizing a population requires meeting these needs. People pursue essential needs until they are met, at any cost and from any source. People support the source that meets their needs. If it is an insurgent source, the population is likely to support the insurgency. If the HN government provides reliable essential services, the population is more likely to support it. Commanders therefore identify who provides essential services to each group within the population.

Economy

3-69. A society's individuals and groups satisfy their economic interests by producing, distributing, and consuming goods and services. How individuals satisfy their economic needs depends on the society's level and type of economic development. For instance, in a rural-based society, land ownership may be a major part of any economic development plan. For a more urban society, public- and private-sector jobs may be the greatest concern.

3-70. Sometimes economic disparities between groups contribute to political instability. Insurgent leadership or traditional authority figures often use real or perceived injustices to drive an insurgency. Perceived injustices may include the following:

- Economic disenfranchisement.
- Exploitative economic arrangements.
- Significant income disparity that creates (or allows for) intractable class distinctions.

3-71. Military operations or insurgent actions can adversely affect the economy. Such disruption can generate resentment against the HN government. Conversely, restoring production and distribution systems can energize the economy, create jobs and growth, and positively influence local perceptions. To determine how to reduce support for insurgency and increase support for the government, commanders determine the following:

- Whether the society has a functioning economy.
- Whether people have fair access to land and property.
- How to minimize the economic grievances of the civilian population.

Political Participation

3-72. Another interest of the population is political participation. Many insurgencies begin because groups within a society believe that they have been denied political rights. Groups may use preexisting cultural narratives and symbols to mobilize for political action. Very often, they rally around traditional or charismatic authority figures. Commanders should investigate whether—

- All members of the civilian population have a guarantee of political participation.
- Ethnic, religious, or other forms of discrimination exist.
- Legal, social, or other policies are creating grievances that contribute to the insurgency.

Commanders should also identify traditional or charismatic authority figures and what narratives mobilize political action.

Grievances

3-73. Unsatisfied interests may become grievances. Table 3-1 lists factors to consider when an interest has become a grievance.

Table 3-1. Factors to consider when addressing grievances

- What are the insurgents' grievances?
- What are the population's grievances?
- Would a reasonable person consider the population's grievances valid? The validity of a grievance should be assessed using both subjective and objective criteria.
- Are the grievances of the population and those of the insurgency the same?
- What does the host-nation government believe to be the population's grievances? Does it consider those grievances valid?
- Are the population's grievances the same as those perceived by the host-nation government?
- Has the host-nation government made genuine efforts to address these grievances?
- Are there practical actions the host-nation government can take to address these grievances? Or is addressing the source of the grievances beyond the host-nation government's capability? (For example, major social and economic dislocations may be caused by globalization.)
- Can U.S. forces act to address these interests or grievances?

Evaluate the Threat

3-74. The purpose of evaluating the insurgency and related threats is to understand the enemy, enemy capabilities, and enemy vulnerabilities. This evaluation also identifies opportunities commanders may exploit.

3-75. Evaluating an insurgency is difficult. Neatly arrayed enemy orders of battle are neither available nor what commanders need to know. Insurgent organizational structures are functionally based and continually adaptive. Attempts to apply traditional order of battle factors and templates can produce oversimplified, misleading conclusions. Commanders require knowledge of difficult-to-measure characteristics. These may include the following:

- Insurgent goals.
- Grievances insurgents exploit.
- Means insurgents use to generate support.
- Organization of insurgent forces.
- Accurate locations of key insurgent leaders.

However, insurgents usually look no different from the general populace and do their best to blend with noncombatants. Insurgents may publicly claim motivations and goals different from what is truly driving their actions. Further complicating matters, insurgent organizations are often rooted in ethnic or tribal groups. They often take part in criminal activities or link themselves to political parties, charities, or religious organizations as well. These conditions and practices make it difficult to determine what and who constitutes the threat. Table 3-2 lists characteristics of an insurgency that can provide a basis for evaluating a threat. Table 3-2 also lists the conventional order of battle element as a supplement for analysts.

Table 3-2. Insurgency characteristics and order of battle factors

Insurgency characteristics	*Conventional order of battle factors*
• Insurgent objectives • Insurgent motivations • Popular support or tolerance • Support activities, capabilities, and vulnerabilities • Information activities, capabilities, and vulnerabilities • Political activities, capabilities, and vulnerabilities • Violent activities, capabilities, and vulnerabilities • Organization • Key leaders and personalities	• Composition • Disposition • Strength • Tactics and operations • Training • Logistics • Operational effectiveness • Electronic technical data • Personalities • Miscellaneous data • Other factors

3-76. The following insurgency characteristics are often the most important intelligence requirements and the most difficult to ascertain:

- Objectives.
- Motivations.
- Means of generating popular support or tolerance.

In particular, the ability to generate and sustain popular support, or at least acquiescence and tolerance, often has the greatest impact on the insurgency's long-term effectiveness. This ability is usually the insurgency's center of gravity. Support or tolerance, provided either willingly or unwillingly, provides the following for an insurgency:

- Safe havens.
- Freedom of movement.
- Logistic support.
- Financial support.
- Intelligence.
- New recruits.

Support or tolerance is often generated using violent coercion and intimidation of the populace. In these cases, even if people do not favor the insurgent cause, they are forced to tolerate the insurgents or provide them material support.

3-77. Understanding the attitudes and perceptions of the society's groups is very important to understanding the threat. It is important to know how the population perceives the insurgents, the host nation, and U.S. forces. In addition, HN and insurgent perceptions of one another and of U.S. forces are also very important. Attitudes and perceptions of different groups and organizations inform decision-making processes and shape popular thinking on the legitimacy of the actors in the conflict.

3-78. As analysts perform IPB, they should focus on insurgent vulnerabilities to exploit and strengths to mitigate. (See chapter 1.) Evaluating threats from insurgents and other armed groups and learning the people's interests and attitudes lets analysts identify divisions between the insurgents and the populace. This analysis also identifies divisions between the HN government and the people. For instance, if the insurgent ideology is unpopular, insurgents may use intimidation to generate support. Another example is discovering that insurgents gain support by providing social services that the HN government neglects or cannot provide. Determining such divisions identifies opportunities to conduct operations that expand splits between the insurgents and the populace or lessen divides between the HN government and the people.

Objective and Motivation Identification

3-79. Insurgents have political objectives and are motivated by an ideology or grievances. The grievances may be real or perceived. Identifying insurgent objectives and motivations lets counterinsurgents address the conflict's underlying causes. Broadly speaking, there are two kinds of insurgencies: national insurgencies and resistance movements. Both can be further classified according to the five approaches explained in paragraphs 1-25 through 1-39.

3-80. In a *national insurgency*, the conflict is between the government and one or more segments of the population. In this type of insurgency, insurgents seek to change the political system, take control of the government, or secede from the country. A national insurgency polarizes the population and is generally a struggle between the government and insurgents for legitimacy and popular support.

3-81. In contrast, a *resistance movement* (sometimes called a liberation insurgency) occurs when insurgents seek to expel or overthrow what they consider a foreign or occupation government. The grievance is foreign rule or foreign intervention. Resistance movements tend to unite insurgents with different objectives and motivations. However, such an insurgency can split into competing factions when foreign forces leave and the focus of resistance is gone. That situation may result in a civil war.

3-82. Identification of insurgent goals and motivations can be difficult for a number of reasons:

- There may be multiple insurgent groups with differing goals and motivations. This case requires separately monitoring each group's goals and motivations.
- Insurgent leaders may change and the movement's goals change with them.
- Movement leaders may have different motivations from their followers. For instance, a leader may want to become a new dictator; followers may be motivated by a combination of political ideology and money.
- Insurgents may hide their true motivations and make false claims. For instance, the differences between insurgents and outsiders generally make resistance movements easier to unify and mobilize. Thus, insurgents may try to portray a national insurgency as a resistance movement.
- The goals of an insurgency may change due to changes in the operational environment. Foreign forces joining a COIN effort can transform a national insurgency into resistance movement. The reverse may happen when foreign forces depart.

For all these reasons, analysts continuously track insurgent actions, internal communications, and public rhetoric to determine insurgent goals and motivations.

Popular Support or Tolerance

3-83. Developing passive support (tolerance or acquiescence) early in an insurgency is often critical to an insurgent organization's survival and growth. Such support often has a great effect on the insurgency's long-term effectiveness. As an insurgent group gains support, its capabilities grow. New capabilities enable the group to gain more support. Insurgents generally view popular support as a zero-sum commodity; that is, a gain for the insurgency is a loss for the government, and a loss for the government is a gain for the insurgency.

Forms of Popular Support

3-84. Popular support comes in many forms. It can originate internally or externally, and it is either active or passive. There are four forms of popular support:

- Active external.
- Passive external.
- Active internal.
- Passive internal.

The relative importance of each form of support varies depending on the circumstances. However, all forms benefit an insurgency.

3-85. *Active external support* includes finance, logistics, training, fighters, and safe havens. These forms of support may be provided by a foreign government or by nongovernmental organizations, such as charities.

3-86. *Passive external support* occurs when a foreign government supports an insurgency through inaction. Forms of passive support include the following:

- Not curtailing the activities of insurgents living or operating within the state's borders.
- Recognizing the legitimacy of an insurgent group.
- Denying the legitimacy of the HN government.

3-87. *Active internal support* is usually the most important to an insurgent group. Forms of active support include the following:

- Individuals or groups joining the insurgency.
- Providing logistic or financial support.
- Providing intelligence.
- Providing safe havens.
- Providing medical assistance.
- Providing transportation.
- Carrying out actions on behalf of the insurgents.

3-88. *Passive internal support* is also beneficial. Passive supporters do not provide material support; however, they do allow insurgents to operate and do not provide information to counterinsurgents. This form of support is often referred to as tolerance or acquiescence.

Methods of Generating Popular Support

3-89. Insurgents use numerous methods to generate popular support. These include the following:

- Persuasion.
- Coercion.
- Encouraging overreaction.
- Apolitical fighters.

3-90. *Persuasion* can be used to obtain either internal or external support. Forms of persuasion include—

- Charismatic attraction to a leader or group.
- Appeal to an ideology.

- Promises to address grievances.
- Demonstrations of potency, such as large-scale attacks or social programs for the poor.

Persuasion through demonstrations of potency can be the most effective technique because it can create the perception that the insurgency has momentum and will succeed.

3-91. Insurgents use *coercion* to force people to support or tolerate insurgent activities. Means of coercion include terrorist tactics, violence, and the threat of violence. Coercion may be used to alter the behavior of individuals, groups, organizations, or governments. Coercion is often very effective in the short term, particularly at the community level. However, terrorism against the general populace and popular leaders or attacks that negatively affect people's way of life can undermine insurgent popularity. Coercion is an easy way for insurgents to generate passive support; however, this support exists only as long as insurgents are able to intimidate.

3-92. *Encouraging overreaction* refers to enticing counterinsurgents to use repressive tactics that alienate the populace and bring scrutiny upon the government. It is also referred to as provocation of a government response.

3-93. *Apolitical fighters* may be attracted by many nonideological means. These means include monetary incentives, the promise of revenge, and the romance of fighting a revolutionary war.

3-94. Although difficult to quantify, analysts evaluate the popular support an insurgent group receives and its ability to generate more support. Open sources and intelligence reporting provide data to support this analysis. Polling data can be a valuable, though imprecise, means of gauging support for the HN government and support for the insurgency. Media and other open-source publications are important at all echelons. Assessing community attitudes, by gauging such things as the reactions of local populace to the presence of troops or government leaders, can be used to estimate popular support at the tactical level. At a minimum, the information in table 3-3 should be known.

Table 3-3. Critical information regarding popular support

• Overall level of popular support to the insurgency relative to that for the government. • Forms of popular support the insurgents receive (active, passive, internal, external) and their relative importance. • Foreign government support. • Support from nongovernmental organizations, including charities and transnational terrorist organizations.	• Criminal network support. • Segments of the populace supporting the insurgency. • Methods used to generate popular support and their relative effectiveness. • Grievances (real or perceived) exploited by insurgents. • Capabilities and vulnerabilities in generating popular support.

Support Activities

3-95. Although noticeable, violence may be only a small part of overall insurgent activity. Unseen insurgent activities include training and logistic actions. These are the support activities that sustain insurgencies. They come from an insurgency's ability to generate popular support. Like conventional military forces, insurgencies usually require more sustainers than fighters. Insurgent support networks may be large, even when violence levels are low. For this reason, it is easy to overlook them early in the development of an insurgency.

3-96. Undermining an insurgency's popular support is the most effective way to reduce insurgent support capabilities. However, identifying support capabilities and vulnerabilities is still important. Doing this lets analysts evaluate potential threat courses of action. Such analysis also lets commanders target vulnerable parts of the insurgents' support network. Table 3-4 lists support activities and capabilities to evaluate.

Table 3-4. Insurgent support activities and capabilities

• Locations of safe havens • Freedom of movement • Logistic support • Financial support • Means of communication	• Ability to train recruits in military or terrorist tactics and techniques • Means of collecting intelligence • Means of maintaining operations security

Information and Media Activities

3-97. Information and media activities can be an insurgency's main effort, with violence used in support. Insurgents use information activities to accomplish the following:

- Undermine HN government legitimacy.
- Undermine COIN forces.
- Excuse insurgent transgressions of national and international laws and norms.
- Generate popular support.

To achieve these effects, insurgents broadcast their successes, counterinsurgent failures, HN government failures, and illegal or immoral actions by counterinsurgents or the HN government. Insurgent broadcasts need not be factual; they need only appeal to the populace. Table 3-5 lists media forms that insurgents commonly use.

Table 3-5. Media forms insurgents use

• Word of mouth • Speeches by elites and key leaders • Flyers and handouts • Newspapers • Journals or magazines • Books • Audio recordings	• Video recordings • Radio broadcasts • Television broadcasts • Web sites • E-mail • Internet • Cellular telephones • Text messaging

3-98. To supplement their own media activities, insurgents take advantage of existing private and public media companies through press releases and interviews. These efforts, in addition to using the Internet, broadcast insurgent messages worldwide. By broadcasting to a global audience, insurgents directly attack public support for the COIN effort. Information and media activities to evaluate include the following:

- Commitment of assets and personnel to information activities.
- Types of media employed.
- Professionalism of products, such as newspaper articles or videos.
- Effectiveness and reach of information activities.

Political Activities

3-99. Insurgents use political activities to achieve their goals and enhance their cause's legitimacy. Political activities are tightly linked to information activities and violent acts. Political parties affiliated with an insurgent organization may negotiate or communicate on behalf of the insurgency, thereby serving as its public face. Insurgencies may grow out of political parties, or political parties may grow out of insurgencies. However, links between insurgents and political parties may be weak or easily broken by disputes between insurgents and politicians. In such cases, political parties may not be able to keep promises to end violent conflict. It is important to understand not only the links between insurgent groups and political organizations but also the amount of control each exerts over the other.

3-100. Understanding insurgent political activities enables effective political engagement of insurgents. Without this knowledge, the wrong political party may be engaged, the wrong messages may be used, or the government may make deals with political parties that cannot deliver on their promises. Political activities to be evaluated include the following:

- Links, if any, between the insurgency and political parties.
- Influence of political parties over the insurgency and vice versa.
- Political indoctrination and recruiting by insurgent groups.

Violent Activities

3-101. Violent actions by insurgents include three major types, which may occur simultaneously:

- Terrorist.
- Guerrilla.
- Conventional.

3-102. Insurgents are by nature an asymmetric threat. They do not use terrorist and guerrilla tactics because they are cowards afraid of a "fair fight"; insurgents use these tactics because they are the best means available to achieve the insurgency's goals. Terrorist and guerrilla attacks are usually planned to achieve the greatest political and informational impact with the lowest amount of risk to insurgents. Thus, commanders need to understand insurgent tactics and targeting as well as how the insurgent organization uses violence to achieve its goals and how violent actions are linked to political and informational actions.

> **Asymmetric Tactics in Ireland**
>
> In 1847, Irish insurgents were advised to engage the British Army in the following way:
>
> "The force of England is entrenched and fortified. You must draw it out of position; break up its mass; break its trained line of march and manoeuvre, its equal step and serried array... nullify its tactic and strategy, as well as its discipline; decompose the science and system of war, and resolve them into their first elements."

3-103. Terrorist tactics employ violence primarily against noncombatants. Terror attacks generally require fewer personnel than guerrilla warfare or conventional warfare. They allow insurgents greater security and have relatively low support requirements. Insurgencies often rely on terrorist tactics early in their formation due to these factors. Terrorist tactics do not involve mindless destruction nor are they employed randomly. Insurgents choose targets that produce the maximum informational and political effects. Terrorist tactics can be effective for generating popular support and altering the behavior of governments.

3-104. Guerrilla tactics, in contrast, feature hit-and-run attacks by lightly armed groups. The primarily targets are HN government activities, security forces, and other COIN elements. Insurgents using guerrilla tactics usually avoid decisive confrontations unless they know they can win. Instead, they focus on harassing counterinsurgents. As with terrorist tactics, guerrilla tactics are neither mindless nor random. Insurgents choose targets that produce maximum informational and political effects. The goal is not to militarily defeat COIN forces but to outlast them while building popular support for the insurgency. Terrorist and guerrilla tactics are not mutually exclusive. An insurgent group may employ both forms of violent action simultaneously.

3-105. Insurgents rarely use conventional tactics. Conventional operations are not always necessary for an insurgency's success. However, insurgents may engage in conventional operations after the insurgency develops extensive popular support and sustainment capabilities. The insurgents can then generate a conventional military force that can engage HN government forces.

3-106. Knowledge of violent capabilities is used to evaluate insurgent courses of action. Commanders use this knowledge to determine appropriate protection measures and tactics to counter insurgent actions. In addition, knowledge of how insurgents conduct attacks provides a baseline that helps determine the effectiveness of COIN operations. The following should be evaluated to determine insurgents' capabilities for violent action:

- Forms of violent action used.
- Weapons available and their capabilities.
- Training.
- Known methods of operating.
  - Frequency of attacks.
  - Timing of attacks.
  - Targets of attacks.
  - Tactics and techniques.
- Known linkages between violent, political, and information actions. How do the insurgents use violence to increase their popular support and undermine counterinsurgents?
- Means of command and control during attacks (including communications means used).

Insurgent Organizational Structure and Key Personalities

3-107. Conducting the preceding activities requires some form of organizational structure and leadership. Insurgencies can be organized in several ways. Each structure has its own strengths and limitations. The structure used balances the following:

- Security.
- Efficiency and speed of action.
- Unity of effort.
- Survivability.
- Geography.
- Social structures and cultures of the society.

Organizations also vary greatly by region and time. Insurgent organizations are often based on existing social networks—familial, tribal, ethnic, religious, professional, or others. Analysts can use social network analysis to determine organizational structure. (See paragraphs B-15 through B-18.)

3-108. An insurgency's structure often determines whether it is more effective to target enemy forces or enemy leaders. For instance, if an insurgent organization is hierarchical with few leaders, removing the leaders may greatly degrade the organization's capabilities. However, if the insurgent organization is nonhierarchical, targeting the leadership may not have much effect. Understanding an insurgent organization's structure requires answers to the following questions:

- Is the organization hierarchical or nonhierarchical?
- Is the organization highly structured or unsystematic?
- Are movement members specialists or generalists?
- Do leaders exercise centralized control or do they allow autonomous action and initiative?
- Are there a few leaders (promotes rapid decision making) or is there redundant leadership (promotes survivability)?
- Does the movement operate independently or does it have links to other organizations and networks (such as criminal, religious, and political organizations)?
- Does the movement place more weight on political action or violent action?

3-109. As explained in paragraphs 1-58 through 1-66, insurgents fall into five overlapping categories: movement leaders, combatants, political cadre, auxiliaries, and the mass base. Movement leaders are important because they choose the insurgency's organization, approach, and tactics. The movement leaders' personalities and decisions often determine whether the insurgency succeeds. Therefore, the movement leaders must be identified and their basic beliefs, intentions, capabilities, and vulnerabilities understood. Important leader characteristics include the following:

- Role in the organization.
- Known activities.
- Known associates.
- Background and personal history.
- Beliefs, motivations, and ideology.
- Education and training.
- Temperament (for example, careful, impulsive, thoughtful, or violent).
- Importance of the organization.
- Popularity outside the organization.

ASSOCIATED THREATS

3-110. When an insurgency has widespread support, it usually means the HN government is weak and losing control. In such situations, other armed groups—particularly criminal organizations, militias, and terrorist groups—can be significant players. Moreover, these groups can support each other's operations.

Criminal Networks

3-111. Criminal networks may not be a part of an insurgency. However, their activities—for example, banditry, hijackings, kidnappings, and smuggling—can further undermine the HN government's authority. Insurgent organizations often link themselves to criminal networks to obtain funding and logistic support. In some cases, insurgent networks and criminal networks become indistinguishable. As commanders work to reassert government control, they need to know the following:

- Which criminal networks are present.
- What their activities are.
- How they interact with insurgents.

Nongovernment Militias

3-112. As the HN government weakens and violence increases, people look for ways to protect themselves. If the government cannot provide protection, people may organize into armed militias to provide that essential service. Examples of this sort of militia include the following:

- Loyalist militias formed in Northern Ireland.
- Right-wing paramilitary organizations formed in Colombia to counter the FARC.
- Militias of various ethnic and political groups formed in Iraq during Operation Iraqi Freedom.

If militias are outside the HN government's control, they can often be obstacles to ending an insurgency. Militias may become more powerful than the HN government, particularly at the local level. They may also fuel the insurgency and a precipitate a downward spiral into full-scale civil war.

3-113. Militias may or may not be an immediate threat to U.S. forces; however, they constitute a long-term threat to law and order. The intelligence staff should track them just like insurgent and other armed groups. Commanders need to understand the role militias play in the insurgency, the role they play in politics, and how they can be disarmed.

DETERMINE THREAT COURSES OF ACTION

3-114. The purpose of this IPB step is to understand insurgent approaches and tactics so they can be effectively countered. The initial determination of threat courses of action focuses on two levels of analysis. The first is determining the overall approach, or combination of approaches, the movement leaders have selected to achieve their goals. The second is determining tactical courses of action used to execute that approach.

3-115. The insurgents' approach is based on their objectives, desired end state, and requirements of the operational environment. The approach and the tactics used to execute it set the conditions for the insurgents to achieve their desired end state. Insurgents can accomplish this goal by maintaining preexisting adverse conditions or by creating those conditions.

INSURGENT APPROACHES

3-116. As indicated in paragraphs 1-25 through 1-39, there are six approaches insurgents may follow:

- Conspiratorial.
- Military-focused.
- Urban.
- Protracted popular war.
- Identity-focused.
- Composite and coalition.

3-117. These approaches may be combined with one another. They may also occur in parallel as different insurgent groups follow different paths, even within a single AO. In addition, insurgents may change approaches over time. The approach pursued affects the insurgents' organization, types of activities, and emphasis placed on different activities.

3-118. Table 3-6 (page 3-22) lists potential indicators of different insurgent approaches. The conspiratorial and identity-focused approaches present distinct collection challenges. Insurgents using a conspiratorial approach execute few overt acts until the conditions appear ripe to seize power. Thus, this approach is difficult to identify without sources within the insurgent organization. Similarly, members of an identity-focused insurgency strongly identify with the insurgent organization. It is difficult to get such people to provide useful information.

Table 3-6. Potential indicators of insurgent approaches

- **Conspiratorial**
  - Absence of overt violent or informational actions.
  - Large cadre relative to the number of combatants in the organization.
  - Small mass base or no mass base at all.
- **Military-focused**
  - Presence of leaders and combatants, but little, if any, cadre or mass base.
- **Urban**
  - Terrorist attacks in urban areas.
  - Infiltration and subversion of host-nation government and security forces in urban areas.
  - Organization composed of small, compartmentalized cells.
  - Cadre and mass base small relative to the number of combatants.
- **Protracted popular war**
  - A large mass base.
  - Overt violence.
  - Heavy use of informational and political activities.
  - Focus on building popular support for the insurgency.
- **Identity-focused**
  - Presence of a resistance movement.
  - Presence of an "us-and-them" gap between the government and one or more ethnic, tribal, or religious groups.
  - Large mass base of passive and active supporters built around preexisting social networks.
  - Many auxiliaries.
  - Small cadre composed primarily of traditional authority figures.
  - Large numbers of part-time combatants.

3-119. It should be noted that insurgents may be inept at the use of a given approach. Alternatively, they may misread the operational environment and use an inappropriate approach. Knowledge of misapplication of approach or the use of different approaches by different insurgent groups may provide opportunities for counterinsurgents to exploit. It is imperative not only to identify insurgent approaches but also to understand their strengths and weaknesses in the context of the operational environment.

Tactical Courses of Action

3-120. Insurgents base their tactical courses of action on their capabilities and intentions. Evaluating the support, information, political, and violent capabilities of insurgent organizations was discussed in paragraphs 3-95 through 3-106. The intentions come from goals, motivations, approach, culture, perceptions, and leadership personalities. Insurgents may pursue many different courses of action in an AO at any time. Their tactical courses of action change with both time and location. People and their attitudes, both within the nation and often outside it, are the ultimate targets of the insurgents. Therefore, commanders pay special attention to the effects insurgent actions have on the populace and how the insurgents achieve those effects. Finally, tactical actions can have strategic effects. This is because insurgent propaganda and media reporting can reach a global audience, multiplying the effects of insurgent tactical actions. Insurgents can employ a wide variety of tactics. (See table 3-7.)

Table 3-7. Examples of insurgent tactics

- **Ambushes**. Guerrilla-style attacks to kill or intimidate counterinsurgents.
- **Assassination**. A term generally applied to the killing of prominent persons and symbolic personnel. It may also pertain to "traitors" who defect from the insurgency, human intelligence sources, and others who work with or for the host-nation government or U.S. forces.
- **Arson**. Less dramatic than most tactics, arson has the advantage of low risk to the perpetrator. It requires only a low level of technical knowledge.
- **Bombing and high explosives**. The improvised explosive device is often the insurgent's weapon of choice. Improvised explosive devices can be inexpensive to produce and, because of the various detonation techniques available, may be a low risk to the perpetrator. However, suicidal bombing cannot be overlooked as an employment method. Another advantage is the publicity that such attacks produce. Yet another is the insurgents' ability to control casualties through timed detonation and careful placement of the device. Attacks are also easily deniable should the action produce undesirable results. From 1983 through 1996, approximately half of all terrorist incidents worldwide involved the use of explosives.
- **Chemical, biological, radiological, or nuclear weapons**. There is a potential for use of both chemical and biological weapons in the future. These types of weapons are relatively cheap and easy to make. They may be used in place of conventional explosives in many situations. The potential for mass destruction and the deep-seated fear most people have for these weapons make them attractive to groups wishing to attract international attention. Although an explosive nuclear device is acknowledged to be beyond the financial and technical reach of most terrorist groups, a chemical or biological weapon, or a radiological dispersion device using nuclear contaminants, is not. The technology is simple and the payoff is potentially higher than that of conventional explosives.
- **Demonstrations**. Demonstrations can be used to incite violent responses by counterinsurgents and also to display the popularity of the insurgent cause.
- **Denial and deception**.
  - *Denial* consists of measures taken by the threat to block, prevent, or impair U.S. intelligence collection. Examples include killing human intelligence sources.
  - *Deception* involves deliberately manipulating information and perceptions in order to mislead. Examples include providing false intelligence.

- **Hijacking and skyjacking**. Sometimes insurgents employ hijacking as a means of escape. However, hijackings are normally executed to produce a spectacular hostage situation. Although trains, buses, and ships have been hijacked, aircraft are the preferred target because of their greater mobility and because they are difficult to enter once they have been seized.
- **Hoaxes**. Any credible insurgent or terrorist group can employ a hoax with considerable success. A threat against a person's life causes that person and those associated with that individual to devote time and efforts to security measures. A bomb threat can close a commercial building, empty a theater, or delay an aircraft flight at no cost to insurgents. False alarms dull the analytical and operational efficiency of key security personnel, thus degrading readiness.
- **Hostage taking**. Hostage taking is an overt seizure of one or more people to gain publicity or other concessions in return for releasing the hostage. While dramatic, hostage and hostage barricade situations are risky for perpetrators.
- **Indirect fire**. Insurgents may use indirect fire, such as mortars, to harass counterinsurgents or cause them to commit forces that can then be attacked by secondary ambushes.
- **Infiltration and subversion**. Insurgents use these tactics to gain intelligence and degrade the effectiveness of host-nation government organizations. These tactics involve getting host-nation government agencies to hire insurgent agents or by convincing members of the government to support the insurgency. Subversion may be achieved through intimidation, indoctrination of sympathetic individuals, or bribes.
- **Kidnapping**. While similar to hostage taking, kidnapping has significant differences. Kidnapping is usually a covert seizure of one or more persons to obtain specific results. It is normally very difficult to execute. Perpetrators may or may not be known for a long time. Media attention is initially intense but decreases over time. Because of the time involved, successful kidnapping requires elaborate planning and logistics. The risk to the perpetrators may be less than in the hostage situation.
- **Propaganda**. Insurgents may disseminate propaganda using any form of media, including face-to-face talks.
- **Raids or attacks on facilities**. Armed attacks on facilities are usually undertaken to—
  - Demonstrate the host-nation government's inability to secure critical facilities or national symbols.
  - Acquire resources (for example, robbery of a bank or armory).
  - Kill host-nation government personnel.
  - Intimidate the host-nation government and the populace.
- **Sabotage**. The objective in most sabotage incidents is to demonstrate how vulnerable the society or host-nation government is to terrorist actions. Industrialized areas are more vulnerable to sabotage than less developed areas. Utility, communication, and transportation systems are interdependent; a serious disruption of any affects them all and gains immediate public attention. Sabotage of industrial and commercial facilities can create significant disruption while making a statement of future intent. Military facilities and installations, information systems, and information infrastructures are possible targets.
- **Seizure**. Seizure usually involves a building or object of value in the eyes of the audience. There is some risk to perpetrators because security forces have time to react.

SECTION III – INTELLIGENCE, SURVEILLANCE, AND RECONNAISSANCE OPERATIONS

3-121. The purpose of intelligence, surveillance, and reconnaissance (ISR) operations during a COIN is to develop the intelligence needed to address the issues driving the insurgency. Several factors are particularly important for ISR operations in COIN environments. These include the following:

- A focus on the local populace.
- Collection occurring at all echelons.
- Localized nature of insurgencies.

- All Soldiers and Marines functioning as potential collectors.
- Insurgent use of complex terrain.

3-122. Intelligence gaps and information requirements determined during IPB may range from insurgent leaders' locations, to the populace's perceptions of insurgents, to HN political parties' status. In general, collection focuses on the populace, insurgents, and host nation.

3-123. The fact that all units collect and report information, combined with the mosaic nature of insurgencies, means that the intelligence flow in COIN is more bottom up than top down. Conducting aggressive ISR operations and pushing intelligence collection assets and analysts to the tactical level, sometimes as far as company level, therefore benefits all echelons. It strengthens local intelligence, enhances regional and national reporting, and bolsters operations at all levels. Two techniques— either attaching a basic intelligence analytical capability down to battalion or company level, or forming a company information management capability from assigned personnel—can help commanders handle the tactical information flow better.

3-124. Collection may occur in any unit and collectors may be pushed to the lowest levels; nonetheless, the overall intelligence synchronization plan (formerly the collection plan) must remain synchronized so that all echelons receive the intelligence they require. There are several means of ensuring this happens. One is to ensure that priority intelligence requirements (PIRs) are "nested" at all echelons. They may be tailored to local or regional circumstances, but tactical and operational collection efforts should support one another. Headquarters monitor requests for information from lower echelons and taskings from higher echelons to get information to requestors when they need it. Commanders ensure their subordinates understand the PIRs. Such understanding helps Soldiers and Marines know when to report something and what they should report.

3-125. Feedback from analysts and intelligence consumers to collectors is important to synchronizing the ISR effort. Responses tell collectors that a report is of interest and that they should follow it up. Such feedback may come from any unit at any echelon.

3-126. Also affecting intelligence synchronization is the requirement to work closely with U.S. Government agencies, HN security and intelligence organizations, and multinational intelligence organizations. Operational-level ISR planning drives the synchronization of these agencies' and organizations' efforts; however, coordination occurs at all echelons. Communication among collection managers and collectors down to the battalion level is important; it can eliminate circular reporting and unnecessary duplicate work. (See section IV.)

3-127. Insurgents often try to use complex terrain and seams between maneuver units to their advantage. (Seams are boundaries between units not adequately covered by any unit.) Collection managers do not ignore areas of complex terrain. They monitor seams to ensure insurgents do not establish undetected bases of operation.

THE INTELLIGENCE-OPERATIONS DYNAMIC

3-128. Intelligence and operations have a dynamic relationship. Even in permissive environments where a great deal is known about the enemy, there is an intelligence aspect to all operations. Intelligence drives operations and successful operations generate additional intelligence. For instance, an operation increasing the security and general happiness of a town often increases the amount of information its inhabitants offer. This information is processed into more intelligence, which results in more effective operations. The reverse is also true. Operations conducted without accurate intelligence may upset the populace and lead them to offer less information. In many cases, newly arrived units have little intelligence on their AO. They have to conduct operations to generate intelligence.

3-129. Because intelligence and operations are so closely related, it is important for collectors to be linked directly to the analysts and operators they support. Analysts must remain responsive to their supported units' intelligence requirements. Further, collectors should not passively wait for operators to submit requirements; rather, they should closely monitor the operational environment and recommend requirements based on their understanding of operators' needs.

HUMAN INTELLIGENCE AND OPERATIONAL REPORTING

3-130. *Human intelligence* (HUMINT) is the collection of information by a trained human intelligence collector from people and their associated documents and media sources to identify elements, intentions, composition, strength, dispositions, tactics, equipment, personnel, and capabilities (FM 2-22.3). (Trained HUMINT collectors are Soldiers holding military occupational specialties 97E, 351Y [formerly 351C], 351M [formerly 351E], 35E, and 35F, and Marines holding the specialty 0251.) HUMINT uses human sources as tools and a variety of collection methods, both passive and active, to gather information to satisfy intelligence requirements and cross-cue other intelligence disciplines. Interrogation is just one of the HUMINT tasks. (FM 2-22.3 provides the authoritative doctrine for HUMINT operations. It also contains policy for interrogation.) HUMINT operations often collect information that is difficult or sometimes impossible to obtain by other, more technical, means. During COIN operations, much intelligence is based on information gathered from people.

3-131. Operational reporting may also have information of intelligence value that originates from the local populace. People may approach Soldiers and Marines during the course of their day-to-day operations and offer information. Soldiers and Marines should take information and report it to the intelligence section. Doing so allows for verification of the information and establishes a means for HUMINT collectors to contact individuals offering information of value.

3-132. The lives of people offering information on insurgents are often in danger. Insurgents continuously try to defeat collection operations. Careless handling of human sources by untrained personnel can result in murder or intimidation of these sources. When this occurs, HUMINT can be dramatically reduced due to the word spreading that U.S. forces are careless or callous about protecting their sources. HUMINT collectors are trained in procedures that limit the risk to sources and handlers.

3-133. Counterinsurgents should not expect people to willingly provide information if insurgents have the ability to violently intimidate sources. HUMINT reporting increases if counterinsurgents protect the populace from insurgents and people begin to believe the insurgency will be defeated.

3-134. People often provide inaccurate and conflicting information to counterinsurgents. They may be spreading rumors or providing inaccurate information purposefully for their own reasons. Examples of reasons include accomplishing the following:

- Using counterinsurgents to settle tribal, ethnic, or business disputes.
- Leading counterinsurgents into ambushes.
- Enticing counterinsurgents into executing operations that upset the populace.
- Learning about U.S. planning time and tactics.
- Stretching COIN forces thin by causing them to react to false reports.

The accuracy of information obtained by Soldiers and Marines should be verified before being used to support operations. This means that information reported to patrols should be verified with all-source intelligence.

MILITARY SOURCE OPERATIONS

3-135. Because of their continuous contact with the populace, Soldiers and Marines regularly identify potential sources for HUMINT personnel to develop. It is therefore imperative that all counterinsurgents know the PIRs and that every patrol is debriefed. These debriefings should be as detailed as possible. Analysts and HUMINT collectors should work closely with operations staffs and other personnel to ensure new sources are properly developed. (Table 3-8) lists some potential HUMINT sources.)

3-136. Establishing a reliable source network is an effective collection method. Military source operations provide the COIN equivalent of the reconnaissance and surveillance conducted by scouts in conventional operations. HUMINT sources serve as "eyes and ears" on the street and provide an early warning system for tracking insurgent activity. Although counterinsurgents regularly get information from "walk-in" or "walk-up" sources, only HUMINT personnel are trained and authorized to work with HUMINT sources. Due to legal considerations, the potential danger sources face if identified, and the potential danger to troops involved, only HUMINT personnel may conduct military source operations. All Soldiers and Marines may record information given to them by walk-up contacts, including liaison relationships, but they may not develop HUMINT sources or networks. (Refer to FM 2-22.3 for more information on military source operations.)

INTERROGATION OF DETAINEES AND DEBRIEFING OF DEFECTORS

3-137. Detainees and insurgent defectors are important HUMINT sources. The information they provide about the internal workings of an insurgency may be better than any other HUMINT source can provide. In addition, defectors can provide otherwise unobtainable insights into an insurgent organization's perceptions, motivations, goals, morale, organization, and tactics. Both detainees and defectors should be thoroughly questioned on all aspects of an insurgency discussed in section II. Their answers should be considered along with information obtained from captured equipment, pocket litter, and documents to build a better understanding of the insurgency. Properly trained Soldiers and Marines can conduct immediate tactical questioning of detainees or defectors. However, only trained HUMINT personnel are legally authorized to conduct interrogations. A trained debriefer should be used for questioning a defector. All questioning of detainees is conducted to comply with U.S. law and regulation, international law, execution orders and other operationally specific guidelines. (FM 2-22.3 provides the authoritative doctrine and policy for interrogation. Chapter 7 and appendix D of this manual also address this subject.)

Table 3-8. Potential sources of human intelligence

- **Patrol debriefings and after-action reviews**. Patrols regularly encounter individuals offering information and observe new enemy tactics and techniques. Patrol debriefings are especially important to units at brigade level and below; however, the information collected can be of higher echelon significance.
- **Civil affairs reports**. These reports are especially useful for gathering information about politics, economy, and infrastructure. Civil affairs personnel also regularly come into contact with individuals offering information.
- **Psychological operations (PSYOP) reports**. PSYOP personnel conduct opinion polls and gather information on community attitudes, perceptions, interests, and grievances. PSYOP personnel also regularly encounter individuals offering information.
- **Special operations forces reporting**. Special operations forces often work closely with local nationals and produce valuable human intelligence reports.
- **Leadership liaison**. Commanders and leaders regularly meet with their counterparts in the host-nation security forces and with community leaders. These meetings produce information or tips.
- **Contracting**. Contracting officers work with theater contractors, both host nation and external, performing support functions or building infrastructure. Contractors may offer information to contracting officers.
- **Multinational operations centers**. These provide a place to share information between host-nation and U.S. personnel.
- **Tips hotlines**. Telephone or e-mail hotlines provide a safe means for people to provide information. They are especially useful for obtaining time-sensitive intelligence, such as warning of an attack or the current location of an insurgent.
- **U.S. persons**. There will be times when U.S. civilians, such as contractors or journalists, offer information to counterinsurgents. For legal reasons, it is important to understand regulations regarding intelligence-related information collected on U.S. persons. (See FM 2-22.3.)

SURVEILLANCE AND RECONNAISSANCE CONSIDERATIONS

3-138. Because all Soldiers and Marines are potential collectors, the ISR plan addresses all day-to-day tactical operations. This means every patrol or mission should be given intelligence collection requirements as well as operations requirements.

3-139. Overt area and zone reconnaissances are excellent means for tactical units to learn more about their AO, especially the terrain, infrastructure, people, government, local leaders, and insurgents. Overt reconnaissance by patrols allows commanders to fill intelligence gaps and develop relationships with local leaders, while simultaneously providing security to the populace.

3-140. Covert reconnaissance and surveillance operations employing scouts or concealed observation posts are often ineffective in places where the populace is alert and suspicious of outsiders. Such places include urban areas, suburban areas, and close-knit communities. In those places, it is very difficult for scouts to conduct reconnaissance or surveillance without being observed by insurgents or people who may tip off insurgents. Reconnaissance of a target may be noticed and cause insurgents to leave the area. Likewise, small groups of scouts may be attractive targets for insurgent attacks if the scouts' location is known. For these reasons, using a HUMINT network or aerial imagery platforms is often preferable to ground reconnaissance and surveillance. Successful ground reconnaissance in populated areas requires leaders to be creative in how they establish observation posts. One technique is for dismounted night patrols to leave a small "stay behind" observation post while the rest of a patrol moves on. Another effective technique is secretly photographing a place of interest while driving by it. However, commanders must weigh the benefits of these operations with the potential cost of insurgents receiving early warning of counterinsurgent intentions.

CONSIDERATIONS FOR OTHER INTELLIGENCE DISCIPLINES

3-141. An *intelligence discipline* is a well-defined area of intelligence collection, processing, exploitation, and reporting using a specific category of technical or human resources (JP 1-02). HUMINT is one of these disciplines. The following discussion addresses COIN-specific considerations for other selected intelligence disciplines and information types. Because of their importance to COIN, counterintelligence is covered separately in section IV and all-source analysis in section V.

SIGNALS INTELLIGENCE

3-142. In conventional environments, signals intelligence (SIGINT) collection is a good source for determining enemy locations, intentions, capabilities, and morale. The same applies in COIN operations. SIGINT is often helpful for confirming or denying HUMINT reporting and may be the primary source of intelligence in areas under insurgent control. Pushing SIGINT collection platforms down to tactical units can therefore improve intelligence collection.

OPEN-SOURCE INTELLIGENCE

3-143. Open-source intelligence (OSINT) is valuable for understanding the operational environment. It is often more useful than any other discipline for understanding public attitudes and public support for insurgents and counterinsurgents. OSINT is also an important means of determining the effectiveness of IO. Monitoring a wide variety of media in multiple languages benefits the COIN effort. If possible, monitoring should occur at every echelon with collection requirements. Each echelon should monitor the media that contain information relevant to operations at that echelon. For instance, reporting by major news networks often matters a lot at the combatant command level; in contrast, local newspapers or radio stations may be more important to tactical units.

IMAGERY INTELLIGENCE

3-144. In COIN operations, imagery intelligence (IMINT) platforms may be used for surveillance of likely insurgent safe houses and other facilities. Further, aerial IMINT platforms are also effective at detecting unusual personnel and supply movements. This information can help commanders determine where best to interdict insurgent lines of communications.

3-145. Static imagery, such as aerial photos of facilities, is useful for detecting long-term changes in structures or activities.

3-146. Real-time video, often from aerial surveillance platforms, is critical to assessing whether particular locations are likely sites of insurgent activity. This capability may also be used to track insurgents during operations. If flown high enough that insurgents cannot hear the platform, real-time video provides surveillance in areas where it is difficult or impossible to use observation posts.

TECHNICAL INTELLIGENCE

3-147. Insurgents often adapt their tactics, techniques, and procedures rapidly. Technical intelligence on insurgent equipment can help understand insurgent capabilities. These may include how insurgents are using improvised explosive devices, homemade mortars, and other pieces of customized military equipment.

MEASUREMENT AND SIGNATURES INTELLIGENCE

3-148. Measurement and signature intelligence (MASINT) sensors can provide remote monitoring of avenues of approach or border regions for smugglers or insurgents. They can also be used to locate insurgent safe havens and cache sites and determining insurgent activities and capabilities. MASINT can also contribute to targeting.

GEOSPATIAL INTELLIGENCE

3-149. Geospatial intelligence (GEOINT) is the exploitation and analysis of imagery and geospatial information to describe, assess, and visually depict physical features and geographically referenced activities on the Earth. GEOINT consists of imagery, IMINT, and geospatial information. GEOINT may have some benefit for identifying smuggling routes and safe havens. Imagery can be very beneficial to operations in urban areas as well. It can help identify structures of interest and aid urban terrain navigation. (Paragraphs B-5 through B-9 contain more GEOINT-related information.)

INTELLIGENCE-RELATED ACTIVITIES

3-150. There are several activities and information sources important to COIN that are not intelligence disciplines but are related to them. Chief among these are target exploitation (TAREX), document exploitation (DOCEX), property ownership records, and financial records.

Target Exploitation and Document Exploitation

3-151. Documents and pocket litter, as well as information found in computers and cell phones, can provide critical information that analysts need to evaluate insurgent organizations, capabilities, and intentions. TAREX and DOCEX are also of great benefit to HUMINT collectors in substantiating what detainees know and whether they are telling the truth.

3-152. TAREX in a COIN environment is like evidence collection in a law enforcement environment. Procedures that ensure captured equipment and documents are tracked accurately and attached to the correct insurgents is necessary. Evidence needs to be enough to justify using operational resources to apprehend the individuals in question; however, it does not necessarily need be enough to convict in a court of law. Pushing HUMINT or law enforcement personnel to the battalion level and below can improve TAREX and DOCEX by tactical units. Procedures for ensuring that tactical units get the results of higher level TAREX and DOCEX are also important. Units must be able to receive intelligence collected from the documents, equipment, and personnel they capture in enough time to exploit it.

Property Ownership Records

3-153. Property ownership records include census records, deeds, and other means of determining ownership of land and buildings. They help counterinsurgents to determine who should or should not be living in a specific area and help them secure the populace. In some cases, it may be necessary for Soldiers and Marines to go door to door and collect census data themselves.

Financial Records

3-154. Information gathered on sources of insurgent funding can be very helpful to the COIN effort. Collection of financial records often requires help from agencies like the Department of the Treasury and financial institutions. It may also require analyzing criminal activities or traditional means of currency transfer.

SECTION IV – COUNTERINTELLIGENCE AND COUNTERRECONNAISSANCE

3-155. Counterintelligence counters or neutralizes intelligence collection efforts through collection, counterintelligence investigations, operations, analysis and production, and functional and technical services. Counterintelligence includes all actions taken to detect, identify, exploit, and neutralize the multidiscipline intelligence activities of friends, competitors, opponents, adversaries, and enemies.

3-156. Insurgents place heavy emphasis on gathering intelligence. They use informants, double agents, reconnaissance, surveillance, open-source media, and open-source imagery. Insurgents can potentially use any person interacting with U.S. or multinational personnel as an informant. These include the same people that U.S. forces use as potential HUMINT sources. Operations security is thus very important; U.S. personnel must carefully screen the contractors, informants, translators, and other personnel working with them. Failure to do so can result in infiltration of U.S. facilities and deaths of U.S. personnel and their partners.

3-157. Background screenings should include collection of personal and biometric data and a search through available reporting databases to determine whether the person is an insurgent. (Biometrics concerns the measurement and analysis of unique physical or behavioral characteristics [as fingerprint or voice patterns].) Identification badges may be useful for providing security and personnel accountability for local people working on U.S. and HN government facilities. Biometric data is preferable, when available, because identification badges may be forged or stolen and insurgents can use them to identify people working with the HN government.

3-158. Insurgents have their own reconnaissance and surveillance networks. Because they usually blend well with the populace, insurgents can execute reconnaissance without easily being identified. They also have an early warning system composed of citizens who inform them of counterinsurgent movements. Identifying the techniques and weaknesses of enemy reconnaissance and surveillance enables commanders to detect signs of insurgent preparations and to surprise insurgents by neutralizing their early warning systems.

3-159. Insurgents may also have a SIGINT capability based on commercially available scanners and radios, wiretaps, or captured counterinsurgent equipment. Counterinsurgents should not use commercial radios or phones because insurgents can collect information from them. If Soldiers and Marines must use commercial equipment or unencrypted communications, they should employ authorized brevity codes to reduce insurgents' ability to collect on them.

SECTION V – ALL-SOURCE INTELLIGENCE

3-160. Joint doctrine defines *all-source intelligence* as products and/or organizations and activities that incorporate all sources of information, most frequently including human resources intelligence, imagery intelligence, measurement and signature intelligence, signals intelligence, and open-source data in the production of finished intelligence (JP 1-02). Intelligence organizations fuse data and information into all-source intelligence products to support COIN operations. Analysis for COIN operations is very challenging, due in part to the—

- Need to understand perceptions and culture.
- Need to track hundreds or thousands of personalities.
- Local nature of insurgencies.
- Tendency of insurgencies to change over time.

3-161. Databases are very important for analyzing insurgent activities and personalities. At a minimum, there should be common searchable combatant command databases of insurgent actions and personnel, as well as another database of all intelligence reporting. These should be accessible by analysts in and out of theater. The common operational picture should include reporting from all units and organizations involved in the effort.

3-162. Because all echelons collect and use intelligence, all staffs are heavily involved in analysis. Units are simultaneously intelligence producers and consumers. This situation is normal at brigade and above; however, battalion staffs often do not have the personnel to collect patrol debriefs, analyze incoming information from multiple sources, produce finished intelligence products, and disseminate products to appropriate consumers. In many cases brigade intelligence sections may also be inadequate for a COIN environment.

3-163. COIN requirements may require pushing analysts to battalion and brigade staffs to give those echelons the required analytical support. There are also instances when analysts can be beneficial at the company level. This is the case when a maneuver company must collect large amounts of information on the local populace and insurgents. An analyst can help collect and process this information and develop an operational picture of the AO. Pushing analysts to brigade level and below places analysts closer to collectors, improves the common operational picture, and helps higher echelon staffs receive answers to their PIRs. Commanders may need to be creative in developing analytical capabilities within their units. Though it is not ideal, commanders have assigned nonintelligence personnel to work in the intelligence section.

3-164. Analysis at brigade and below is the basis for operational-level intelligence. This is due to the bottom-up flow of intelligence in COIN. Battalions and brigades develop intelligence for their AOs; higher echelons fuse it into combatant-command-wide intelligence of the insurgency. Operational-level intelligence adds information about national and international politics and their effects on the operational environment.

3-165. Analysis of enemy actions and comprehensive insurgency analysis are done at battalion level and above. These processes build on IPB and use the tools discussed in appendix B. Analysis of enemy actions, commonly called current operations, focuses on what the enemy is doing now. Comprehensive insurgency analysis focuses on the people in the AO. It develops information about relationships among them and the ideas and beliefs driving their actions. Comprehensive insurgency analysis brings together all other forms of analysis.

CURRENT OPERATIONS

3-166. Current operations intelligence supports a commander's understanding of what insurgents are currently doing. The basic tasks of analysts working in current operations are to—

- Analyze past and current enemy actions (event analysis and pattern analysis) to look for changes in the insurgents' approach or tactics.
- Track the effects of friendly operations on the populace and insurgents.
- Provide intelligence support to ongoing operations.
- Disseminate immediate threat warnings to appropriate consumers.

3-167. Intelligence for current operations comes from a variety of sources, but operations reports are particularly important. This is because current enemy activities are more often reported by patrols, units conducting raids, or observation posts than they are by dedicated intelligence collectors. OSINT is important for tracking IO effects. Current operations analysis depends on the insurgent actions database for determining changes in insurgent tactics and techniques.

COMPREHENSIVE INSURGENCY ANALYSIS

3-168. Accurate and thorough intelligence on insurgent organizations, leadership, financial support networks, and the operational environment contribute to more effective friendly operations. Comprehensive insurgency analysis integrates a range of analytic tools to develop this intelligence. (These tools include social network analysis and socio-cultural factors analysis.) Comprehensive insurgency analysis provides information upon which commanders and staffs base their understanding of the following:

- Insurgent organization.
- Insurgent leadership.
- Key nodes in the insurgent organization.
- Insurgents' approach, capabilities, and motivations.
- Insurgents' support base.
- Insurgent links to the community.

Effectively developing and integrating information from a range of intelligence and operations sources provides the detailed knowledge and insights required to exploit insurgents' vulnerabilities and mitigate their strengths. Table 3-9 lists key tasks associated with comprehensive insurgency analysis.

Table 3-9. Comprehensive insurgency analysis tasks

• Identify insurgent strategic, operational, and tactical goals, objectives, and imperatives. • Identify motivations, fears, concerns, and perceptions that shape the actions of insurgents and their supporters. • Identify grievances, fears, and concerns that the insurgents exploit. • Determine how culture, interests, and history inform insurgent and host-nation decision making. • Understand links among political, religious, tribal, criminal, and other social networks. • Determine how social networks, key leaders, and groups interact with insurgent networks. • Determine the structure and function of insurgent organizations. • Identify key insurgent activities and leaders. • Understand popular and insurgent perceptions of the host-nation, insurgency, and counterinsurgents —and how these affect the insurgency.

3-169. Developing knowledge and using network analytic tools requires an unusually large investment of time compared to conventional analytic problem-solving methods. Comprehensive insurgency analysis may not provide immediate usable intelligence. Analysts may have to spend weeks or months analyzing numerous all-source intelligence reports before providing an accurate picture of insurgent groups, leaders and activities. It is essential that commanders designate a group of analysts to perform comprehensive insurgency analysis. This team must be insulated from the short-term demands of current operations and day-to-day intelligence demands. These analysts focus on long-term intelligence development. It is ultimately the commander's responsibility to ensure that comprehensive and basic insurgent network analysis still occurs despite high-profile demands and time sensitive requirements.

3-170. Comprehensive insurgency analysis examines interactions among individuals, groups, and beliefs within the operational environment's historic and cultural context. One of the more important products of this analysis is an understanding of how local people think. This knowledge allows predictive analysis of enemy actions. It also contributes to the ability to develop effective IO and civil-military operations.

REACHBACK

3-171. *Reachback* refers to the process of obtaining products, services, and applications, or forces, or equipment, or material from organizations that are not forward deployed (JP 1-02). Deployed or deploying units should use reach capabilities to "outsource" time-intensive aspects of analysis. Reachback is particularly useful when deployments occur with little warning and when organizations used for reach have a great deal of expertise available on a given subject. Analysts may receive reach assistance from higher echelons or external sources. Most organizations affiliated with DOD regard assisting field commanders as one of their primary missions.

ANALYTIC CONTINUITY

3-172. The complexity and difficulty of analyzing an insurgency means it often takes analysts months to understand the operational environment and the insurgency. The most productive analysts and action officers generally have more than a year focused on an aspect of the insurgency problem; therefore, commanders should try to maintain continuity among their analysts. Intelligence and other staff sections should track operations from their home station and immerse themselves in related intelligence before deploying. This flattens the learning curve of units rotating in to an AO and increases their effectiveness during the deployment.

3-173. Part of unit transition should include exchange of relevant databases and knowledge of the AO. Effective intelligence handover saves time and effort for an incoming unit and ensures consistency of approach.

SECTION VI – INTELLIGENCE COLLABORATION

3-174. Effective intelligence collaboration organizes the collection and analysis actions of various units and organizations into a coherent, mutually supportive intelligence effort. Collaboration and synchronization of the effort between lower and higher echelon units and organizations reduces the likelihood of gaps in the intelligence effort. Some of the important operational- and strategic-level analytic support is called "tactical overwatch." The intelligence portion of the common operational picture and other supporting intelligence for COIN operations is complex. Insurgencies do not normally lend themselves to generalizations like "if this leader is removed, the insurgency is over" or "this group drives the movement." It is important not to oversimplify an insurgency. However, analysts and commanders still require a commonly understood means of defining and describing the enemy. One such means is using the following categories to track and report the insurgency:

- Region.
- Insurgent organization.
- Key personalities.
- Insurgent goals and motivations.

The mutual support that various intelligence units and organizations provide across all echelons facilitates timely and relevant intelligence.

3-175. Insurgencies are often localized; however, most have national or international aspects to them. This characteristic complicates intelligence collaboration between adjacent units and among various echelons. For instance, if numerous insurgent groups operate in one country, adjacent battalions within the country may face very different threats. Higher echelon analysts must then understand multiple insurgent organizations and determine the links, if any, among them. Usually, battalions focus on the population and insurgents in their AO. Higher echelon analysts determine links and interactions among the populace and insurgents across unit boundaries. Combatant-command-level analysts determine the major linkages within the area of responsibility and internationally. Based on these requirements, a common database based on intelligence reporting is a prerequisite for effective intelligence fusion.

3-176. Also complicating collaboration is the fact that COIN operations involve many government agencies and foreign security forces. Analysts must establish good working relationships with various agencies and elements to ensure they can fuse intelligence.

INTELLIGENCE CELLS AND WORKING GROUPS

3-177. Intelligence community assets operating in an AO work in or coordinate with the intelligence cell in one of the unit's command posts. They are under the staff supervision of the unit intelligence officer. Table 3-10 (page 3-34) lists examples of intelligence community assets that may operate in a division AO.

3-178. As necessary, intelligence officers form working groups or boards to synchronize collection, analysis, and targeting efforts. Cells and working groups conduct regular meetings to accomplish the following:

- Establish and maintain shared situational awareness.
- Share collection priorities.
- Deconflict activities and operations.

- Discuss target development.
- Share results of operations.

These meetings build mutual trust and understanding of each member's mission, capabilities, and limitations. Meetings should be coordinated with meetings of other staff cells, working groups, and boards (for example, the targeting board) as part of the command post's battle rhythm.

3-179. An effective intelligence cell enhances the commander's knowledge of the enemy, local populace, and friendly forces and agencies operating in the unit's AO. Incorporating HN representatives (for example, intelligence services, military forces, and local government officials) and multinational partners into the intelligence cell should also be considered to foster teamwork, gain insight into the local society, and prepare the host nation to assume the COIN mission when multinational forces depart.

Table 3-10. Possible intelligence community assets in a division area of operations

• Defense Intelligence Agency ▪ Defense human intelligence case officers ▪ Reports officers ▪ Document exploitation teams • Central Intelligence Agency ▪ Chief of base ▪ Case officers ▪ Reports officers • National intelligence support team (from the U.S. Army Intelligence and Security Command —INSCOM) • National Geospatial-Intelligence Agency • Department of the Treasury analytic support teams	• U.S. Special Operations Command ▪ Special forces operational detachment-alpha teams ▪ Special mission units ▪ Civil affairs teams ▪ Psychological operations teams • Immigration and Customs Enforcement/ Department of Homeland Security agents • Air Force Office of Special Investigations special agents • Federal Bureau of Investigation agents • Department of State political advisor • National Security Agency

Protecting Sources

3-180. Protecting sources is another important consideration when sharing intelligence. Organizations may sometimes choose not to share information because acting on intelligence can compromise its sources. Using the targeting process to synchronize targeting decisions is usually a good way to protect sources. (See paragraphs 5-100 through 5-112.)

HOST-NATION INTEGRATION

3-181. COIN operations require U.S. personnel to work closely with the host nation. Sharing intelligence with HN security forces and government personnel is an important and effective means of supporting their COIN efforts. However, HN intelligence services may not be well developed and their personnel may not be well trained. Thus, HN intelligence should be considered useful but definitely not the only intelligence available. Usually, HN services are not the most important intelligence source. It is essential for U.S. personnel to evaluate HN intelligence capabilities and offer training as required.

3-182. In addition, infiltration of HN security forces by insurgents or foreign intelligence services can create drawbacks to intelligence sharing. Insurgents may learn what is known about them, gain insight into COIN intelligence sources and capabilities, and get early warning of targeting efforts.

3-183. When sharing intelligence with the host nation, it is important to understand the level of infiltration by insurgents or foreign intelligence services. Insofar as possible, intelligence should be tailored so required intelligence still gets to HN consumers but does not give away information about sources and capabilities. In addition, care is needed when providing targeting information; it should be done such that insurgents do not receive early warning of an upcoming operation. As trust develops between HN and U.S. personnel, the amount of intelligence shared should grow. This will make the COIN effort more effective.

SECTION VII – SUMMARY

3-184. What makes intelligence analysis for COIN so distinct and so challenging is the amount of sociocultural information that must be gathered and understood. However, truly grasping the operational environment requires commanders and staffs to devote at least as much effort to understanding the people they support as they do to understanding the enemy. All this information is essential to get at the root causes of the insurgency and to determine the best ways to combat it. Identifying the real problem and developing solutions is the essence of operational design, which is discussed in the next chapter.

This page intentionally left blank.

Chapter 4

Designing Counterinsurgency Campaigns and Operations

The first, the supreme, the most far-reaching act of judgment that the statesman and commander have to make is to establish...the kind of war on which they are embarking; neither mistaking it for, nor trying to turn it into, something that is alien to its nature. This is the first of all strategic questions and the most comprehensive.

Carl von Clausewitz, *On War*

This chapter describes considerations for designing counterinsurgency campaigns and operations. For Army forces, this chapter applies aspects of command and control doctrine and planning doctrine to counterinsurgency campaign planning. While campaign design is most often associated with a joint force command, all commanders and staffs need to understand it.

THE IMPORTANCE OF CAMPAIGN DESIGN

4-1. In chapter 1, insurgency is described as an organized, protracted politico-military struggle designed to weaken government control and legitimacy while increasing insurgent control. Ultimately, the long-term objective for both sides in that struggle remains acceptance by the people of the state or region of the legitimacy of one side's claim to political power. The reason an insurgency forms to challenge the existing order is different in each case. The complexity of insurgency presents problems that have incomplete, contradictory, and changing requirements. The solutions to these intensely challenging and complex problems are often difficult to recognize as such because of complex interdependencies. While attempting to solve an intensely complex problem, the solution of one of its aspects may reveal or create another, even more complex, problem. The purpose of design is to achieve a greater understanding, a proposed solution based on that understanding, and a means to learn and adapt. For a U.S. military commander directed to counter an insurgency, knowing why an insurgent movement has gained support and the purpose of American involvement is essential in designing a counterinsurgency (COIN) campaign. Failure to understand both factors can have disastrous consequences, as illustrated by Napoleon's experience in Spain.

Campaign Assessment and Reassessment

During Napoleon's occupation of Spain in 1808, it seems little thought was given to the potential challenges of subduing the Spanish populace. Conditioned by the decisive victories at Austerlitz and Jena, Napoleon believed the conquest of Spain would be little more than a "military promenade." Napoleon's campaign included a rapid conventional military victory but ignored the immediate requirement to provide a stable environment for the populace.

The French failed to analyze the Spanish people, their history, culture, motivations, and potential to support or hinder the achievement of French political objectives. The Spanish people were accustomed to hardship, suspicious of foreigners and constantly involved in skirmishes with security forces. Napoleon's cultural miscalculation resulted in a protracted occupation struggle that lasted nearly six years and ultimately required approximately three-fifths of the Empire's total armed strength, almost four times the force of 80,000 Napoleon originally designated.

> The Spanish resistance drained the resources of the French Empire. It was the beginning of the end for Napoleon. At the theater level, a complete understanding of the problem and a campaign design that allowed the counterinsurgency force to learn and adapt was lacking.

4-2. Design and planning are qualitatively different yet interrelated activities essential for solving complex problems. While planning activities receive consistent emphasis in both doctrine and practice, discussion of design remains largely abstract and is rarely practiced. Presented a problem, staffs often rush directly into planning without clearly understanding the complex environment of the situation, purpose of military involvement, and approach required to address the core issues. This situation is particularly problematic with insurgencies. Campaign design informs and is informed by planning and operations. It has an intellectual foundation that aids continuous assessment of operations and the operational environment. Commanders should lead the design process and communicate the resulting framework to other commanders for planning, preparation, and execution.

THE RELATIONSHIP BETWEEN DESIGN AND PLANNING

4-3. It is important to understand the distinction between design and planning. (See figure 4-1.) While both activities seek to formulate ways to bring about preferable futures, they are cognitively different. Planning applies established procedures to solve a largely understood problem within an accepted framework. Design inquires into the nature of a problem to conceive a framework for solving that problem. In general, planning is problem solving, while design is problem setting. Where planning focuses on generating a plan—a series of executable actions—design focuses on learning about the nature of an unfamiliar problem.

Design	Planning
Problem-setting	• Problem-solving
Conceptual—blank sheet	• Physical and detailed
Questions assumptions and methods	• Procedural
Develops understanding	• Develops products
Paradigm-setting	• Paradigm-accepting
Complements planning, preparation, execution, and assessment	• Patterns and templates activity
Commander-driven dialog	• Staff-centered process

Figure 4-1. Design and planning continuum

4-4. When situations do not conform to established frames of reference—when the hardest part of the problem is figuring out what the problem is—planning alone is inadequate and design becomes essential. In these situations, absent a design process to engage the problem's essential nature, planners default to doctrinal norms; they develop plans based on the familiar rather than an understanding of the real situation. Design provides a means to conceptualize and hypothesize about the underlying causes and dynamics that explain an unfamiliar problem. Design provides a means to gain understanding of a complex problem and insights towards achieving a workable solution.

4-5. This description of design at the tactical level is a form of what Army doctrine calls commander's visualization. Commanders begin developing their design upon receipt of a mission. Design precedes and forms the foundation for staff planning. However, design is also continuous throughout the operation. As part of assessment, commanders continuously test and refine their design to ensure the relevance of military action to the situation. In this sense, design guides and informs planning, preparation, execution, and assessment. However, a plan is necessary to translate a design into execution. (FM 6-0, paragraphs 4-17 through 4-25, discusses commander's visualization.)

4-6. Planning focuses on the physical actions intended to directly affect the enemy or environment. Planners typically are assigned a mission and a set of resources; they devise a plan to use those resources to accomplish that mission. Planners start with a design (whether explicit or implicit) and focus on generating a plan—a series of executable actions and control measures. Planning generally is analytic and reductionist. It breaks the design into manageable pieces assignable as tasks, which is essential to transforming the design into an executable plan. Planning implies a stepwise process in which each step produces an output that is the necessary input for the next step. (FM 5-0 contains Army planning doctrine. MCDP 5 contains Marine Corps planning doctrine.)

THE NATURE OF DESIGN

4-7. Given the difficult and multifaceted problems of insurgencies, dialog among the commander, principal planners, members of the interagency team, and host-nation (HN) representatives helps develop a coherent design. This involvement of all participants is essential. The object of this dialog is to achieve a level of situational understanding at which the approach to the problem's solution becomes clear. The underlying premise is this: when participants achieve a level of understanding such that the situation no longer appears complex, they can exercise logic and intuition effectively. As a result, design focuses on framing the problem rather than developing courses of action.

4-8. COIN design must be iterative. By their nature, COIN efforts require repeated assessments from different perspectives to see the various factors and relationships required for adequate understanding. Assessment and learning enable incremental improvements to the design. The aim is to rationalize the problem—to construct a logical explanation of observed events and subsequently construct the guiding logic that unravels the problem. The essence of this is the mechanism necessary to achieve success. This mechanism may not be a military activity—or it may involve military actions in support of nonmilitary activities. Once commanders understand the problem and what needs to be accomplished to succeed, they identify the means to assess effectiveness and the related information requirements that support assessment. This feedback becomes the basis for learning, adaptation, and subsequent design adjustment.

CONSIDERATIONS FOR DESIGN

4-9. Key design considerations include the following:

- Critical discussion.
- Systems thinking.
- Model making.
- Intuitive decision making.
- Continuous assessment.
- Structured learning.

4-10. Rigorous and structured *critical discussion* provides an opportunity for interactive learning. It deepens shared understanding and leverages the collective intelligence and experiences of many people.

4-11. *Systems thinking* involves developing an understanding of the relationships within the insurgency and the environment. It also concerns the relationships of actions within the various logical lines of operations (LLOs). This element is based on the perspective of the systems sciences that seeks to understand the interconnectedness, complexity, and wholeness of the elements of systems in relation to one another.

4-12. In *model making*, the model describes an approach to the COIN campaign, initially as a hypothesis. The model includes operational terms of reference and concepts that shape the language governing the conduct (planning, preparation, execution, and assessment) of the operation. It addresses questions like these: Will planning, preparation, execution, and assessment activities use traditional constructs like center of gravity, decisive points, and LLOs? Or are other constructs—such as leverage points, fault lines, or critical variables—more appropriate to the situation?

4-13. The Army and Marine Corps define *intuitive decision making* as the act of reaching a conclusion which emphasizes pattern recognition based on knowledge, judgment, experience, education, intelligence, boldness, perception, and character. This approach focuses on assessment of the situation vice comparison of multiple options (FM 6-0/MCRP 5-12A). An operational design emerges intuitively as understanding of the insurgency deepens.

4-14. *Continuous assessment* is essential as an operation unfolds because of the inherent complexity of COIN operations. No design or model completely matches reality. The object of continuous assessment is to identify where and how the design is working or failing and to consider adjustments to the design and operation.

4-15. The objective of *structured learning* is to develop a reasonable initial design and then learn, adapt, and iteratively and continuously improve that design as more about the dynamics of the COIN problem become evident.

DESIGN FOR COUNTERINSURGENCY

4-16. Through design commanders gain an understanding of the problem and the COIN operation's purpose within the strategic context. Communicating this understanding of the problem, purpose, and context to subordinates allows them to exercise subordinates' initiative. *Subordinates' initiative* is assumption of responsibility for deciding and initiating independent actions when the concept of operations or order no longer applies or when an unanticipated opportunity leading to the accomplishment of the commander's intent presents itself (FM 6-0). (Subordinates' initiative is discussed in FM 6-0, paragraphs 2-83 through 2-92.) It facilitates decentralized execution and continuous assessment of operations at all levels throughout the campaign. While traditional aspects of campaign design as expressed in joint and Service doctrine remain relevant, they are not adequate for a discussion of the broader design construct for a COIN environment. Inherent in this construct is the tension created by understanding that military capabilities provide only one component of an overall approach to a COIN campaign. Design of a COIN campaign must be viewed holistically. Only a comprehensive approach employing all relevant design components, including the other instruments of national power, is likely to reach the desired end state.

4-17. As noted above, this description of campaign design is a form that Army doctrine calls commander's visualization. Design begins with identification of the end state, as derived from the policy aim. (Joint doctrine defines the *end state* as the set of required conditions that defines achievement of the commander's objectives [JP 1-02]). The end state provides context and logic for operational and tactical decision making. Consequently, strategic goals must be communicated clearly to commanders at every level. While strategy drives design, which in turn drives tactical actions, the reverse is also true. The observations of tactical actions result in learning and greater understanding that may generate modifications to the design, which in turn may have strategic implications. The COIN imperative to "Learn and Adapt" is essential in making the design process work correctly. Figure 4-2 illustrates the iterative nature of COIN campaign design and the large number of factors involved.

COMMANDER'S INTENT AND VISION OF RESOLUTION

4-18. Guided by the campaign's purpose, commanders articulate an operational logic for the campaign that expresses in clear, concise, conceptual language a broad vision of what they plan to accomplish. The operational logic is the commander's assessment of the problem and approach toward solving it. Commanders express it as the commander's intent. Ideally, the operational logic is expressed clearly and simply but in comprehensive terms, such as what the commander envisions achieving with various components or particular LLOs. This short statement of the operational logic helps subordinate commanders and planners, as well as members of other agencies and organizations, see the campaign's direction. It provides a unifying theme for interagency planning.

4-19. In addition, commanders also issue a form of planning guidance called the vision of resolution. The vision of resolution is usually expressed in the form of LLOs. LLOs for a counterinsurgency may include the following:

- Conduct information operations.
- Conduct combat operations/civil security operations.
- Train and employ HN security forces.

- Establish or restore essential services.
- Support development of better governance.
- Support economic development.

This list is an example only. Commanders determine the LLOs appropriate to the situation based on their assessment and their dialog with the leaders of other participating organizations.

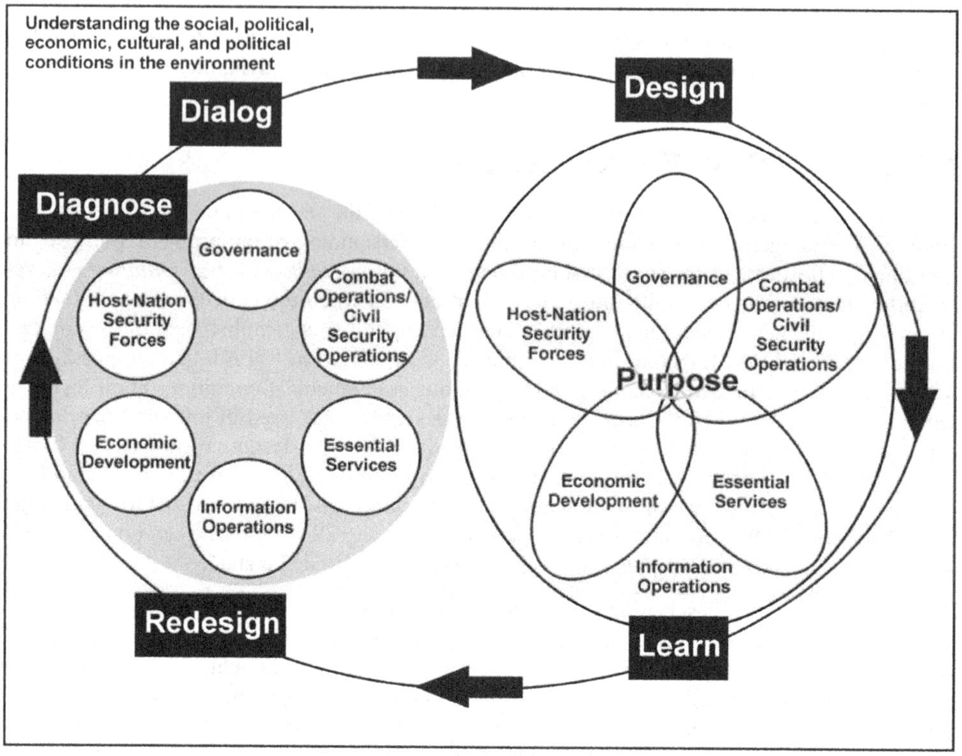

Figure 4-2. Iterative counterinsurgency campaign design

4-20. LLOs like those listed in paragraph 4-19 are not intended as a "success template." Selecting and applying them requires judgment. The mosaic nature of insurgencies and the shifting circumstances within each area of operations (AO) requires a different emphasis on and interrelationship among the various lines. The situation may also require that military forces closely support, or temporarily assume responsibility for, tasks normally accomplished by other government agencies and private organizations. By broadly describing how the LLOs interact to achieve the end state, commanders provide the operational logic to link the various components in a comprehensive framework. This framework guides the initiative of subordinate commanders as they establish local conditions that support achieving the overall end state. It also promotes unity of effort among joint, interagency, multinational, and HN partners.

LOCAL RELEVANCE

4-21. Informed by the commander's intent—including the end state and vision of resolution—subordinate commanders tailor and prioritize their actions within the LLOs based on the distinct and evolving circumstances within their respective AOs. Military forces are accustomed to unity of command; however, the interagency and multinational nature of COIN operations usually makes such arrangements unlikely. All participating organizations do share attitudes and goals. General cooperation on matters of mutual concern, established through informal agreements, may be the most practicable arrangement. Therefore, effective commanders empower subordinate leaders to perform the coordination, cooperation, and innovation needed to achieve unity of effort and execute operations in the manner best suited to local conditions. The design—consisting of the commander's intent, vision of resolution and other guidance issued as the campaign unfolds, and end state—provides the framework within which subordinates exercise this form of initiative.

LEARNING IN EXECUTION

4-22. Before commanders deploy their units, they make every effort to mentally prepare their Soldiers or Marines for the anticipated challenges, with a particular focus on situational awareness of the anticipated AO. *Situational awareness* is knowledge of the immediate present environment, including knowledge of the factors of METT-TC (FMI 5-0.1). COIN operations require a greater focus on civil considerations—the C in METT-TC—than conventional operations do. This situational awareness is only the beginning of an understanding of the AO that will mature as operations progress. However, commanders use it to begin to establish a common frame of reference.

4-23. Design begins based on this initial awareness. Aspects of the problem and means of resolving them do not remain static. Conditions are seldom consistent throughout any AO and continue to change based on actions by friendly, enemy, neutral, and other involved organizations. Rather than being uniform in character, the operational environment is likely to display a complex, shifting mosaic of conditions. To be effective, commanders—and indeed all personnel—continually develop and enhance their understanding of the mosaic peculiar to their AO. Observing tactical actions and the resulting changing conditions deepens understanding of the environment and enables commanders to relearn and refine their design and implementation actions.

4-24. Initially, situational awareness will probably be relatively low and the design will, by necessity, require a number of assumptions, especially with respect to the populace and the force's ability to positively influence their perception of events. The design can be viewed as an experiment that tests the operational logic, with the expectation of a less-than-perfect solution. As the experiment unfolds, interaction with the populace and insurgents reveals the validity of those assumptions, revealing the strengths and weaknesses of the design.

4-25. *Assessment* is the continuous monitoring and evaluation of the current situation and progress of an operation (FMI 5-0.1). Effective assessment is necessary for commanders to recognize changing conditions and determine their meaning. It is crucial to successful adaptation and innovation by commanders within their respective AOs. A continuous dialog among commanders at all echelons provides the feedback the senior commander needs to refine the design. The dialog is supported by formal assessment techniques and red-teaming to ensure commanders are fully cognizant of the causal relationships between their actions and the insurgents' adaptations. Accordingly, assessment is a learning activity and a critical aspect of design. This learning leads to redesign. Therefore, design can be viewed as a perpetual design-learn-redesign activity, with the commander's intent, vision of resolution, and end state providing the unifying themes.

4-26. The critical role of assessment necessitates establishing measures of effectiveness during planning. Commanders should choose these carefully so that they align with the design and reflect the emphasis on and interrelationship among the LLOs. Commanders and staffs revise their assessment and measures of effectiveness during the operation in order to facilitate redesign and stay abreast of the current situation. Sound assessment blends qualitative and quantitative analysis with the judgment and intuition of all leaders. Great care must be applied here, as COIN operations often involve complex societal issues that may not lend themselves to quantifiable measures of effectiveness. Moreover, bad assumptions and false data can undermine the validity of both assessments and the conclusions drawn from them. Data and metrics can inform a commander's assessment. However they must not be allowed to dominate it in uncertain situations. Subjective and intuitive assessment must not be replaced by an exclusive focus on data or metrics. Commanders must exercise their professional judgment in determining the proper balance.

GOALS IN COUNTERINSURGENCY

4-27. In an ideal world, the commander of military forces engaged in COIN operations would enjoy clear and well-defined goals for the campaign from the very beginning. However, the reality is that many goals emerge only as the campaign develops. For this reason, counterinsurgents usually have a combination of defined and emerging goals toward which to work. Likewise, the complex problems encountered during COIN operations can be so difficult to understand that a clear design cannot be developed initially. Often, the best choice is to create iterative solutions to better understand the problem. In this case, these iterative solutions allow the initiation of intelligent interaction with the environment. The experiences of the 1st Marine Division during Operation Iraqi Freedom II illustrate this situation.

Iterative Design During Operation Iraqi Freedom II

During Operation Iraqi Freedom II (2004-2005), the 1st Marine Division employed an operational design similar to that used during the Philippine Insurrection (circa 1902). The commanding general, Major General James N. Mattis, USMC, began with an assessment of the people that the Marines, Soldiers, and Sailors would encounter within the division's area of operations. The area of operations was in western Iraq/Al Anbar Province, which had a considerably different demographic than the imam-led Shia areas in which the division had operated during Operation Iraqi Freedom I.

Major General Mattis classified provincial constituents into three basic groups: the tribes, former regime elements, and foreign fighters. The tribes constituted the primary identity group in western Iraq/Al Anbar Province. They had various internal tribal affiliations and looked to a diverse array of sheiks and elders for leadership. The former regime elements were a minority that included individuals with personal, political, business, and professional ties to the Ba'ath Party. These included civil servants and career military personnel with the skills needed to run government institutions. Initially, they saw little gain from a democratic Iraq. The foreign fighters were a small but dangerous minority of transnational Islamic subversives.

To be successful, U.S. forces had to apply a different approach to each of these groups within the framework of an overarching plan. As in any society, some portion of each group included a criminal element, further complicating planning and interaction. Major General Mattis's vision of resolution comprised two major elements encompassed in an overarching "bodyguard" of information operations. (See figure 4-3, page 4-8.)

The first element and main effort was diminishing support for insurgency. Guided by the maxims of "first do no harm" and "no better friend–no worse enemy," the objective was to establish a secure local environment for the indigenous population so they could pursue their economic, social, cultural, and political well-being and achieve some degree of local normalcy. Establishing a secure environment involved both offensive and defensive combat operations with a heavy emphasis on training and advising the security forces of the fledgling Iraqi government. It also included putting the populace to work. Simply put, an Iraqi with a job was less likely to succumb to ideological or economic pressure to support the insurgency. Other tasks included the delivery of essential services, economic development, and the promotion of governance. All were geared towards increasing employment opportunities and furthering the establishment of local normalcy. Essentially, diminishing support for insurgency entailed gaining and maintaining the support of the tribes, as well as converting as many of the former regime members as possible. "Fence-sitters" were considered a winnable constituency and addressed as such.

The second element involved neutralizing the bad actors, a combination of irreconcilable former regime elements and foreign fighters. Offensive combat operations were conducted to defeat recalcitrant former regime members. The task was to make those who were not killed outright see the futility of resistance and give up the fight. With respect to the hard-core extremists, who would never give up, the task was more straightforward: their complete and utter destruction. Neutralizing the bad actors supported the main effort by improving the local security environment. Neutralization had to be accomplished in a discrete and discriminate manner, however, in order to avoid unintentionally increasing support for insurgency.

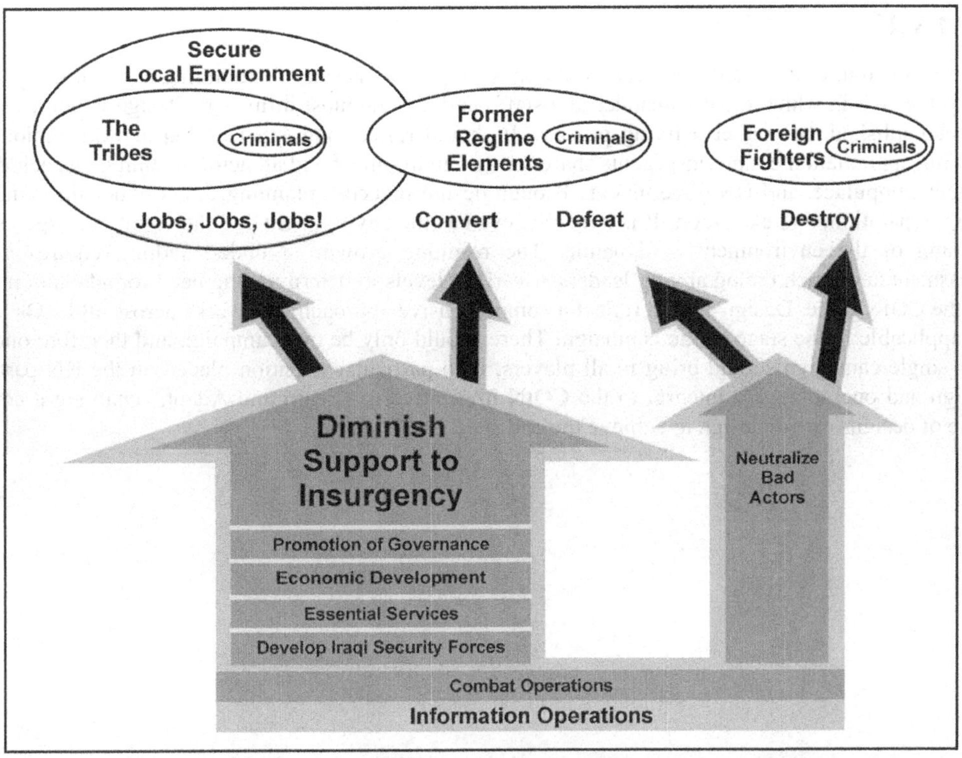

Figure 4-3. 1st Marine Division's operational design for Operation Iraqi Freedom II

Both elements described above were wrapped in an overarching "bodyguard" of information operations. Information operations, both proactive and responsive, were aggressively employed to favorably influence the populace's perception of all coalition actions while simultaneously discrediting the insurgents. These tasks were incredibly difficult for a number of reasons. Corruption had historically been prevalent among Iraqi officials, generating cynicism toward any government. Additionally, decades of Arab media mischaracterization of U.S. actions had instilled distrust of American motives. The magnitude of that cynicism and distrust highlighted the critical importance of using information operations to influence every situation.

In pursuing this vision of resolution, the 1st Marine Division faced an adaptive enemy. Persistent American presence and interaction with the populace threatened the insurgents and caused them to employ more open violence in selected areas of Al Anbar province. This response resulted in learning and adaptation within the 1st Marine Division. The design enabled 1st Marine Division to adjust the blend of "diminishing support for insurgents" and "neutralizing bad actors" to meet the local challenges. Throughout the operation, 1st Marine Division continued learning and adapting with the vision of resolution providing a constant guide to direct and unify the effort.

SUMMARY

4-28. Campaign design may very well be the most important aspect of countering an insurgency. It is certainly the area in which the commander and staff can have the most influence. Design is not a function to be accomplished, but rather a living process. It should reflect ongoing learning and adaptation and the growing appreciation counterinsurgents share for the environment and all actors within it, especially the insurgents, populace, and HN government. Though design precedes planning, it continues throughout planning, preparation, and execution. It is dynamic, even as the environment and the counterinsurgents' understanding of the environment is dynamic. The resulting growth in understanding requires integrated assessment and a rich dialog among leaders at various levels to determine the need for adaptation throughout the COIN force. Design should reflect a comprehensive approach that works across all LLOs in a manner applicable to the stage of the campaign. There should only be one campaign and therefore one design. This single campaign should bring in all players, with particular attention placed on the HN participants. Design and operations are integral to the COIN imperative to "Learn and Adapt," enabling a continuous cycle of design-learn-redesign to achieve the end state.

This page intentionally left blank.

Chapter 5

Executing Counterinsurgency Operations

It is a persistently methodical approach and steady pressure which will gradually wear the insurgent down. The government must not allow itself to be diverted either by countermoves on the part of the insurgent or by the critics on its own side who will be seeking a simpler and quicker solution. There are no short-cuts and no gimmicks.

Sir Robert Thompson,
Defeating Communist Insurgency: The Lessons of Malaya and Vietnam, 1966 [2]

This chapter addresses principles and tactics for executing counterinsurgency (COIN) operations. It begins by describing the different stages of a COIN operation and logical lines of operations that commanders can use to design one. It continues with discussions of three COIN approaches and how to continuously assess a COIN operation. The chapter concludes by describing lethal and nonlethal targeting in a COIN environment.

THE NATURE OF COUNTERINSURGENCY OPERATIONS

5-1. Counterinsurgency (COIN) operations require synchronized application of military, paramilitary, political, economic, psychological, and civic actions. Successful counterinsurgents support or develop local institutions with legitimacy and the ability to provide basic services, economic opportunity, public order, and security. The political issues at stake are often rooted in culture, ideology, societal tensions, and injustice. As such, they defy nonviolent solutions. Military forces can compel obedience and secure areas; however, they cannot by themselves achieve the political settlement needed to resolve the situation. Successful COIN efforts include civilian agencies, U.S. military forces, and multinational forces. These efforts purposefully attack the basis for the insurgency rather than just its fighters and comprehensively address the host nation's core problems. Host-nation (HN) leaders must be purposefully engaged in this effort and ultimately must take lead responsibility for it.

5-2. There are five overarching requirements for successful COIN operations:

- U.S. and HN military commanders and the HN government together must devise the plan for attacking the insurgents' strategy and focusing the collective effort to bolster or restore government legitimacy.
- HN forces and other counterinsurgents must establish control of one or more areas from which to operate. HN forces must secure the people continuously within these areas.
- Operations should be initiated from the HN government's areas of strength against areas under insurgent control. The host nation must retain or regain control of the major population centers to stabilize the situation, secure the government's support base, and maintain the government's legitimacy.
- Regaining control of insurgent areas requires the HN government to expand operations to secure and support the population. If the insurgents have established firm control of a region, their military apparatus there must be eliminated and their politico-administrative apparatus rooted out.
- Information operations (IO) must be aggressively employed to accomplish the following:

- Favorably influence perceptions of HN legitimacy and capabilities.
- Obtain local, regional, and international support for COIN operations.
- Publicize insurgent violence.
- Discredit insurgent propaganda and provide a more compelling alternative to the insurgent ideology and narrative.

5-3. COIN operations combine offensive, defensive, and stability operations to achieve the stable and secure environment needed for effective governance, essential services, and economic development. The focus of COIN operations generally progresses through three indistinct stages that can be envisioned with a medical analogy:

- Stop the bleeding.
- Inpatient care—recovery.
- Outpatient care—movement to self-sufficiency.

Understanding this evolution and recognizing the relative maturity of the operational environment are important to the conduct (planning, preparation, execution, and assessment) of COIN operations. This knowledge allows commanders to ensure that their activities are appropriate to the current situation.

Initial Stage: "Stop the Bleeding"

5-4. Initially, COIN operations are similar to emergency first aid for the patient. The goal is to protect the population, break the insurgents' initiative and momentum, and set the conditions for further engagement. Limited offensive operations may be undertaken, but are complemented by stability operations focused on civil security. During this stage, friendly and enemy information needed to complete the common operational picture is collected and initial running estimates are developed. Counterinsurgents also begin shaping the information environment, including the expectations of the local populace.

Middle Stage: "Inpatient Care—Recovery"

5-5. The middle stage is characterized by efforts aimed at assisting the patient through long-term recovery or restoration of health—which in this case means achieving stability. Counterinsurgents are most active here, working aggressively along all logical lines of operations (LLOs). The desire in this stage is to develop and build resident capability and capacity in the HN government and security forces. As civil security is assured, focus expands to include governance, provision of essential services, and stimulation of economic development. Relationships with HN counterparts in the government and security forces and with the local populace are developed and strengthened. These relationships increase the flow of human and other types of intelligence. This intelligence facilitates measured offensive operations in conjunction with the HN security forces. The host nation increases its legitimacy through providing security, expanding effective governance, providing essential services, and achieving incremental success in meeting public expectations.

Late Stage: "Outpatient Care—Movement to Self-sufficiency"

5-6. Stage three is characterized by the expansion of stability operations across contested regions, ideally using HN forces. The main goal for this stage is to transition responsibility for COIN operations to HN leadership. In this stage, the multinational force works with the host nation in an increasingly supporting role, turning over responsibility wherever and whenever appropriate. Quick reaction forces and fire support capabilities may still be needed in some areas, but more functions along all LLOs are performed by HN forces with the low-key assistance of multinational advisors. As the security, governing, and economic capacity of the host nation increases, the need for foreign assistance is reduced. At this stage, the host nation has established or reestablished the systems needed to provide effective and stable government that sustains the rule of law. The government secures its citizens continuously, sustains and builds legitimacy through effective governance, has effectively isolated the insurgency, and can manage and meet the expectations of the nation's entire population.

LOGICAL LINES OF OPERATIONS IN COUNTERINSURGENCY

5-7. Commanders use LLOs to visualize, describe, and direct operations when positional reference to enemy forces has little relevance. (See figure 5-1.) LLOs are appropriate for synchronizing operations against enemies that hide among the populace. A plan based on LLOs unifies the efforts of joint, interagency, multinational, and HN forces toward a common purpose. Each LLO represents a conceptual category along which the HN government and COIN force commander intend to attack the insurgent strategy and establish HN government legitimacy. LLOs are closely related. Successful achievement of the end state requires careful coordination of actions undertaken along all LLOs.

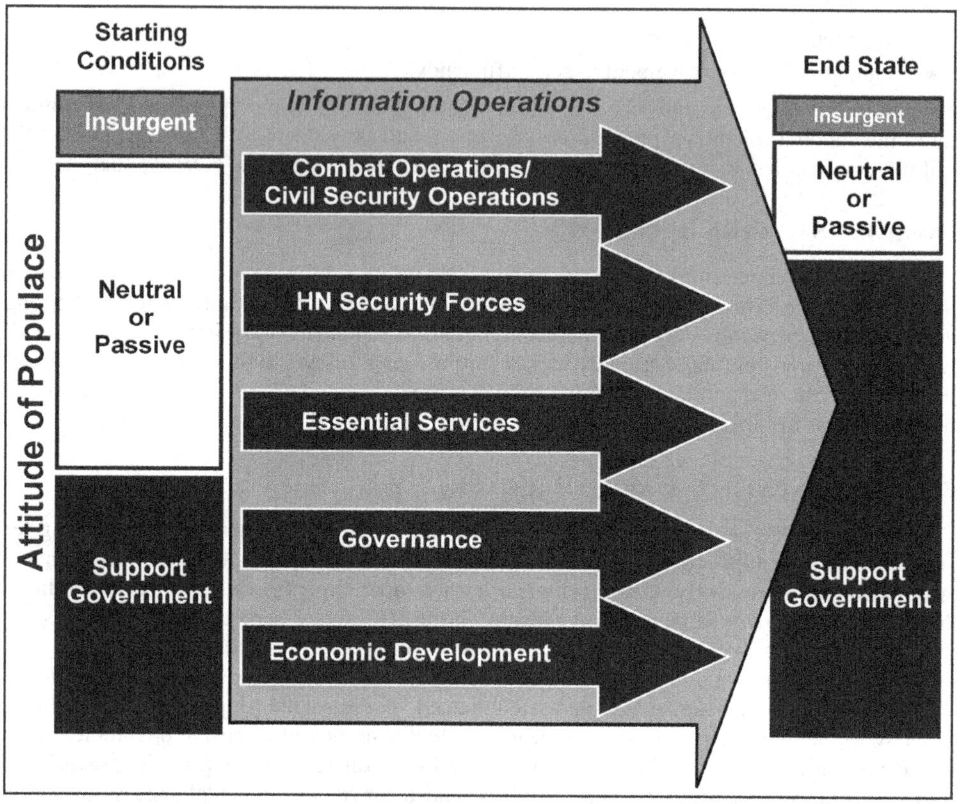

Figure 5-1. Example logical lines of operations for a counterinsurgency

5-8. Success in one LLO reinforces successes in the others. Progress along each LLO contributes to attaining a stable and secure environment for the host nation. Stability is reinforced by popular recognition of the HN government's legitimacy, improved governance, and progressive, substantive reduction of the root causes of the insurgency. There is no list of LLOs that applies in all cases. Commanders select LLOs based on their understanding of the nature of the insurgency and what the COIN force must do to counter it. Commanders designate LLOs that best focus counterinsurgent efforts against the insurgents' subversive strategy.

5-9. Commanders at all echelons can use LLOs. Lower echelon operations are nested within the higher echelon's operational design and LLOs; however, lower echelon operations are conducted based on the operational environment in each unit's area of operations (AO).

5-10. The commander's intent and vision of resolution, expressed as LLOs, describe the design for a COIN operation. Commanders and staffs synchronize activities along all LLOs to gain unity of effort. This approach ensures the LLOs converge on a well-defined, commonly understood end state.

5-11. LLOs are directly related to one another. They connect objectives that, when accomplished, support achieving the end state. Operations designed using LLOs typically employ an extended, event-driven timeline with short-, mid-, and long-term goals. These operations combine the effects of long-term operations, such as neutralizing the insurgent infrastructure, with cyclic and short-term events, like regular trash collection and attacks against insurgent bases. (See figure 5-2.)

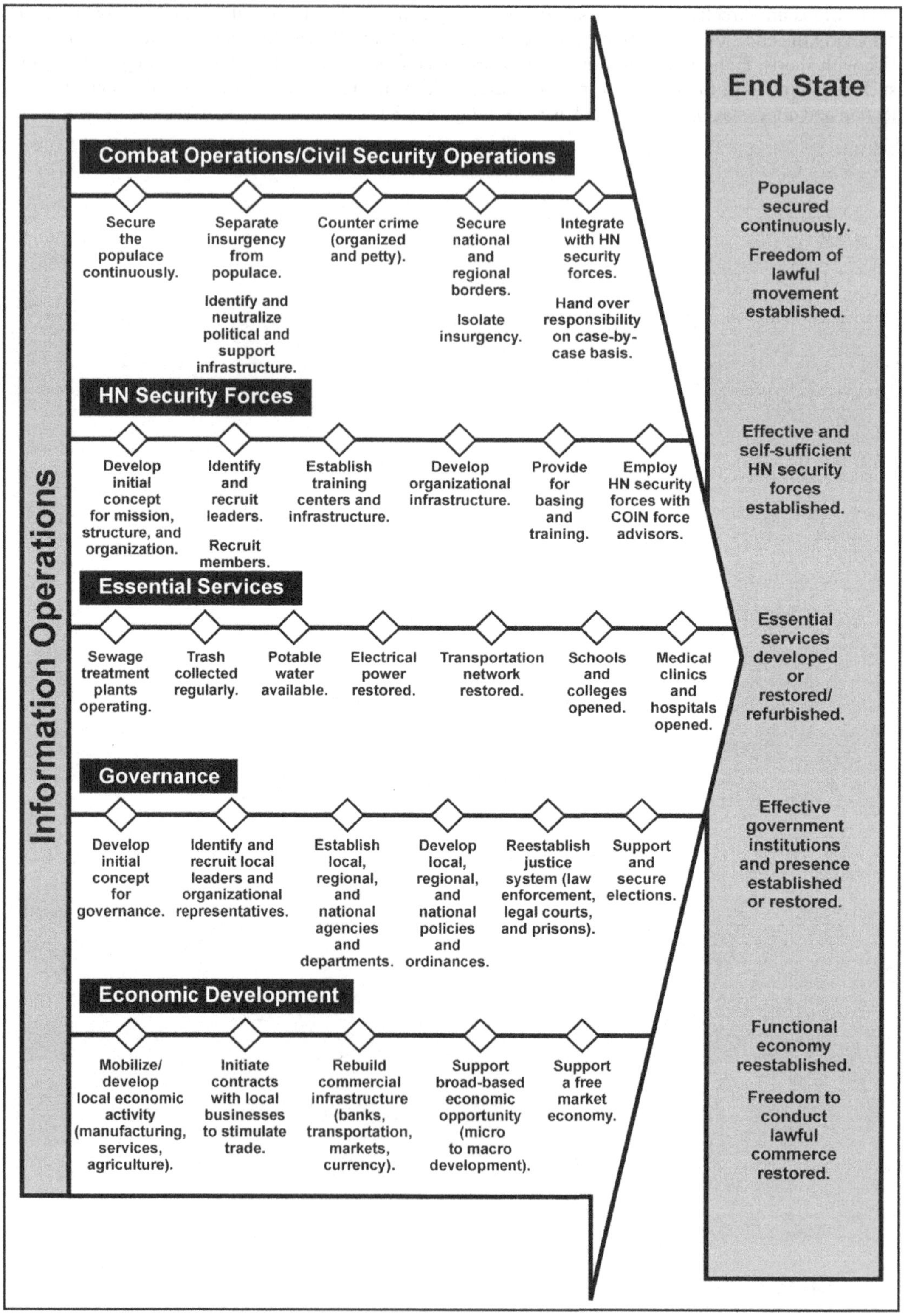

Figure 5-2. Example goals and objectives along logical lines of operations

5-12. Commanders determine which LLOs apply to their AO and how the LLOs connect with and support one another. For example, commanders may conduct offensive and defensive operations to form a shield behind which simultaneous stability operations can maintain a secure environment for the populace. Accomplishing the objectives of combat operations/civil security operations sets the conditions needed to achieve essential services and economic development objectives. When the populace perceives that the environment is safe enough to leave families at home, workers will seek employment or conduct public economic activity. Popular participation in civil and economic life facilitates further provision of essential services and development of greater economic activity. Over time such activities establish an environment that attracts outside capital for further development. Neglecting objectives along one LLO risks creating vulnerable conditions along another that insurgents can exploit. Achieving the desired end state requires linked successes along all LLOs.

5-13. The relationship of LLOs to the overall operation is similar to the stands of a rope. (See figure 5-3.) Each LLO is a separate string. Operations along it cannot accomplish all objectives required for success in a COIN operation. However, a strong rope is created when strands are woven together. The overall COIN effort is further strengthened through IO, which support and enhance operations along all LLOs by highlighting the successes along each one.

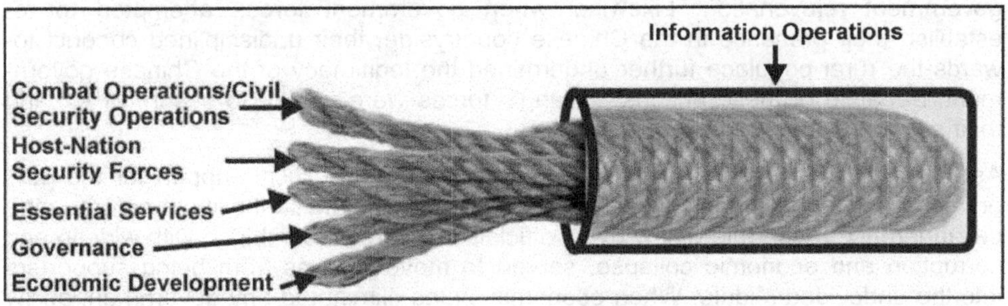

Figure 5-3. The strengthening effect of interrelated logical lines of operations

5-14. LLOs help commanders identify missions, assign tasks, allocate resources, and assess operations. Commanders specify the LLO that is the decisive operation; others shape the operational environment for the decisive operation's success. This prioritization usually changes as COIN operations create insurgent vulnerabilities, insurgents react or adjust their activities, or the environment changes. In this sense, commanders adapt their operations not only to the state of the insurgency, but also to the environment's overall condition. Greater stability indicates progress toward the end state.

5-15. Well-designed operations are based on LLOs that are mutually supportive between echelons and adjacent organizations. For example, similar LLOs among brigade combat teams produce complementary effects, while brigade-level accomplishments reinforce achievement of division objectives. LLOs are normally used at brigade and higher levels, where the staff and unit resources needed to use them are available; however, battalions can use LLOs. Commanders at various levels may expect subordinates to describe their operations in these terms.

5-16. Commanders at all levels should select the LLOs that relate best to achieving the desired end state in accordance with the commander's intent. The following list of possible LLOs is not all inclusive. However, it gives commanders a place to start:

- Conduct information operations.
- Conduct combat operations/civil security operations.
- Train and employ HN security forces.
- Establish or restore essential services.
- Support development of better governance.
- Support economic development.

The Importance of Multiple Lines of Operations in COIN

The Chinese Civil War illustrates the importance of pursuing and linking multiple logical lines of operations. Chiang Kai-shek's defeat in 1949 resulted from his failure to properly establish security, good governance, the rule of law, essential services, and economic stability. Failures in each undermined his government's position in the others.

In China during 1945, Chiang Kai-shek adopted a strategy to secure and defend the coastal financial and industrial centers from the communist insurgency led by Mao Zedong. These areas had been the Chinese government's prewar core support areas. Although a logical plan, this strategy suffered in implementation. Republic of China administration and military forces were often corrupt. They neither provided good governance and security nor facilitated provision of essential services. Furthermore, the government, because of an insufficient number of soldiers, relied on warlord forces, which lacked quality and discipline. Their actions undermined the legitimacy of and popular support for the government within the very core areas vital to government rejuvenation. Likewise, when government forces attempted to reestablish their presence in the Chinese countryside, their undisciplined conduct towards the rural populace further undermined the legitimacy of the Chinese government. Because of these actions, Chiang's forces were unable to secure or expand their support base.

As a result, there was increasing lack of material and political support for the government, whose legitimacy was undercut. The government's inability to enforce ethical adherence to the rule of law by its officials and forces, combined with widespread corruption and economic collapse, served to move millions from being supporters into the undecided middle. When economic chaos eliminated any government ability to fund even proper and justified efforts, an insurgent victory led by the Chinese Communist Party became inevitable.

As government defeat followed defeat, a collapse of morale magnified the impact of material shortages. Chiang's defeat in Manchuria, in particular, created a psychological loss of support within China It caused economic dislocation due to substantial price inflation of foodstuffs and sowed discord and dissension among government allies. As the regime lost moral authority, it also faced a decreasing ability to govern. All these factors served to create a mythical yet very powerful psychological impression that the success of the Chinese Communist Party was historically inevitable. The failure of the leaders of the Republic of China to address the requirements of logical lines of operations like good governance, economic development, and essential services magnified their military shortcomings and forced their abandonment of the Chinese mainland.

5-17. These lines can be customized, renamed, changed altogether, or simply not used. Commanders may combine two or more of the listed LLOs or split one LLO into several. For example, IO are integrated into all LLOs; however, commanders may designate a separate LLO for IO if necessary to better describe their intent. Likewise, some commanders may designate separate LLOs for combat operations and civil security operations.

5-18. LLOs should be used to isolate the insurgents from the population, address and correct the root causes of the insurgency, and create or reinforce the societal systems required to sustain the legitimacy of the HN government. The following discussion addresses six LLOs common during COIN operations. The IO LLO may be the most important one. However, IO are interwoven throughout all LLOs and shape the information environment in which COIN operations are executed.

Conduct Information Operations

5-19. The IO LLO may often be the decisive LLO. By shaping the information environment, IO make significant contributions to setting conditions for the success of all other LLOs. (See JP 3-13 and FM 3-13 for IO doctrine. IO include elements not addressed here.) By publicizing government policies, the actual situation, and counterinsurgent accomplishments, IO, synchronized with public affairs, can neutralize insurgent propaganda and false claims. Major IO task categories include the following:

- Ensure that IO are synchronized at all levels and nested within the interagency strategic communications operation.
- Identify all the audiences (local, regional, and international), the various news cycles, and how to reach them with the HN government's message.
- Manage the local populace's expectations regarding what counterinsurgents can achieve.
- Develop common, multiechelon themes based on and consistent with HN government policies and the operation's objectives. Sustain unity of the message.
- Coordinate and provide a comprehensive assessment of the information environment, incorporating the activities of all other LLOs.
- Remember actions always speak louder than words—every Soldier and Marine is an integral part of IO communications. IO are executed every day through the actions of firm, fair, professional, and alert Soldiers and Marines on the streets among the populace.
- Work to establish and sustain transparency that helps maintain HN government legitimacy.

5-20. Commanders and staffs synchronize IO with operations along all other LLOs. IO address and manage the public's expectations by reporting and explaining HN government and counterinsurgent actions. When effectively used, IO address the subject of root causes that insurgents use to gain support. (Table 5-1) lists considerations for developing the IO LLO.)

5-21. IO are tailored to address the concerns of the populace of specific areas. IO should inform the public of successfully completed projects and improvements, including accomplishments in security, infrastructure, essential services, and economic development. This publicity furthers popular acceptance of the HN government's legitimacy.

5-22. Effective IO use consistent themes based on policy, facts, and deeds—not claims or future plans, because these can be thwarted. Themes must be reinforced by actions along all LLOs. Making unsubstantiated claims can undermine the long-term credibility and legitimacy of the HN government. Counterinsurgents should never knowingly commit themselves to an action that cannot be completed. However, to reduce the negative effects of a broken promise, counterinsurgents should publicly address the reasons expectations cannot be met before insurgents can take advantage of them.

5-23. Command themes and messages based on policy should be distributed simultaneously or as soon as possible using all available media. Radio, television, newspapers, flyers, billboards, and the Internet are all useful dissemination means. Polling and analysis should be conducted to determine which media allow the widest dissemination of themes to the desired audiences at the local, regional, national, and international levels.

5-24. Insurgents are not constrained by truth; they create propaganda that furthers their aims. Insurgent propaganda may include lying, deception, and creating false causes. Historically, as the environment changes, insurgents change their message to address the issues that gain them support. IO should point out the insurgency's propaganda and lies to the local populace. Doing so creates doubt regarding the viability of the insurgents' short- and long-term intentions among the uncommitted public and the insurgency's supporters.

Table 5-1. Considerations for developing the information operations LLO

- Consider word choices carefully. Words are important—they have specific meanings and describe policy. For example, are counterinsurgents liberators or occupiers? Occupiers generate a "resistance," whereas liberators may be welcomed for a time. Soldiers and Marines can be influenced likewise. In a conflict among the people, terms like "battlefield" influence perceptions and confuse the critical nature of a synchronized approach. Refrain from referring to and considering the area of operations as a "battlefield" or it may continue to be one.
- Publicize insurgent violence and use of terror to discredit the insurgency. Identify barbaric actions by extremists and the insurgents' disregard for civilian losses.
- Admit mistakes (or actions perceived as mistakes) quickly. Explain these mistakes and actions as fully as possible—including mistakes committed by U.S. military forces. However, do not attempt to explain actions by the host-nation government. Instead encourage host-nation officials to handle such information themselves. They know the cultural implications of their actions better, and honesty should help to build legitimacy.
- Highlight successes of the host-nation government and counterinsurgents promptly. Positive results speak loudly and resonate with people. Do not delay announcements while waiting for all results. Initiate communications immediately to let people know what counterinsurgents are doing and why. Delaying announcements creates "old news" and misses news cycles.
- Respond quickly to insurgent propaganda. Delaying responses can let the insurgent story dominate several news cycles. That situation can lead to the insurgents' version of events becoming widespread and accepted. This consideration may require giving increased information assets and responsibilities to lower level leaders.
- Shape the populace's expectations. People generally expect too much too soon. When the host-nation government or counterinsurgency force is slow to deliver, people become easily and perhaps unfairly disgruntled.
- Give the populace some way to voice their opinions and grievances, even if that activity appears at first to cause friction. Such opportunities are important to both the formal political process and to informal, local issues (where government touches people directly). Develop a feedback mechanism from the populace to the local government to identify needs and align perceptions.
- Keep Soldiers and Marines engaged with the populace. Presence patrols facilitate Soldiers and Marines mingling with the people. As the populace and counterinsurgents learn to know each other better, two-way communication develops, building trust and producing intelligence.
- Conduct ongoing perception assessments. Identify leaders who influence the people at the local, regional, and national levels. Determine a population's relevant lines of loyalty as accurately as possible.
- Treat detainees professionally and publicize their treatment. Arrange for host-nation leaders to visit and tour your detention facility. Consider allowing them to speak to detainees and eat the same food detainees receive. If news media or host-nation government representatives visit your detention facility, allow them as much access as prudent. Provide a guided tour and explain your procedures.
- Consider encouraging host-nation leaders to provide a forum for initiating a dialog with the opposition. This does not equate to "negotiating with terrorists." It is an attempt to open the door to mutual understanding. There may be no common ground and the animosity may be such that nothing specifically or directly comes of the dialog. However, if counterinsurgents are talking with their adversaries, they are using a positive approach and may learn something useful. If the host nation is reluctant to communicate with insurgents, other counterinsurgents may have to initiate contact. Consider adopting a "We understand why you fight" attitude and stating this position to the insurgents.

- Work to convince insurgent leaders that the time for resistance has ended and that other ways to accomplish what they desire exist.
- Turn the insurgents' demands on the insurgents. Examine the disputed issues objectively; then work with host-nation leaders to resolve them where possible. Portray any success as a sign of responsiveness and improvement.
- Portray the counterinsurgency force as robust, persistent, and willing to help the population through the present difficulty.
- Learn the insurgents' messages or narratives. Develop countermessages and counter-narratives to attack the insurgents' ideology. Understanding the local culture is required to do this. Host-nation personnel can play a key role.
- Remember that the media's responsibility is to report the news. The standard against which the media should be judged is accuracy complemented by the provision of context and proper characterization of overall trends, not whether it portrays the actions of counterinsurgents, host-nation forces, and host-nation officials positively or negatively.
- Conduct town meetings to assess and address areas where counterinsurgents can make things better.
- When insurgents follow an ideology based on religious extremism, information operations should encourage, strengthen, and protect the society's moderating elements. Command themes need to portray a credible, publicly attractive vision that resonates with local culture. At the same time, commanders should avoid the appearance of interfering in the society's internal religious affairs.

5-25. Impartiality is a common theme for information activities when there are political, social, and sectarian divisions in the host nation. Counterinsurgents should avoid taking sides, when possible. Perceived favoritism can exacerbate civil strife and make counterinsurgents more desirable targets for sectarian violence.

5-26. Effective commanders directly engage in a dialog with the media and communicate command themes and messages personally. The worldwide proliferation of sophisticated communication technologies means that media coverage significantly affects COIN operations at all echelons. Civilian and military media coverage influences the perceptions of the political leaders and public of the host nation, United States, and international community. The media directly influence the attitude of key audiences toward counterinsurgents, their operations, and the opposing insurgency. This situation creates a war of perceptions between insurgents and counterinsurgents conducted continuously using the news media.

5-27. Commanders often directly engage the local populace and stakeholders through face-to-face meetings, town meetings, and community events highlighting counterinsurgent community improvements. These engagements give commanders additional opportunities to assess their efforts' effects, address community issues and concerns, and personally dispel misinformation. These events often occur in the civil-military operations center.

5-28. The media are a permanent part of the information environment. Effective media/public affairs operations are critical to successful military operations. All aspects of military operations are subject to immediate scrutiny. Well-planned, properly coordinated, and clearly expressed themes and messages can significantly clarify confusing situations. Clear, accurate portrayals can improve the effectiveness and morale of counterinsurgents, reinforce the will of the U.S. public, and increase popular support for the HN government. The right messages can reduce misinformation, distractions, confusion, uncertainty, and other factors that cause public distress and undermine the COIN effort. Constructive and transparent information enhances understanding and support for continuing operations against the insurgency.

5-29. There are several methods for working with the media to facilitate accurate and timely information flow. These include the following:

- Embedded media.
- Press conferences.
- Applying resources.
- Network with media outlets.

5-30. Embedded media representatives experience Soldiers' and Marines' perspectives of operations in the COIN environment. Media representatives should be embedded for as long as practicable. Representatives embedded for weeks become better prepared to present informed reports. Embedding for days rather than weeks risks media representatives not gaining a real understanding of the context of operations. Such short exposure may actually lead to unintended misinformation. The media should be given access to Soldiers and Marines in the field. These young people nearly always do a superb job of articulating the important issues for a broad audience. Given a chance, they can share their courage and sense of purpose with the American people and the world.

5-31. Commanders may hold weekly press conferences to explain operations and provide transparency to the people most affected by COIN efforts. Ideally, these sessions should include the HN media and HN officials. Such events provide opportunities to highlight the accomplishments of counterinsurgents and the HN government.

5-32. Commanders should apply time, effort, and money to establish the proper combination of media outlets and communications to transmit the repetitive themes of HN government accomplishments and insurgent violence against the populace. This might require counterinsurgents to be proactive, alerting the media to news opportunities and perhaps providing transportation or other services to ensure proper coverage. Helping establish effective HN media is another important COIN requirement. A word of caution: the populace and HN media must never perceive that counterinsurgents and HN forces are manipulating the media. Even the slightest appearance of impropriety can undermine the credibility of the COIN force and the host nation.

5-33. Good working relationships between counterinsurgent leaders and members of the U.S. media are in the Nation's interest. Similar relationships can be established with international media sources. When they do not understand COIN efforts, U.S. media representatives portray the situation to the American public based on what they do know. Such reports can be incomplete, if not incorrect. Through professional relationships, military leaders can ensure U.S. citizens better understand what their military is doing in support of the Nation's interests.

5-34. The media are ever present and influence perceptions of the COIN environment. Therefore, successful leaders engage the media, create positive relationships, and help the media tell the story. Operations security must always be maintained; however, security should not be used as an excuse to create a media blackout. In the absence of official information, some media representatives develop stories on their own that may be inaccurate and may not include the COIN force perspective. (See JP 3-61, FM 46-1, FM 3-61.1 for public affairs doctrine.)

Conduct Combat Operations/Civil Security Operations

5-35. This LLO is the most familiar to military forces. Care must be taken not to apply too many resources to this LLO at the expense of other LLOs that facilitate the development or reinforcement of the HN government's legitimacy. Commanders may describe actions related to combat operations and civil security operations as a single LLO or as multiple LLOs. Commanders base their decision on conditions in the AO and their objectives. (Table 5-2 [page 5-12] lists considerations for developing the combat operations/civil security LLO.)

5-36. Under full spectrum operations, forces conduct simultaneous offensive, defensive, and stability operations. Offensive and defensive operations focus on defeating enemy forces. Security operations, including area security, pertain to actions taken to protect the force. They are associated with offensive and defensive operations. In contrast, stability operations focus on security and control of areas, resources, and populations. Civil security and civil control are types of stability operations. Army commanders expect a mission of protecting and providing security for a population to be expressed in terms of civil security or civil control.

5-37. Within the COIN context, Marine Corps doctrine does not draw a distinction in this manner; rather, it places tasks related to civil security and area security under combat operations. A Marine force assigned an area security mission during a COIN operation executes it as a combat operation. The force establishes and maintains measures to protect people and infrastructure from hostile acts or influences while actively seeking out and engaging insurgent forces.

Table 5-2. Considerations for developing the combat operations/civil security operations LLO

• Develop cultural intelligence, which assumes a prominent role. Make every effort to learn as much about the environment as possible. Human dynamics tend to matter the most.
• Ensure that rules of engagement adequately guide Soldiers and Marines engaged in combat while encouraging the prudent use of force commensurate with mission accomplishment and self-defense.
• Consider how the populace might react when planning tactical situations, even for something as simple as a traffic control point. Anticipate how people might respond to each operation.
• Identify tasks the host-nation government and populace generally perceive to be productive and appropriate for an outside force. Focus counterinsurgents on them.
• Win over, exhaust, divide, capture, or eliminate the senior- and mid-level insurgent leaders as well as network links. (Appendix B discusses networks and links.)
• Frustrate insurgent recruiting.
• Disrupt base areas and sanctuaries.
• Deny outside patronage (external support). Make every effort to stop insurgents from bringing materiel support across international and territorial borders.
• When patrolling in or occupying an area, clear only what the unit intends to hold. Otherwise, the effort will be wasted as the insurgents reoccupy the area. An exception to this policy is when commanders deem disruption of enemy strongholds necessary.
• When Soldiers and Marines interact with the populace, encourage them to treat people with respect to avoid alienating anyone.
• Support efforts to disarm, demobilize, and reintegrate into society members of armed groups outside of government control, such as militias and paramilitary organizations. Also include insurgents who are captured, surrender, or accept amnesty.
• Take a census as soon as is practicable. Help the host-nation government do this. This information can be helpful for learning about the people and meeting their needs. The census also establishes a necessary database for civil security operations.
• Help the host-nation government produce and distribute identification cards. Register all citizens—or at least those nearing a predetermined, adult age. Identification cards may help to track people's movements. This information is useful in identifying illicit activity and also contributes to civil security.

5-38. Insurgents use unlawful violence to weaken the HN government, intimidate people into passive or active support, and murder those who oppose the insurgency. Measured combat operations are always required to address insurgents who cannot be co-opted into operating inside the rule of law. These operations may sometimes require overwhelming force and the killing of fanatic insurgents. However, COIN is "war amongst the people." Combat operations must therefore be executed with an appropriate level of restraint to minimize or avoid injuring innocent people. Not only is there a moral basis for the use of restraint or measured force; there are practical reasons as well. Needlessly harming innocents can turn the populace against the COIN effort. Discriminating use of fires and calculated, disciplined response should characterize COIN operations. Kindness and compassion can often be as important as killing and capturing insurgents.

5-39. Battalion-sized and smaller unit operations are often most effective for countering insurgent activities. Counterinsurgents need to get as close as possible to the people to secure them and glean the maximum amount of quality information. Doing this helps counterinsurgents gain a fluidity of action equal or superior to that of the enemy. This does not mean larger unit operations are not required. Brigades are usually synchronizing headquarters. Divisions shape the environment to set conditions and facilitate brigade and battalion success. The sooner counterinsurgents can execute small-unit operations effectively, the better.

Train and Employ Host-Nation Security Forces

5-40. Most societal and government functions require a secure environment. Although U.S. and multinational forces can provide direct assistance to establish and maintain security, this situation is at best a provisional solution. Ultimately, the host nation must secure its own people. (Table 5-3 lists considerations for developing the HN security forces LLO. Chapter 6 addresses this LLO in detail.)

Table 5-3. Considerations for developing the host-nation security forces LLO

- Understand the security problem. The function, capabilities, and capacities required for host-nation security (military and police) forces should align with the theater strategy, host-nation culture, and the threat these forces face.
- Take a comprehensive approach, beginning with planning. Consult local representatives to determine local needs. Include host-nation government and military authorities as partners. Consult with multinational partners and intergovernmental organizations that may be involved. Share leadership with local authorities to achieve legitimacy in the eyes of the populace.
- Avoid mirror-imaging (trying to make host-nation forces look like the U.S. military). That solution fits few cultures or situations.
- Establish separate training academies for military and police forces. Staff them with multinational personnel. (Tap into the talents of as many nations as possible for this.) Maintaining public order, enforcing laws, and preventing and detecting crime requires a trained and capable host-nation police force. Likewise, an operational penal system providing adequate and humane confinement is needed to support host-nation police and judicial processes.
- Establish mobile training teams.
- Train the host-nation cadre first. Focus on identifying leaders. Use trained leaders to establish new units, staff training academies, and staff mobile training teams (where appropriate).
- Create special-purpose forces based on threats facing the host nation.
  - For police, consider special reaction teams, a counterespionage organization, port security, and public figure security.
  - For the host-nation military, consider riverine operations forces, explosive ordinance disposal, and other specialized units.
- Put host-nation personnel in charge of as much as possible as soon as possible.
- Conduct operations with host-nation forces. Show that you respect their partnership. Once host-nation forces are ready to work with the counterinsurgency force, include them in planning. Encourage host-nation leaders to take ownership of plans and operations as they move toward self-sufficiency.
- Respect host-nation security force leaders in public and private. Show the populace that their security forces have earned counterinsurgents' respect. However, do not tolerate abuses. Base respect on generally upright comportment by host-nation security forces.
- Provide advisors for host-nation units under development. Place liaison officers with trained host-nation units. Make sure all involved understand how an advisor differs from a liaison officer.
- Establish competent military and police administrative structures early. Provision and pay host-nation forces on time. Pay should come from the host-nation organization, not the counterinsurgency force.
- Encourage insurgents to change sides—welcome them in with an "open-arms" policy. However, identify insurgents seeking to join the security forces under false pretext. Vetting repatriated insurgents is a task for the host-nation government in partnership with the country team.
- Encourage the host nation to establish a repatriation or amnesty program to allow insurgents an alternative to the insurgency.

5-41. The U.S. military can help the host nation develop the forces required to establish and sustain stability within its borders. This task usually involves other government agencies and multinational partners. This assistance can include developing, equipping, training, and employing HN security forces. It may extend to operations in which multinational military units fight alongside the newly formed, expanded, or reformed HN forces.

Establish or Restore Essential Services

5-42. Essential services address the life support needs of the HN population. The U.S. military's primary task is normally to provide a safe and secure environment. HN or interagency organizations can then develop the services or infrastructure needed. In an unstable environment, the military may initially have the leading role. Other agencies may not be present or might not have enough capability or capacity to meet HN needs. Therefore, COIN military planning includes preparing to perform these tasks for an extended period. (Table 5-4 lists considerations for developing the essential services LLO.)

Table 5-4. Considerations for developing the essential services LLO

- Make this effort a genuine partnership between counterinsurgents and host-nation authorities. Use as much local leadership, talent, and labor as soon as possible.
- Plan for a macro and a micro assessment effort. Acknowledge early what is known and not known—and honestly appraise what needs to be accomplished. The macro assessment concerns national-level needs; it is long term in focus. The micro assessment focuses on the local level; it determines, by region, specific short-term needs.
- Appreciate local preferences. An accurate needs assessment reflects cultural sensitivity; otherwise, great time and expense can be wasted on something the populace considers of little value. Ask, How do I know this effort matters to the local populace? If there is no answer, the effort may not be important. Host-nation authorities are a good place to start with this question. A local perception assessment may also be useful. (See appendix B.)
- Establish realistic, measurable goals. Establish ways to assess their achievement.
- Form interagency planning teams to discuss design, assessment, and redesign. Recognize and understand other agencies' institutional cultures.
- Meet with representatives from organizations beyond the host-nation government team. Many nongovernmental organizations do not want to appear closely aligned with the counterinsurgency effort. Encourage their participation in planning, even if it means holding meetings in neutral areas. When meeting with these organizations, help them understand mutual interests in achieving local security, stability, and relief objectives.
- Be as transparent as possible with the local populace. Do your best to help people understand what counterinsurgents are doing and why they are doing it.
- Consider the role of women in the society and how this cultural factor may influence these activities.
- Discourage the attitude that counterinsurgents have arrived to "save the day"—or that their arrival will only cause greater problems. Helping the populace understand what is possible avoids frustrations based on unrealized high expectations.

5-43. Counterinsurgents should work closely with the host nation in establishing achievable goals. If lofty goals are set and not achieved, both counterinsurgents and the HN government can lose the populace's respect. The long-term objective is for the host nation to assume full responsibility and accountability for these services. Establishing activities that the HN government is unable to sustain may be counterproductive. IO nested within this LLO manage expectations and ensure that the public understands the problems involved in providing these services, for example, infrastructure sabotage by insurgents. Figure 5-4 shows an example of common essential services categories. Accomplishing objectives in each category contributes to achieving the higher commander's desired end state.

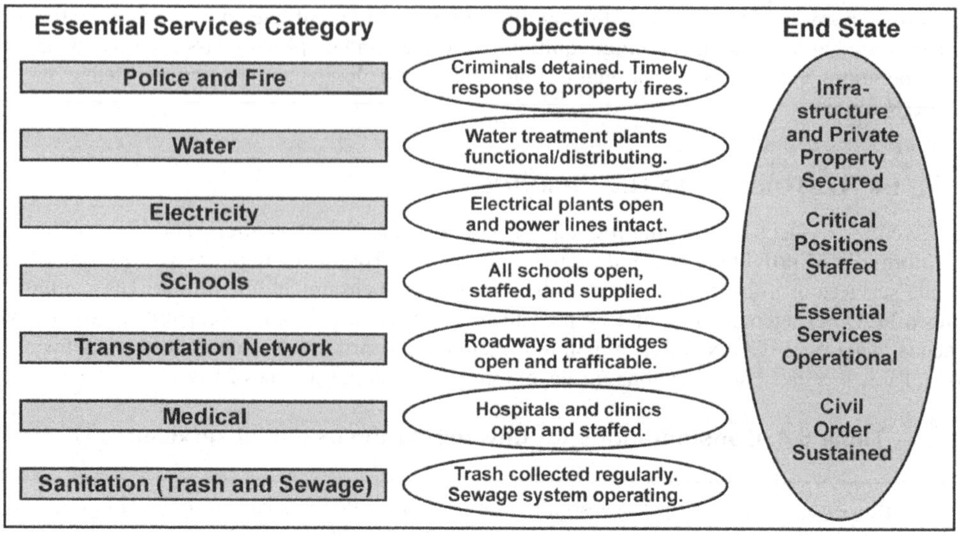

Figure 5-4. Example of essential services categories and objectives

SUPPORT DEVELOPMENT OF BETTER GOVERNANCE

5-44. This LLO relates to the HN government's ability to gather and distribute resources while providing direction and control for society. These include regulation of public activity; taxation; maintenance of security, control, and essential services; and normalizing the means of succession of power. Good governance is normally a key requirement to achieve legitimacy for the HN government. Activities related to it include the following:

- Controlling military and police activities.
- Establishing and enforcing the rule of law.
- Public administration.
- Justice (a judiciary system, prosecutor/defense representation, and corrections).
- Property records and control.
- Public finance.
- Civil information.
- Historical, cultural, and recreational services.
- An electoral process for representative government.
- Disaster preparedness and response.

5-45. Sometimes no HN government exists or the government is unable or unwilling to assume full responsibility for governance. In those cases, this LLO may involve establishing and maintaining a military government or a civil administration while creating and organizing a HN capability to govern. In the long run, developing better governance will probably affect the lives of the populace more than any other COIN activities. When well executed, these actions may eliminate the root causes of the insurgency. Governance activities are among the most important of all in establishing lasting stability for a region or nation. (Table 5-5 (page 5-16) lists considerations for developing the governance LLO.)

Table 5-5. Considerations for developing the governance LLO

- Encourage community leaders to participate in local governance. If no local council exists, encourage the populace to create one. Ask teachers, businesspeople, and others whom the community respects to form a temporary council until a more permanent organization can be formed.
- Help (or encourage) the host-nation government to remove or reduce genuine grievances, expose imaginary ones, and resolve contradictions, immediately where possible. Accomplishing these tasks may be difficult because—
  - Genuine grievances may be hard to separate from unreasonable complaints.
  - Host-nation leaders may be unable or unwilling to give up the necessary power to local governments.
- Make only commitments that can be fulfilled in the foreseeable future.
- Help the host nation develop and empower competent and responsive leaders and strengthen their civil service and security forces. Doing this is often difficult; however, backing an incompetent (or worse) host-nation leader can backfire. Do not be afraid to step in and make a bold change where necessary. A corrupt official, such as a chief of police who is working for both sides, can be doing more harm than good. You may be forced to replace him. If so, move decisively. Arrange the removal of all officials necessary to solve the problem. The pain of the affair may be acute, but it will be brief and final. Wherever possible, have host-nation authorities conduct the actual removal.
- Be accessible to the populace to facilitate two-way communication. Establish rapport for the host-nation government and counterinsurgents.
- Encourage the host nation to grant local demands and meet acceptable aspirations. Some of these might be driving the insurgency.
- Emphasize the national perspective in all host-nation government activities. Downplay sectarian divides.
- Provide liaison officers to host-nation government ministries or agencies. When possible, use an interagency team approach. Structure teams based on function.
- Once the legal system is functioning, send someone to observe firsthand a person moving through the legal process (arrest by police, trial, and punishment by confinement to a correctional facility). Ask to see the docket of the judges at the provincial courthouse. If there is no one on the docket or if it is full and there are no proceedings, there may be a problem.
- Create a system for citizens to pursue redress for perceived wrongs by authorities. Rule of law includes means for a citizen to petition the government for redress of government wrongs. The ability to petition the counterinsurgency force for redress of wrongs perpetrated by that force (intentionally or otherwise) is also required.
- Build on existing capabilities wherever possible. Host nations often have some capability; counterinsurgents may only need to help develop greater capacity.

SUPPORT ECONOMIC DEVELOPMENT

5-46. The economic development LLO includes both short- and long-term aspects. The short-term aspect concerns immediate problems, such as large-scale unemployment and reestablishing an economy at all levels. The long-term aspect involves stimulating indigenous, robust, and broad economic activity. The stability a nation enjoys is often related to its people's economic situation and its adherence to the rule of law. However, a nation's economic health also depends on its government's ability to continuously secure its population.

5-47. Planning economic development requires understanding the society, culture, and operational environment. For example, in a rural society, land ownership and the availability of agricultural equipment, seed, and fertilizer may be the chief parts of any economic development plan. In an urban, diversified society, the availability of jobs and the infrastructure to support commercial activities may be more important. Except for completely socialist economies, governments do not create jobs other than in the public bureaucracy. However, the micro economy can be positively stimulated by encouraging small businesses development. Jump-starting small businesses requires micro finance in the form of some sort of banking activities. So then, supporting economic development requires attention to both the macro economy and the micro economy.

5-48. Without a viable economy and employment opportunities, the public is likely to pursue false promises offered by insurgents. Sometimes insurgents foster the conditions keeping the economy stagnant. Insurgencies attempt to exploit a lack of employment or job opportunities to gain active and passive support for their cause and ultimately undermine the government's legitimacy. Unemployed males of military age may join the insurgency to provide for their families. Hiring these people for public works projects or a local civil defense corps can remove the economic incentive to join the insurgency. The major categories of economic activity include the following:

- Fossil fuels, mining, and related refining infrastructure.
- Generation and transmission of power and energy.
- Transportation and movement networks.
- Stock and commodities exchange.
- Banking.
- Manufacturing and warehousing.
- Building trades and services.
- Agriculture, food processing, fisheries, and stockyard processing.
- Labor relations.
- Education and training.

5-49. Table 5-6 lists considerations for developing the economic development LLO.

Table 5-6. Considerations for the economic development LLO

• Work with the host-nation government to strengthen the economy and quality of life. In the long run, success depends on supporting people's livelihoods.
• Create an environment where business can thrive. In every state (except perhaps a completely socialist one), business drives the economy. To strengthen the economy, find ways to encourage and support legitimate business activities. Even providing security is part of a positive business environment.
• Work with the host-nation government to reduce unemployment to a manageable level.
• Seek to understand the effects of military operations on business activities and vice versa. Understand the effects of outsourcing and military support on the local economy and employment level.
• Use economic leverage to enter new areas and reach new people. Remember that in many societies, monies are distributed though tribal or clan networks. For instance, giving a clan leader a contracting job may lead to employing many local men. Those employees are less likely to join the insurgency. It may be necessary to pay more than seems fair for a job; however, the money is well spent if it keeps people from supporting the insurgency.
• Ensure that noncompliance with government policies has an economic price. Likewise, show that compliance with those policies is profitable. In the broadest sense, counterinsurgency operations should reflect that "peace pays."
• Program funds for commanders to use for economic projects in their area of operations from the beginning of any operation. No one appreciates the situation better than those "on the ground." Creating these funds may require congressional action. (Appendix D contains a description of relevant funding sources.)

COUNTERINSURGENCY APPROACHES

5-50. There are many approaches to achieving success in a COIN effort. The components of each approach are not mutually exclusive. Several are shared by multiple approaches. The approaches described below are not the only choices available and are neither discrete nor exclusive. They may be combined, depending on the environment and available resources. The following methods and their components have proven effective. However, they must be adapted to the demands of the local environment. Three examples of approaches are—

- Clear-hold-build.
- Combined action.
- Limited support.

CLEAR-HOLD-BUILD

5-51. A clear-hold-build operation is executed in a specific, high-priority area experiencing overt insurgent operations. It has the following objectives:

- Create a secure physical and psychological environment.
- Establish firm government control of the populace and area.
- Gain the populace's support.

Popular support can be measured in terms of local participation in HN programs to counter the insurgency and whether people give counterinsurgents usable information about insurgent locations and activities.

5-52. COIN efforts should begin by controlling key areas. Security and influence then spread out from secured areas. The pattern of this approach is to clear, hold, and build one village, area, or city—and then reinforce success by expanding to other areas. This approach aims to develop a long-term, effective HN government framework and presence that secures the people and facilitates meeting their basic needs. Success reinforces the HN government's legitimacy. The primary tasks to accomplish during clear-hold-build are—

- Provide continuous security for the local populace.
- Eliminate insurgent presence.
- Reinforce political primacy.
- Enforce the rule of law.
- Rebuild local HN institutions.

5-53. To create success that can spread, a clear-hold-build operation should not begin by assaulting the main insurgent stronghold. However, some cases may require attacks to disrupt such strongholds, even if counterinsurgents cannot clear and hold the area. "Disrupt and leave" may be needed to degrade the insurgents' ability to mount attacks against cleared areas. Clear-hold-build objectives require lots of resources and time. U.S. and HN commanders should prepare for a long-term effort. All operations require unity of effort by civil authorities, intelligence agencies, and security forces. Coherent IO are also needed.

5-54. Clear-hold-build operations should expand outward from a secure base. An example is an urban industrial complex whose population supports the government effort and where security forces are in firm control. No population subjected to the intense organizational efforts of an insurgent organization can be won back until certain conditions are created:

- The counterinsurgent forces are clearly superior to forces available to the insurgents.
- Enough nonmilitary resources are available to effectively carry out all essential improvements needed to provide basic services and control the population.
- The insurgents are cleared from the area.
- The insurgent organizational infrastructure and its support have been neutralized or eliminated.
- A HN government presence is established to replace the insurgents' presence, and the local populace willingly supports this HN presence.

5-55. The following discussion describes some examples of activities involved in the clear-hold-build approach. Its execution involves activities across all LLOs. There can be overlap between steps—especially between hold and build, where relevant activities are often conducted simultaneously.

Clearing the Area

5-56. *Clear* is a tactical mission task that requires the commander to remove all enemy forces and eliminate organized resistance in an assigned area (FM 3-90). The force does this by destroying, capturing, or forcing the withdrawal of insurgent combatants. This task is most effectively initiated by a clear-in-zone or cordon-and-search operation. This operation's purpose is to disrupt insurgent forces and force a reaction by major insurgent elements in the area. Commanders employ a combination of offensive small-unit operations. These may include area saturation patrolling that enables the force to defeat insurgents in the area, interdiction ambushes, and targeted raids.

5-57. These offensive operations are only the beginning, not the end state. Eliminating insurgent forces does not remove the entrenched insurgent infrastructure. While their infrastructure exists, insurgents continue to recruit among the population, attempt to undermine the HN government, and try to coerce the populace through intimidation and violence. After insurgent forces have been eliminated, removing the insurgent infrastructure begins. This should be done so as to minimize the impact on the local populace. Rooting out such infrastructure is essentially a police action that relies heavily on military and intelligence forces until HN police, courts, and legal processes can assume responsibility for law enforcement within the cleared area.

5-58. If insurgent forces are not eliminated but instead are expelled or have broken into smaller groups, they must be prevented from reentering the area or reestablishing an organizational structure inside the area. Once counterinsurgents have established their support bases, security elements cannot remain static. They should be mobile and establish a constant presence throughout the area. Use of special funds should be readily available for all units to pay compensation for damages that occur while clearing the area of insurgent forces. Offensive and stability operations are continued to maintain gains and set the conditions for future activities. These include—

- Isolating the area to cut off external support and to kill or capture escaping insurgents.
- Conducting periodic patrols to identify, disrupt, eliminate, or expel insurgents.
- Employing security forces and government representatives throughout the area to secure the populace and facilitate follow-on stages of development.

5-59. Operations to clear an area are supplemented by IO focused on two key audiences: the local populace and the insurgents. The message to the populace focuses on gaining and maintaining their overt support for the COIN effort. This command theme is that the continuous security provided by U.S. and HN forces is enough to protect the people from insurgent reprisals for their cooperation. Conversely, the populace should understand that actively supporting the insurgency will prolong combat operations, creating a risk to themselves and their neighbors. The command message to the insurgents focuses on convincing them that they cannot win and that the most constructive alternatives are to surrender or cease their activities.

Holding with Security Forces

5-60. Ideally HN forces execute this part of the clear-hold-build approach. Establishment of HN security forces in bases among the population furthers the continued disruption, identification, and elimination of the local insurgent leadership and infrastructure. The success or failure of the effort depends, first, on effectively and continuously securing the populace and, second, on effectively reestablishing a HN government presence at the local level. Measured offensive operations continue against insurgents as opportunities arise, but the main effort is focused on the population.

5-61. Key infrastructure must be secured. Since resources are always limited, parts of the infrastructure vital for stability and vulnerable to attack receive priority for protection. These critical assets should be identified during planning. For instance, a glassmaking factory may be important for economic recovery, but it may not be at risk of insurgent attack and therefore may not require security.

5-62. There are four key target audiences during the hold stage:

- Population.
- Insurgents.
- COIN force.
- Regional and international audiences.

5-63. Command themes and messages to the population should affirm that security forces supporting the HN government are in the area to accomplish the following:

- Protect the population from insurgent intimidation, coercion, and reprisals.
- Eliminate insurgent leaders and infrastructure.
- Improve essential services where possible.
- Reinstate HN government presence.

IO should also emphasize that U.S. and HN security forces will remain until the current situation is resolved or stated objectives are attained. This message of a persistent presence can be reinforced by making long-term contracts with local people for supply or construction requirements.

5-64. The commander's message to the insurgents is to surrender or leave the area. It emphasizes the permanent nature of the government victory and presence. The HN government might try to exploit success by offering a local amnesty. Insurgent forces will probably not surrender, but they may cease hostile actions against the HN government agencies in the area.

5-65. The commander's message to the COIN force should explain changes in missions and responsibilities associated with creating or reinforcing the HN government's legitimacy. The importance of protecting the populace, gaining people's support by assisting them, and using measured force when fighting insurgents should be reinforced and understood.

5-66. Operations during this stage are designed to—

- Continuously secure the people and separate them from the insurgents.
- Establish a firm government presence and control over the area and populace.
- Recruit, organize, equip, and train local security forces.
- Establish a government political apparatus to replace the insurgent apparatus.
- Develop a dependable network of sources by authorized intelligence agents.

5-67. Major actions occurring during this stage include—

- Designating and allocating area-oriented counterinsurgent forces to continue offensive operations. Other forces that participated in clearing actions are released or assigned to other tasks.
- A thorough population screening to identify and eliminate remaining insurgents and to identify any lingering insurgent support structures.
- Conducting area surveys to determine available resources and the populace's needs. Local leaders should be involved.
- Environmental improvements designed to convince the populace to support the HN government, participate in securing their area, and contribute to the reconstruction effort.
- Training of local paramilitary security forces, including arming them and integrating them into successful operations against the insurgents.
- Establishing a communications system that integrates the area into the HN communications grid and system.

Building Support and Protecting the Population

5-68. Progress in building support for the HN government requires protecting the local populace. People who do not believe they are secure from insurgent intimidation, coercion, and reprisals will not risk overtly supporting COIN efforts. The populace decides when it feels secure enough to support COIN efforts.

5-69. To protect the populace, HN security forces continuously conduct patrols and use measured force against insurgent targets of opportunity. Contact with the people is critical to the local COIN effort's success. Actions to eliminate the remaining covert insurgent political infrastructure must be continued; an insurgent presence will continue to threaten and influence people.

5-70. Tasks that provide an overt and direct benefit for the community are key, initial priorities. Special funds (or other available resources) should be available to pay wages to local people to do such beneficial work. Accomplishing these tasks can begin the process of establishing HN government legitimacy. Sample tasks include—

- Collecting and clearing trash from the streets.
- Removing or painting over insurgent symbols or colors.
- Building and improving roads.
- Digging wells.
- Preparing and building an indigenous local security force.
- Securing, moving, and distributing supplies.
- Providing guides, sentries, and translators.
- Building and improving schools and similar facilities.

Population Control Measures

5-71. Population control includes determining who lives in an area and what they do. This task requires determining societal relationships—family, clan, tribe, interpersonal, and professional. Establishing control normally begins with conducting a census and issuing identification cards. A census must be advertised and executed systematically. Census tasks include establishing who resides in which building and each household's family head. Those heads of households are required to report any changes to the appropriate agencies. Census records provide information regarding real property ownership, relationships, and business associations.

5-72. Insurgents may try to force people to destroy their identification cards. The benefits of retaining identification cards must be enough to motivate people to resist losing them. Insurgents may participate in the census to obtain valid identification cards. Requiring applicants to bring two men from outside their family to swear to their identity can reduce this probability. Those who affirm the status of an applicant are accountable for their official statements made on behalf of the applicant. Identification cards should have a code that indicates where the holders live.

5-73. Other population control measures include—

- Curfews.
- A pass system (for example, one using travel permits or registration cards) administered by security forces or civil authorities.
- Limits on the length of time people can travel.
- Limits on the number of visitors from outside the area combined with a requirement to register them with local security forces or civil authorities.
- Checkpoints along major routes to monitor and enforce compliance with population control measures.

5-74. The HN government should explain and justify new control measures to the affected population. People need to understand what is necessary to protect them from insurgent intimidation, coercion, and reprisals. Once control measures are in place, the HN government should have an established system of punishments for offenses related to them. These should be announced and enforced. The host nation should establish this system to ensure uniform enforcement and conformity with the rule of law throughout its territory. The HN government must be able to impose fines and other punishments for such civil infractions.

Increasing Popular Support

5-75. Counterinsurgents should use every opportunity to help the populace and meet its needs and expectations. Projects to improve economic, social, cultural, and medical needs can begin immediately. Actions speak louder than words. Once the insurgent political infrastructure is destroyed and local leaders begin to establish themselves, necessary political reforms can be implemented. Other important tasks include the following:

- Establishing HN government agencies to perform routine administrative functions and begin improvement programs.
- Providing HN government support to those willing to participate in reconstruction. Selection for participation should be based on need and ability to help. People should also be willing to secure what they create.
- Beginning efforts to develop regional and national consciousness and rapport between the population and its government. Efforts may include participating in local elections, making community improvements, forming youth clubs, and executing other projects.

5-76. Commanders can use IO to increase popular support. Command messages are addressed to the populace, insurgents, and counterinsurgents.

5-77. The IO message to the population has three facets:

- Obtaining the understanding or approval of security force actions that affect the populace, such as control measures or a census. Tell the people what forces are doing and why they are doing it.
- Establishing human intelligence sources that lead to identification and destruction of any remaining insurgent infrastructure in the area.
- Winning over passive or neutral people by demonstrating how the HN government is going to make their life better.

5-78. The IO message to insurgents should aim to create divisions between the movement leaders and the mass base by emphasizing failures of the insurgency and successes of the government. Success is indicated when insurgents abandon the movement and return to work with the HN government.

5-79. Commanders should emphasize that counterinsurgents must remain friendly towards the populace while staying vigilant against insurgent actions. Commanders must ensure Soldiers and Marines understand the rules of engagement, which become more restrictive as peace and stability return.

5-80. The most important activities during the build stage are conducted by nonmilitary agencies. HN government representatives reestablish political offices and normal administrative procedures. National and international development agencies rebuild infrastructure and key facilities. Local leaders are developed and given authority. Life for the area's inhabitants begins to return to normal. Activities along the combat operations/civil security operations LLO and HN security force LLO become secondary to those involved in essential services, good governance, and essential services LLOs.

Clear-Hold-Build in Tal Afar

In early 2005, the city of Tal Afar in northern Iraq had become a focal point for Iraqi insurgent efforts. The insurgents tried to assert control over the population. They used violence and intimidation to inflame ethnic and sectarian tensions. They took control of all schools and mosques, while destroying police stations. There were frequent abductions and executions. The insurgents achieved some success as the populace divided into communities defined by sectarian boundaries. Additionally, Tal Afar became an insurgent support base and sanctuary for launching attacks in the major regional city of Mosul and throughout Nineveh province.

During the summer of 2005, the 3d Armored Cavalry Regiment (ACR) assumed the lead for military efforts in and around Tal Afar. In the months that followed, the 3d ACR applied a clear-hold-build approach to reclaim Tal Afar from the insurgents.

Destruction or Expulsion of Insurgent Forces (Clear)

In August 2005, the 3d ACR and Iraqi forces began the process of destroying the insurgency in Tal Afar. Their first step was to conduct reconnaissance to understand the enemy situation; understand the ethnic, tribal, and sectarian dynamics; and set the conditions for effective operations. Iraqi security forces and U.S. Soldiers isolated the insurgents from external support by controlling nearby border areas and creating an eight-foot-high berm around the city. The berm's purpose was to deny the enemy freedom of movement and safe haven in outlying communities. The berm prevented free movement of fighters and weapons and forced all traffic to go through security checkpoints manned by U.S. and Iraqi forces. Multinational checkpoints frequently included informants who could identify insurgents. Multinational forces supervised the movement of civilians out of contentious areas. Forces conducted house-to-house searches. When they met violent resistance, they used precision fires from artillery and aviation. Targets were chosen through area reconnaissance operations, interaction with the local populace, and information from U.S. and Iraqi sources. Hundreds of insurgents were killed or captured during the encirclement and clearing of the city. Carefully controlled application of violence limited the cost to residents.

Deployment of Security Forces (Hold)

Following the defeat of enemy fighters, U.S. and Iraqi forces established security inside Tal Afar. The security forces immediately enhanced personnel screening at checkpoints based on information from the local population. To enhance police legitimacy in the people's eyes, multinational forces began recruiting Iraqi police from a more diverse, representative mix comprising city residents and residents of surrounding communities. Police recruits received extensive training in a police academy. U.S. forces and the Iraqi Army also trained Iraqi police in military skills. Concurrently, the local and provincial government dismissed or prosecuted Iraqi police involved in offenses against the populace. The government assigned new police leaders to the city from Mosul and other locations. U.S. forces assisted to ensure Iraqi Army, police, and their own forces shared common boundaries and were positioned to provide mutual support to one another. At the same time, U.S. forces continued to equip and train a border defense brigade, which increased the capability to interdict the insurgents' external support. Among its successes, the multinational force destroyed an insurgent network that included a chain of safe houses between Syria and Tal Afar.

Improving Living Conditions and Restoring Normalcy (Build)

With insurgents driven out of their city, the local population accepted guidance and projects to reestablish control by the Iraqi government. The 3d ACR commander noted, "The people of Tal Afar understood that this was an operation for them—an operation to bring back security to the city."

With the assistance of the Department of State and the U.S. Agency for International Development's Office of Transition Initiatives, efforts to reestablish municipal and economic systems began in earnest. These initiatives included providing essential services (water, electricity, sewage, and trash collection), education projects, police stations, parks, and reconstruction efforts. A legal claims process and compensation program to address local grievances for damages was also established.

As security and living conditions in Tal Afar improved, citizens began providing information that helped eliminate the insurgency's infrastructure. In addition to information received on the streets, multinational forces established joint coordination centers in Tal Afar and nearby communities that became multinational command posts and intelligence-sharing facilities with the Iraqi Army and the Iraqi police.

Unity of effort by local Iraqi leaders, Iraqi security forces, and U.S. forces was critical to success. Success became evident when many families who had fled the area returned to the secured city.

COMBINED ACTION

5-81. Combined action is a technique that involves joining U.S. and HN troops in a single organization, usually a platoon or company, to conduct COIN operations. This technique is appropriate in environments where large insurgent forces do not exist or where insurgents lack resources and freedom of maneuver. Combined action normally involves joining a U.S. rifle squad or platoon with a HN platoon or company, respectively. Commanders use this approach to hold and build while providing a persistent counterinsurgent presence among the populace. This approach attempts to first achieve security and stability in a local area, followed by offensive operations against insurgent forces now denied access or support. Combined action units are not designed for offensive operations themselves and rely on more robust combat units to perform this task. Combined action units can also establish mutual support among villages to secure a wider area.

5-82. A combined action program can work only in areas with limited insurgent activity. The technique should not be used to isolate or expel a well-established and supported insurgent force. Combined action is most effective after an area has been cleared of armed insurgents.

5-83. The following geographic and demographic factors can also influence the likelihood of success:

- Towns relatively isolated from other population centers are simpler to secure continuously.
- Towns and villages with a limited number of roads passing through them are easier to secure than those with many routes in and out. All approaches must be guarded.
- Existing avenues of approach into a town should be observable from the town. Keeping these areas under observation facilitates interdiction of insurgents and control of population movements.
- The local populace should be small and constant. People should know one another and be able to easily identify outsiders. In towns or small cities where this is not the case, a census is the most effective tool to establish initial accountability for everyone.
- Combined action or local defense forces must establish mutual support with forces operating in nearby towns. Larger reaction or reserve forces as well as close air support, attack aviation, and air assault support should be quickly available. Engineer and explosive ordnance disposal assets should also be available.

5-84. Combined action unit members must develop and build positive relationships with their associated HN security forces and with the town leadership. By living among the people, combined action units serve an important purpose. They demonstrate the commitment and competence of counterinsurgents while sharing experiences and relationships with local people. These working relationships build trust and enhance the HN government's legitimacy. To build trust further, U.S. members should ask HN security forces for training on local customs, key terrain, possible insurgent hideouts, and relevant cultural dynamics. HN forces should also be asked to describe recent local events.

5-85. Combined action units are integrated into a regional scheme of mutually supporting security and influence; however, they should remain organic to their parent unit. Positioning reinforced squad-sized units (13 to 15 Soldiers or Marines) among HN citizens creates a dispersal risk. Parent units can mitigate this risk with on-call reserve and reaction forces along with mutual support from adjacent villages and towns.

5-86. Thoroughly integrating U.S. and HN combined action personnel supports the effective teamwork critical to the success of each team and the overall program. U.S. members should be drawn from some of the parent unit's best personnel. Designating potential members before deployment facilitates the training and team building needed for combined action unit success in theater. Preferably, team members should have had prior experience in the host nation. Other desirable characteristics include—

- The ability to operate effectively as part of a team.
- Strong leadership qualities, among them—
  - Communicating clearly.
  - Maturity.
  - Leading by example.
  - Making good decisions.
- The ability to apply the commander's intent in the absence of orders.
- Possession of cultural awareness and understanding of the HN environment.
- The absence of obvious prejudices.

- Mutual respect when operating with HN personnel.
- Experience with the HN language, the ability to learn languages, or support of reliable translators.
- Patience and tolerance when dealing with language and translation barriers.

5-87. Appropriate tasks for combined action units include, but are not limited to, the following:

- Helping HN security forces maintain entry control points.
- Providing reaction force capabilities through the parent unit.
- Conducting multinational, coordinated day and night patrols to secure the town and area.
- Facilitating local contacts to gather information in conjunction with local HN security force representatives. (Ensure information gathered is made available promptly and on a regular basis to the parent unit for timely fusion and action.)
- Training HN security forces in leadership and general military subjects so they can secure the town or area on their own.
- Conducting operations with other multinational forces and HN units, if required.
- Operating as a team with HN security forces to instill pride, leadership, and patriotism.
- Assisting HN government representatives with civic action programs to establish an environment where the people have a stake in the future of their town and nation.
- Protecting HN judicial and government representatives and helping them establish the rule of law.

Combined Action Program

Building on their early 20th-century counterinsurgency experiences in Haiti and Nicaragua, the Marine Corps implemented an innovative program in South Vietnam in 1965 called the Combined Action Program. This program paired teams of about 15 Marines led by a noncommissioned officer with approximately 20 host-nation security personnel. These combined action platoons operated in the hamlets and villages in the northern two provinces of South Vietnam adjacent to the demilitarized zone. These Marines earned the trust of villagers by living among them while helping villagers defend themselves. Marines trained and led the local defense forces and learned the villagers' customs and language. The Marines were very successful in denying the Viet Cong access to areas under their control. The Combined Action Program became a model for countering insurgencies. Many lessons learned from it were used in various peace enforcement and humanitarian assistance operations that Marines conducted during the 1990s. These operations included Operations Provide Comfort in northern Iraq (1991) and Restore Hope in Somalia (1992 through 1993).

Limited Support

5-88. Not all COIN efforts require large combat formations. In many cases, U.S. support is limited, focused on missions like advising security forces and providing fire support or sustainment. The longstanding U.S. support to the Philippines is an example of such limited support. The limited support approach focuses on building HN capability and capacity. Under this approach, HN security forces are expected to conduct combat operations, including any clearing and holding missions.

Pattern of Transition

5-89. COIN efforts may require Soldiers and Marines to create the initial secure environment for the populace. Ideally HN forces hold cleared areas. As HN military and civil capabilities are further strengthened, U.S. military activity may shift toward combined action and limited support. As HN forces assume internal and external security requirements, U.S. forces can redeploy to support bases, reduce force strength, and eventually withdraw. Special operations forces and conventional forces continue to provide support as needed to achieve internal defense and development objectives.

ASSESSMENT OF COUNTERINSURGENCY OPERATIONS

The two best guides, which can not be readily reduced to statistics or processed through a computer, are an improvement in intelligence voluntarily given by the population and a decrease in the insurgents' recruiting rate. Much can be learnt merely from the faces of the population in villages that are subject to clear-and-hold operations, if these are visited at regular intervals. Faces which at first are resigned and apathetic, or even sullen, six months or a year later are full of cheerful welcoming smiles. The people know who is winning.

Sir Robert Thompson
Defeating Communist Insurgency: The Lessons of Malaya and Vietnam, 1966 [3]

5-90. *Assessment* is the continuous monitoring and evaluation of the current situation and progress of an operation (FMI 5-0.1). Assessment precedes and is integrated into every operations-process activity and entails two tasks:

- Continuously monitoring the current situation (including the environment) and progress of the operation.
- Evaluating the operation against established criteria.

Commanders, assisted by the staff, continuously compare the operation's progress with their commander's visualization and intent. Based on their assessments, commanders adjust the operation and associated activities to better achieve the desired end state. (See FM 6-0, paragraphs 6-90 through 6-92 and 6-110 through 6-121.)

DEVELOPING MEASUREMENT CRITERIA

5-91. Assessment requires determining why and when progress is being achieved along each LLO. Traditionally, commanders use discrete quantitative and qualitative measurements to evaluate progress. However, the complex nature of COIN operations makes progress difficult to measure. Subjective assessment at all levels is essential to understand the diverse and complex nature of COIN problems. It is also needed to measure local success or failure against the overall operation's end state. Additionally, commanders need to know how actions along different LLOs complement each other; therefore, planners evaluate not only progress along each LLO but also interactions among LLOs.

ASSESSMENT TOOLS

5-92. Assessment tools help commanders and staffs determine—

- Completion of tasks and their impact.
- Level of achievement of objectives.
- Whether a condition of success has been established.
- Whether the operation's end state has been attained.
- Whether the commander's intent was achieved.

For example, planning for transition of responsibility to the host nation is an integral part of COIN operational design and planning. Assessment tools may be used to assess the geographic and administrative transfer of control and responsibility to the HN government as it develops its capabilities. Assessments differ for every mission, task, and LLO, and for different phases of an operation. Leaders adjust assessment methods as insurgents adapt to counterinsurgent tactics and the environment changes.

5-93. The two most common types of assessment measures are measures of effectiveness (MOEs) and measures of performance (MOPs).

5-94. A *measure of effectiveness* is a criterion used to assess changes in system behavior, capability, or operational environment that is tied to measuring the attainment of an end state, achievement of an objective, or creation of an effect (JP 1-02). MOEs focus on the results or consequences of actions. MOEs answer the question, Are we achieving results that move us towards the desired end state, or are additional or alternative actions required?

5-95. A *measure of performance* is a criterion to assess friendly actions that is tied to measuring task accomplishment (JP 1-02). MOPs answers the question, Was the task or action performed as the commander intended?

5-96. Leaders may use observable, quantifiable, objective data as well as subjective indicators to assess progress measured against expectations. A combination of both types of indicators is recommended to reduce the chance of misconstruing trends.

5-97. All MOEs and MOPs for assessing COIN operations should be designed with the same characteristics. These four characteristics are—

- **Measurable**. MOEs and MOPs should have quantitative or qualitative standards against which they can be measured. The most effective measurement would be a combination of quantitative and qualitative measures to guard against an inaccurate view of results.
- **Discrete**. Each MOE and MOP must measure a separate, distinct aspect of the task, purpose, or condition.
- **Relevant**. MOEs and MOPs must be relevant to the measured task, outcome, and condition. HN local, regional, and national leaders, and nongovernmental organization personnel, may provide practical, astute, and professional ideas and feedback to craft relevant MOPs and MOEs.
- **Responsive**. Assessment tools must detect environmental and situational changes quickly and accurately enough to facilitate developing an effective response or counter.

Broad Indicators of Progress

5-98. Numerical and statistical indicators have limits when measuring social environments. For example, in South Vietnam U.S. forces used the body count to evaluate success or failure of combat operations. Yet, the body count only communicated a small part of the information commanders needed to assess their operations. It was therefore misleading. Body count can be a partial, effective indicator only when adversaries and their identities can be verified. (Normally, this identification is determined through a uniform or possession of an insurgent identification card.) Additionally, an accurate appreciation of what insurgent casualty numbers might indicate regarding enemy strength or capability requires knowing the exact number of insurgent armed fighters initially present. In addition, this indicator does not measure several important factors: for example, which side the local populace blames for collateral damage, whether this fighting and resultant casualties damaged the insurgent infrastructure and affected the insurgency strategy in that area, and where families of dead insurgents reside and how they might react. For another example, within the essential services LLO the number of schools built or renovated does not equate to the effective operation of an educational system.

5-99. Planners should start with broad measures of social and economic health or weakness when assessing environmental conditions. (Table 5-7 [page 5-28] lists possible examples of useful indicators in COIN.)

Table 5-7. Example progress indicators

- **Acts of violence** (numbers of attacks, friendly/host-nation casualties).
- **Dislocated civilians**. The number, population, and demographics of dislocated civilian camps or the lack thereof are a resultant indicator of overall security and stability. A drop in the number of people in the camps indicates an increasing return to normalcy. People and families exiled from or fleeing their homes and property and people returning to them are measurable and revealing.[1]
- **Human movement and religious attendance**. In societies where the culture is dominated by religion, activities related to the predominant faith may indicate the ease of movement and confidence in security, people's use of free will and volition, and the presence of freedom of religion. Possible indicators include the following:
  - Flow of religious pilgrims or lack thereof.
  - Development and active use of places of worship.
  - Number of temples and churches closed by a government.
- **Presence and activity of small- and medium-sized businesses**. When danger or insecure conditions exist, these businesses close. Patrols can report on the number of businesses that are open and how many customers they have. Tax collections may indicate the overall amount of sales activity.
- **Level of agricultural activity**.
  - Is a region or nation self-sustaining, or must life-support type foodstuffs be imported?
  - How many acres are in cultivation? Are the fields well maintained and watered?
  - Are agricultural goods getting to market? Has the annual need increased or decreased?
- **Presence or absence of associations.** The formation and presence of multiple political parties indicates more involvement of the people in government. Meetings of independent professional associations demonstrate the viability of the middle class and professions. Trade union activity indicates worker involvement in the economy and politics.
- **Participation in elections**, especially when insurgents publicly threaten violence against participants.
- **Government services available**. Examples include the following:
  - Police stations operational and police officers present throughout the area.
  - Clinics and hospitals in full operation, and whether new facilities sponsored by the private sector are open and operational.
  - Schools and universities open and functioning.
- **Freedom of movement of people, goods, and communications**. This is a classic measure to determine if an insurgency has denied areas in the physical, electronic, or print domains.
- **Tax Revenue**. If people are paying taxes, this can be an indicator of host-nation government influence and subsequent civil stability.
- **Industry exports**.
- **Employment/unemployment rate**.
- **Availability of electricity**.
- **Specific attacks on infrastructure**.

[1] *Dislocated civilian* is a broad term that includes a displaced person, an evacuee, an expellee, an internally displaced person, a migrant, a refugee, or a stateless person (JP 1-02). Dislocated civilians are a product of the deliberate violence associated with insurgencies and their counteraction. (See FM 3-05.40/MCRP 3-33.1A for additional information on dislocated civilians.)

TARGETING

5-100. The targeting process focuses operations and the use of limited assets and time. Commanders and staffs use the targeting process to achieve effects that support the LLOs in a COIN campaign plan. It is important to understand that targeting is done for all operations, not just attacks against insurgents. The targeting process can support IO, civil-military operations (CMO), and even meetings between commanders and HN leaders, based on the commander's desires. The targeting process occurs in the targeting cell of the appropriate command post. (See JP 3-60, FM 3-09.31/MCRP 3-16C, and FM 6-20-10 for joint and Army targeting doctrine. FM 3-13, appendix E, describes how to apply the targeting process to IO-related targets.)

5-101. Targeting in a COIN environment requires creating a targeting board or working group at all echelons. The intelligence cell provides representatives to the targeting board or working group to synchronize targeting with intelligence sharing and intelligence, surveillance, and reconnaissance operations. The goal is to prioritize targets and determine the means of engaging them that best supports the commander's intent and the operation plan.

5-102. The focus for targeting is on people, both insurgents and noncombatants. There are several different approaches to targeting in COIN. For example, all of the following are potential targets that can link objectives with effects:

- Insurgents (leaders, combatants, political cadre, auxiliaries, and the mass base).
- Insurgent internal support structure (bases of operations, finance base, lines of communications, and population).
- Insurgent external support systems (sanctuaries, media, and lines of communications).
- Legitimate government and functions (essential services, promotion of governance, development of security forces, and institutions).

5-103. Effective targeting identifies the targeting options, both lethal and nonlethal, to achieve effects that support the commander's objectives. Lethal targets are best addressed with operations to capture or kill; nonlethal targets are best engaged with CMO, IO, negotiation, political programs, economic programs, social programs and other noncombat methods. Nonlethal targets are usually more important than lethal targets in COIN; they are never less important. (Table 5-8 [page 5-30] lists examples of lethal and nonlethal targets.)

5-104. The targeting process comprises the following four activities:

- Decide which targets to engage.
- Detect the targets.
- Deliver (conduct the operation).
- Assess the effects of the operation.

5-105. Commanders issue targeting guidance during the "decide" activity. The commander's guidance drives subsequent targeting-process activities. Actions during the "detect" activity may give commanders the intelligence needed to refine the guidance. It may be difficult to identify targets when a COIN campaign begins. The focus during the "decide" activity should be on decisive points commanders can engage.

DECIDE

5-106. The decide activity draws on a detailed intelligence preparation of the battlefield and continuous assessment of the situation. Intelligence personnel, with the commander and other staff members, decide when a target is developed well enough to engage. Continuous staff integration and regular meetings of the intelligence cell and targeting board enable this activity. Staff members consider finished intelligence products in light of their understanding of the AO and advise commanders on targeting decisions. Intelligence personnel provide information on the relative importance of different target personalities and areas and the projected effects of lethal or nonlethal engagement. Specifically, the intelligence analysts need to identify individuals and groups to engage as potential COIN supporters, targets to isolate from the population, and targets to eliminate.

5-107. During the decide activity, the targeting board produces a prioritized list of targets and a recommended course of action associated with each. Executing targeting decisions may require the operations section to issue fragmentary orders. Each of these orders is a task that should be nested within the higher headquarters' plan and the commander's intent. Targeting decisions may require changing the intelligence synchronization plan.

Table 5-8. Examples of lethal and nonlethal targets

Personality targets
• **Lethal** ▪ Insurgent leaders to be captured or killed. • **Nonlethal** ▪ People like community leaders and those insurgents who should be engaged through outreach, negotiation, meetings, and other interaction. ▪ Corrupt host-nation leaders who may have to be replaced.
Area targets
• **Lethal** ▪ Insurgent bases and logistic depots or caches. ▪ Smuggling routes. • **Lethal and nonlethal mix** ▪ Populated areas where insurgents commonly operate. ▪ Populated areas controlled by insurgents where the presence of U.S. or host-nation personnel providing security could undermine support to insurgents. • **Nonlethal** ▪ Populations potentially receptive to civil-military operations or information operations.

DETECT

5-108. The detect activity is performed continuously. It requires much analytical work by intelligence personnel. They analyze large quantities of all-source intelligence reporting to determine the following:

- Threat validity.
- Actual importance of potential targets.
- Best means to engage the target.
- Expected effects of engaging the targets (which will guide actions to mitigate negative effects).
- Any changes required to the exploitation plan.

As mentioned in paragraph 3-152 target exploitation in a COIN environment is similar to that in law enforcement. An exploitation plan not only facilitates gathering evidence but also may lead to follow-on targets after successful exploitation. This requires a detailed understanding of social networks, insurgent networks, insurgent actions, and the community's attitude toward counterinsurgents. (See appendix B.)

5-109. Intelligence regarding the perceptions and interests of the populace requires particular attention. This intelligence is crucial to IO and CMO targeting. It is also important for developing political, social, and economic programs.

DELIVER

5-110. The deliver activity involves executing the missions decided upon by the commander.

ASSESS

5-111. The assess activity occurs continuously throughout an operation. During assessment, collectors and analysts evaluate the operation's progress. They adjust the intelligence synchronization plan and analyses based on this evaluation. In addition to assessing changes to their own operations, intelligence personnel look for reports indicating effects on all aspects of the operational environment, including insurgents and civilians. Relevant reporting can come from any intelligence discipline, open sources, or operational reporting. Commanders adjust an operation based on its effects. They may expand the operation, continue it as is, halt it, execute a branch or sequel, or take steps to correct a mistake's damage. Therefore, an accurate after-action assessment is very important. Metrics often include the following:

- Changes in local attitudes (friendliness towards U.S. and HN personnel).
- Changes in public perceptions.
- Changes in the quality or quantity of information provided by individuals or groups.
- Changes in the economic or political situation of an area.
- Changes in insurgent patterns.
- Captured and killed insurgents.
- Captured equipment and documents.

5-112. As indicated in chapter 3, detainees, captured documents, and captured equipment may yield a lot of information. Its exploitation and processing into intelligence often adds to the overall understanding of the enemy. This understanding can lead to more targeting decisions. In addition, the assessment of the operation should be fed back to collectors. This allows them to see if their sources are credible. In addition, effective operations often cause the local populace to provide more information, which drives future operations.

LEARNING AND ADAPTING

5-113. When an operation is executed, commanders may develop the situation to gain a more thorough situational understanding. This increased environmental understanding represents a form of operational learning and applies across all LLOs. Commanders and staffs adjust the operation's design and plan based on what they learn. The result is an ongoing design-learn-redesign cycle.

5-114. COIN operations involve complex, changing relations among all the direct and peripheral participants. These participants adapt and respond to each other throughout an operation. A cycle of adaptation usually develops between insurgents and counterinsurgents; both sides continually adapt to neutralize existing adversary advantages and develop new (usually short-lived) advantages of their own. Victory is gained through a tempo or rhythm of adaptation that is beyond the other side's ability to achieve or sustain. Therefore, counterinsurgents should seek to gain and sustain advantages over insurgents by emphasizing the learning and adaptation that this manual stresses throughout.

5-115. Learning and adapting in COIN is very difficult due to the complexity of the problems commanders must solve. Generally, there is not a single adversary that can be singularly classified as the enemy. Many insurgencies include multiple competing groups. Success requires the HN government and counterinsurgents to adapt based on understanding this very intricate environment. But the key to effective COIN design and execution remains the ability to adjust better and faster than the insurgents.

SUMMARY

5-116. Executing COIN operations is complex, demanding, and tedious. There are no simple, quick solutions. Success often seems elusive. However, contributing to the complexity of the problem is the manner in which counterinsurgents view the environment and how they define success. The specific design of the COIN operation and the manner in which it is executed must be based on a holistic treatment of the environment and remain focused on the commander's intent and end state. Success requires unity of effort across all LLOs to achieve objectives that contribute to the desired end state—establishing legitimacy and gaining popular support for the HN government. Operational design and execution cannot really be separated. They are both part of the same whole.

This page intentionally left blank.

Chapter 6
Developing Host-Nation Security Forces

[H]elping others to help themselves is critical to winning the long war.

Quadrennial Defense Review Report, 2006

This chapter addresses aspects of developing host-nation security forces. It begins with a discussion of challenges involved and resources required. It provides a framework for organizing the development effort. It concludes with a discussion of the role of police in counterinsurgency operations.

OVERVIEW

6-1. Success in counterinsurgency (COIN) operations requires establishing a legitimate government supported by the people and able to address the fundamental causes that insurgents use to gain support. Achieving these goals requires the host nation to defeat insurgents or render them irrelevant, uphold the rule of law, and provide a basic level of essential services and security for the populace. Key to all these tasks is developing an effective host-nation (HN) security force. In some cases, U.S. forces might be actively engaged in fighting insurgents while simultaneously helping the host nation build its own security forces.

6-2. Just as insurgency and COIN are defined by a complex array of factors, training HN security forces is also affected by a variety of determinants. These include whether sovereignty in the host nation is being exercised by an indigenous government or by a U.S. or multinational element. The second gives counterinsurgents more freedom of maneuver, but the first is important for legitimate governance, a key goal of any COIN effort. If the host nation is sovereign, the quality of its governance also has an impact. The scale of the effort is another factor; what works in a small country might not work in a large one. Terrain and civil considerations are also important. A nation compartmentalized by mountains, rivers, or ethnicity presents different challenges for the COIN effort. A large "occupying" force or international COIN effort can facilitate success in training HN security forces; however, it also complicates the situation. Other factors to consider include the following:

- Type of security forces that previously existed.
- Whether the effort involves creating a completely new security force or changing an existing one.
- Existence of sectarian divisions within the forces.
- Resources available.
- Popular support.

Commanders must adapt these doctrinal foundations to the situation in the area of operations (AO).

6-3. The term "security forces" includes all HN forces with the mission of protecting against internal and external threats. Elements of the security forces include, but are not limited to, the following:

- Military forces.
- Police.
- Corrections personnel.
- Border guards (including the coast guard).

Elements of the HN security forces exist at the local through national levels. Only in unusual cases will COIN forces experience a situation where the host nation has no security force. With this in mind, this chapter addresses methods to develop security forces, realizing that the range of assistance varies depending on the situation.

6-4. JP 3-07.1 contains foreign internal defense (FID) doctrine. JP 3-07.1 addresses the legal and fiscal regulations and responsibilities concerning the planning, development, and administration of FID programs. It also discusses command and supervisory relationships of U.S. diplomatic missions, geographic combatant commands, and joint task forces in applying military aid, support, and advisory missions. The tenets presented in this chapter reinforce and supplement those in JP 3-07.1.

CHALLENGES, RESOURCES, AND END STATE

6-5. Each instance of developing security forces is as unique as each insurgency. In Vietnam, the United States committed thousands of advisors for South Vietnamese units and hundreds of thousands of combat troops but ultimately failed to achieve its strategic objectives. In El Salvador, a relative handful of American advisors were enough enable the HN government to execute a successful counterinsurgency, even though that situation had evolved into a recognized, full-blown civil war. Many factors influence the amount and type of aid required. These are discussed in more detail later, but include the following:

- Existing HN security force capabilities.
- Character of the insurgency.
- Population and culture.
- Level of commitment and sovereignty of the host nation.
- Level of commitment from the United States and other nations.

6-6. U.S. and multinational forces may need to help the host nation in improve security; however, insurgents can use the presence of foreign forces as a reason to question the HN government's legitimacy. A government reliant on foreign forces for internal security risks not being recognized as legitimate. While combat operations with significant U.S. and multinational participation may be necessary, U.S. combat operations are secondary to enabling the host nation's ability to provide for its own security.

CHALLENGES TO DEVELOPING EFFECTIVE SECURITY FORCES

6-7. Many common problems and issues arose in training missions U.S. forces undertook after World War II. These problems generally fall under differing national perspectives in one of four broad categories:

- Resources.
- Leadership.
- Exercising power.
- Organizational structures.

6-8. Governments must properly balance national resources to meet the people's expectations. Funding for services, education, and health care can limit resources available for security forces. The result HN spending priorities may be a security force capable of protecting only the capital and key government facilities, leaving the rest of the country unsecured. Undeveloped countries often lack resources to maintain logistic units. This situation results in chronic sustainment problems. Conducting effective COIN operations requires allocating resources to ensure integration of efforts to develop all aspects of the security force. Recognizing the interrelationship of security and governance, the HN government must devote adequate resources to meeting basic needs like health care, clean water, and electricity.

6-9. Counterinsurgents may need to adjust the existing HN approach to leadership. HN leaders may be appointed and promoted based on family ties or membership in a party or faction, rather than on demonstrated competence or performance. Leaders may not seek to develop subordinates. The need to ensure the welfare of subordinates may not be a commonly shared trait. In some cases, leaders enforce the subordinates' obedience by fear and use their leadership position to exploit them. Positions of power can lead to corruption, which can also be affected by local culture.

6-10. The behavior of HN security force personnel is often a primary cause of public dissatisfaction. Corrupting influences of power must be guarded against. Cultural and ethnic differences within a population may lead to significant discrimination within the security forces and by security forces against minority groups. In more ideological struggles, discrimination may be against members of other political parties, whether in a minority cultural group or not. Security forces that abuse civilians do not win the populace's trust and confidence; they may even be a cause of the insurgency. A comprehensive security force development program identifies and addresses biases as well as improper or corrupt practices.

6-11. Perhaps the biggest hurdle for U.S. forces is accepting that the host nation can ensure security using practices that differ from U.S. practices. Commanders must recognize and continuously address that this "The American way is best" bias is unhelpful. While relationships among U.S. police, customs, and military organizations works for the United States, those relationships may not exist in other nations that have developed differently.

RESOURCES

6-12. For Soldiers and Marines, the mission of developing HN security forces goes beyond a task assigned to a few specialists. The scope and scale of training programs today and the scale of programs likely to be required in the future have grown. While FID has been traditionally the primary responsibility of the special operations forces (SOF), training foreign forces is now a core competency of regular and reserve units of all Services. Multinational partners are often willing to help a nation against an insurgency by helping to train HN forces. Partner nations may develop joint training teams or assign teams to a specific element of the security force or a particular specialty. Training resources may be received from the following organizations and programs:

- Special operations forces.
- Ground forces.
- Joint forces.
- Interagency resources.
- Multinational resources.
- International Military Education and Training (IMET) Program.
- Contractor support.

Special Operations Forces

6-13. SOF focus on specific regions of the world and the study of languages and cultures. SOF have long been the lead organization in training and advising foreign armed forces. (FM 31-20-3 outlines Army special forces training programs and tactics, techniques, and procedures.) While SOF personnel may be ideal for some training and advisory roles, their limited numbers restrict their ability to carry out large-scale missions to develop HN security forces. In a low-level COIN, SOF personnel may be the only forces assigned; at the higher end of the spectrum, SOF may train only their counterparts in the HN forces.

Ground Forces

6-14. Large-scale training and advisory missions need to use large numbers of Soldiers and Marines who may not have language training or regional expertise to levels common in SOF. However, such conventional forces may have some advantages in training HN counterparts with similar missions. SOF and conventional ground forces need linguist augmentation and additional cultural training. (Appendix C discusses linguist support.) Commanders must assign the best qualified Soldiers and Marines to training and advisory missions. Those personnel normally come from active-duty forces, but large-scale efforts require using Reserve Component personnel. All land forces assigned to this high-priority mission need thorough training, both before deploying and in theater.

Joint Forces

6-15. Although other Services often play smaller roles, they can still make significant contributions because of their considerable experience in training foreign forces. For example, the Navy and Air Force can train their HN counterparts. The Coast Guard may also be of value, since its coastal patrol, fisheries oversight, and port security missions correlate with the responsibilities of navies in developing countries. To minimize the burden on land forces, specialists—such as lawyers and medical personnel—from other Services participate in HN training wherever possible.

Interagency Resources

6-16. Interagency resources can support training HN security forces. Perhaps most important is training nonmilitary security forces. The Departments of Justice and State can send law enforcement specialists overseas to train and advise HN police forces. Police are best trained by other police. The quick reaction capability of these agencies is limited, although they can attain necessary levels when given time. Such forces are also expensive. During intensive counterinsurgencies, the environment's high-threat nature limits the effectiveness of civilian police advisors and trainers. These forces work more effectively when operating in a benign environment or when security is provided separately. Many legal restrictions about training nonmilitary forces exist. Normally the Department of State takes the lead in such efforts. However, the President occasionally assigns military forces to these missions.

Multinational Resources

6-17. Although their support frequently plays more of a legitimizing role, multinational partners also assist materially in training HN security forces. Some nations more willingly train HN forces, especially police forces, than provide troops for combat operations. Some multinational forces come with significant employment restrictions. Each international contribution is considered on its own merits, but such assistance is rarely declined. Good faith efforts to integrate multinational partners and achieve optimum effectiveness are required.

International Military Education and Training Program

6-18. For more than 50 years, the U.S. military has run the IMET program to provide opportunities for foreign personnel to attend U.S. military schools and courses. Most of these commissioned officers and noncommissioned officers (NCOs) receive English language training before attending the U.S. courses. In the case of Latin American armed forces, the United States operates courses in Spanish.

Contractor Support

6-19. In some cases, additional training support from contractors enables commanders to use Soldiers and Marines more efficiently. Contractor support can provide HN training and education, including the following:

- Institutional training.
- Developing security ministries and headquarters.
- Establishing administrative and logistic systems.

Contracted police development capabilities through the Department of State's Bureau of International Narcotics and Law Enforcement Affairs can provide expertise not resident in the uniformed military.

Organizing U.S. Forces to Develop Host-Nation Security Forces

6-20. Developing HN security forces is a complex and challenging mission. The United States and multinational partners can only succeed if they approach the mission with the same deliberate planning and preparation, energetic execution, and appropriate resourcing as the combat aspects of the COIN operation. Accordingly, COIN force commanders and staffs need to consider the task of developing HN security forces during their initial mission analysis. They must make that task an integral part of all assessments, planning, coordination, and preparation.

6-21. As planning unfolds, mission requirements should drive the initial organization for the unit charged with developing security forces. To achieve unity of effort, a single organization should receive this responsibility.

6-22. For small-scale COIN efforts, SOF may be the only forces used. SOF organizations may be ideally suited for developing security forces through the FID portion of their doctrinal mission.

6-23. If only a single component (land, maritime, air, or special operations) is being developed, commanders can assign the mission to a single-Service task force. For example, if the host nation requires a maritime capability to guard oil distribution platforms, a Navy task force may receive the mission.

6-24. In an area in which COIN operations are already underway, developing security forces can be assigned to a specific unit, such as a brigade combat team, division, or Marine air-ground task force.

6-25. For large, multi-Service, long-duration missions, a separate organization with the sole responsibility of developing security forces and subordinate to the COIN force commander may be required. Such an organization may be multi-Service, multinational, and interagency.

Table 6-1. Staff functions required when training host-nation security forces

- **Financial manager** for managing the significant monetary resources required for training, equipping, and building security forces. A separate internal auditor may be required as a check to ensure host-nation resources are safeguarded and effectively managed.
- **Staff judge advocate** with specific specialties and a robust capability for contract law, military justice, and the law of land warfare.
- **Construction engineer** management to oversee and manage the construction of security forces infrastructure, such as the following:
  - Bases, ranges, and training areas.
  - Depots and logistic facilities.
  - Police stations.
- **Political-military advisors** to ensure development of security forces is integrated with development of civilian ministries and capabilities.
- **Public affairs**, with a focused capability to build the populace's confidence in the host-nation security forces and to develop the host-nation forces' public affairs capability.
- **Force protection and focused intelligence staff** to address the challenge of and threats to the relatively small teams that may be embedded with host-nation security forces and not co-located with U.S. or multinational forces.
- **Materiel management** until such a capability is developed in the host-nation forces. The equipping and supplying of new security forces is critical to their development and employment. It may not be able to wait until the host-nation develops that capability.
- **Health affairs**, since most developing countries have poor health care systems. Host-nation personnel are more likely to stay in new units and fight when they believe that they will be properly treated if wounded. Additionally, disease is a significant threat that must be addressed with preventive medicine and robust care.
- **Security assistance** (IMET) to manage the external training efforts and foreign military sales, and to employ well-developed procedures for purchasing weapons, equipment, goods, and services. In counterinsurgencies, these functions are probably performed by higher headquarters staff elements rather than a stand-alone office (such as an office of military cooperation). U.S. security assistance programs normally try to sell U.S.-manufactured equipment. The organization responsible for equipping host-nation forces should not be constrained to purchase U.S. equipment. It requires the flexibility to procure equipment where time, cost, and quality are appropriate for host-nation needs.
- **Civilian law enforcement**. Staff officers with a civilian law enforcement background or actual civilian law enforcement personnel can play a vitally important role in advising the commander. Traditionally officers from the Reserve Components have done this.

6-26. The internal structure of the organization charged with developing security forces must reflect the desired end state of those security forces. For example, if army, police, air, naval, and special operations capabilities are being developed, the organization in charge of those programs requires teams charged specifically with each of those tasks. If civilian security components, such as a ministry of defense or interior, are being developed, then ministerial teams are needed. Developing security forces in terms of professionalism and ethics is important; a separate element focused on training those values may be needed.

6-27. The U.S. or multinational force responsible for these programs requires a headquarters and staff task-organized for the functions required to support all aspects of developing the security forces. (See paragraph 6-31.) In addition to traditional staff functions, some or all of functions listed in table 6-1 (page 6-5) may require augmentation.

6-28. An effective security force development organization is flexible and adaptive. Requirements for developing the type, character, composition, and quantity of security forces change as security forces grow and the COIN operation matures. The organization must anticipate such changes, since joint manning document procedures and requests for forces have limited responsiveness. Temporary duty and contract personnel may provide support to fill gaps until more permanent individuals or units arrive.

Desired End State

6-29. Training HN security forces is a slow and painstaking process. It does not lend itself to a "quick fix." Real success does not appear as a single decisive victory. To ensure long-term success, commanders clarify their desired end state for training programs early. This end state consists of a set of military characteristics common to all militaries. (See table 6-2.) Those characteristics have nuances in different countries, but well-trained HN security forces should—

- Provide reasonable levels of security from external threats while not threatening regional security.
- Provide reasonable levels of internal security without infringing upon the populace's civil liberties or posing a coup threat.
- Be founded upon the rule of law.
- Be sustainable by the host nation after U.S. and multinational forces depart.

6-30. When dealing with insurgents, HN military and police forces may perform functions not normally considered conventional. The military may fill an internal security role usually reserved for the police. Police may have forces so heavily armed that they would normally be part of the military. In the near term, the HN security forces should—

- Focus on COIN operations, integrating military capabilities with those of local, regional, and national police.
- Maintain the flexibility to transition to more conventional roles of external and internal defense, based on long-term requirements.

To meet both near- and long-term objectives, trainers remember the cumulative effects of training. Effective training programs have short-, mid-, and long-term effects.

6-31. To achieve this end state and intermediate objectives, the host nation should develop a plan—with multinational assistance when necessary. The plan should address all aspects of force development. U.S. doctrine divides force development into domains: doctrine, organization, training, materiel, leadership and education, personnel and facilities (DOTMLPF). Doctrine is listed first. However, these elements are tightly linked, simultaneously pursued, and difficult to prioritize. Commanders monitor progress in all domains. There is always a temptation for Soldiers and Marines involved in such programs to impose their own doctrine and judgment on the host nation. The first U.S. advisors and trainers working with the South Vietnamese Army aimed to create a conventional force to fight another Korean War. They did not recognize their allies' abilities or the real nature of the threat. The organization and doctrine adopted did not suit the South Vietnamese situation and proved vulnerable to North Vietnamese guerrilla tactics. HN security force doctrine, like the remaining DOTMLPF domains discussed in this chapter, must be appropriate to HN capabilities and requirements.

FRAMEWORK FOR DEVELOPMENT

6-32. The mission to develop HN security forces can be organized around these tasks:

- Assess.
- Organize.
- Build or rebuild facilities.

- Train.
- Equip.
- Advise.

These incorporate all DOTMLPF requirements. Although described sequentially, these tasks are normally performed concurrently. For example, training and equipping operations must be integrated and, as the operation progresses, assessments will lead to changes. If U.S. forces are directly involved in operations against insurgents, the development program requires a transition period during which major COIN operations are handed over to HN security forces.

Table 6-2. Characteristics of effective host-nation security forces

- **Flexible**. Forces capable of accomplishing the broad missions required by the host nation—not only to defeat insurgents or defend against outside aggression but also to increase security in all areas. This requires an effective command and organizational structure that makes sense for the host nation.
- **Proficient**.
  - Security forces capable of working effectively in close coordination with each other to suppress lawlessness and insurgency.
  - Military units tactically and technically proficient, capable of ensuring their aspect of national security and capable of integrating their operations with those of multinational partners.
  - Nonmilitary security forces competent in maintaining civil order, enforcing laws, controlling borders, securing key infrastructure (such as power plants), and detaining criminal suspects.
  - Nonmilitary security forces thoroughly trained in modern police ethos and procedures, and who understand the basics of investigation, evidence collection, and proper court and legal procedures.
- **Self-sustained**. Forces capable of managing their own equipment throughout its life cycle (procurement to disposal) and performing administrative support.
- **Well led**. Leaders at all levels who possess sound professional standards and appropriate military values, and are selected and promoted based on competence and merit.
- **Professional**.
  - Security forces that are honest, impartial, and committed to protecting and serving the entire population, operating under the rule of law, and respecting human rights.
  - Security forces that are loyal to the central government and serving national interests, recognizing their role as the people's servants and not their masters.
- **Integrated into society**. Forces that represent the host nation's major ethnic groups and are not seen as instruments of just one faction. Cultural sensitivities toward the incorporation of women must be observed, but efforts should also be made to include women in police and military organizations.

ASSESS

6-33. Commanders assess the situation at the start of every major military operation. The assessment is one part of the comprehensive program of analyzing the insurgency. It includes looking at the society and the economy. The analysis is performed in close collaboration with the U.S. country team, the host nation, and multinational partners. These partners continually assess the security situation and its influence on other logical lines of operations. From the assessment, planners develop short-, mid-, and long-range goals and programs. As circumstances change, so do the goals and programs. A raging insurgency might require the early employment of HN forces at various stages of development. Some existing security forces may be so dysfunctional or corrupt that the organizations must be disbanded rather than rehabilitated. In some cases, commanders will need to replace some HN leaders before their units will become functional.

6-34. While every situation is different, leaders of the development program should assess the following factors throughout planning, preparation, and execution of the operation:

- Social structure, organization, demographics, interrelationships, and education level of security force elements.
- Methods, successes, and failures of HN COIN efforts.
- State of training at all levels, and the specialties and education of leaders.
- Equipment and priority placed on maintenance.
- Logistic and support structure, and its ability to meet the force's requirements.
- Level of sovereignty of the HN government.
- Extent of acceptance of ethnic and religious minorities.
- Laws and regulations governing the security forces and their relationship to national leaders.

6-35. A comprehensive mission analysis should provide a basis for establishing the scope of effort required. This analysis includes a troop-to-task analysis that determines the type and size of forces needed. The HN security forces may require complete reestablishment, or they may only require assistance to increase capacity. They may completely lack a capability (for example, internal affairs, federal investigative department, corrections, logistics for military forces, formal schools for leaders), or they may only require temporary reinforcement. As with other military operations, efforts to assist security forces should reinforce success. For example, instead of building new police stations in every town, improve the good stations and use them as models for weaker organizations.

6-36. Leaders need decisions on what shortfalls to address first. The extent of the insurgency combined with resource limitations inevitably forces commanders to set priorities. Follow-on assessments should start by reviewing areas with restricted resources, determining where resources should be committed or redirected, and deciding whether to request additional resources. If the U.S. or another multinational partner or international entity exercises sovereignty, such as during an occupation or regime change, decisions about security force actions can be imposed on a host nation; however, it is always better to take efforts to legitimize the HN leaders by including them in decisions.

6-37. Developing a strategic analysis and outlining a strategic plan for training the forces of a country facing an insurgency is not necessarily a long process. In fact, situations that include a security vacuum or very active insurgency often require starting programs as soon as possible. Assessment is continuous; initial assessments and the programs they inspire must be adjusted as more experience and information are gained. In 1981, when El Salvador faced a major insurgency, a team of ten U.S. officers visited there. The team consulted with the HN command and the U.S. military assistance advisory group (referred to as the MILGRP) for ten days. In that time, the team outlined a five-year comprehensive plan to rebuild, reorganize, train, and reequip the Salvadoran armed forces to counter the insurgency. The U.S. plan became part of the foundation of a successful national COIN strategy. A team with a similar mission today should include specialists from the Departments of State, Justice, and Homeland Security (in particular, border security and customs experts) to assess the security force requirements.

ORGANIZE

6-38. Organizing HN forces depends on the host nation's social and economic conditions, cultural and historical factors, and security threat. The development program's aim is to create an efficient organization with a command, intelligence, logistic, and operations structure that makes sense for the host nation. Conventional forces with limited special purpose teams (such as explosive ordnance disposal and special weapons and tactics [SWAT]) are preferred. Elite units tend to divert a large share of the best leadership and remove critical talent from the regular forces. Doctrine should be standard across the force, as should unit structures. The organization must facilitate the collection, processing, and dissemination of intelligence across and throughout all security forces.

6-39. Another organizational approach is establishing home guard units. In many COIN operations, these units have effectively provided increased security to the populace. Home guards are part-time, lightly armed, local security forces under HN government control. Often, career military and police officers supervise home guards at the provincial and national levels. Home guards provide point security. They guard vital installations that insurgents will likely target, such as government buildings and businesses. Home guards can also provide security for small villages and man gates and checkpoints. While home guards are not trained to conduct offensive operations, their constant presence reminds the populace that the HN government can provide security. Effective home guards can free police and military forces from stationary guard duties.

General Considerations

6-40. As much as possible, the host nation should determine the security force organization's structure. The host nation may be open to proposals from U.S. and multinational forces but should at least approve all organizational designs. As the HN government strengthens, U.S. leaders and trainers should expect increasingly independent organizational decisions. These may include changing the numbers of forces, types of units, and internal organizational designs. Culture and conditions might result in security forces given what U.S. experience considers nontraditional roles and missions. HN police may be more paramilitary than their U.S. counterparts, and the military may have a role in internal security. Eventually, police and military roles should clearly differ. Police should counter crime while the military should address external threats. However, the exact nature of these missions depends on the HN situation. In any event, police and military roles should be clearly delineated.

6-41. Organized units should include all appropriate warfighting functions (formerly, battlefield operating systems) or some adaptation for police forces. Some systems may be initially deployed in limited numbers or excluded for various reasons, for example, cost, relevance, or training requirements. However, organizational plans should include eventually establishing all appropriate capabilities.

6-42. Organization should address all security force elements, from the ministerial level to the patrolling police officer and soldier. Figure 6-1 (page 6-10) illustrates the complex matrix of simultaneous development programs. Building a competent HN civilian infrastructure—including civilian command and control systems—is critical for success in COIN. The COIN force commander works with HN ministries responsible for national and internal security, including the ministry of national defense, the interior ministry, and the justice ministry. The commander assesses strengths and weaknesses of the ministerial organization as well as training requirements of civilian defense officials and employees. The U.S. and multinational advisory team at the ministry level help the host nation develop a procurement and management system that effectively meets its requirements.

6-43. A thorough review of HN military and police doctrine is a necessary first step in setting up a training program. Advisors review security force regulations to ensure they provide clear, complete instructions for discipline, acquisitions, and support activities. Advisors review and refine doctrine (including tactics, techniques, and procedures) to address COIN operations. Regulations should fit security force personnel's level of education and sophistication. Treatment of prisoners, detainees, and suspected persons should be clear and consistent with the norms of international and military law.

Personnel Considerations

6-44. Organizing a security force requires resolving issues related to the following areas:

- Recruiting.
- Promotion screening and selection.
- Pay and benefits.
- Leader recruiting and selection.
- Personnel accountability.
- Demobilization of security force personnel.

Recruiting

6-45. Recruiting is critical when establishing security forces. The host nation designs the recruiting program, considering local culture and themes that resonate with the populace. The program should ensure that security forces include members from all major demographic groups. U.S. and multinational partners should encourage and support HN efforts to recruit from minority populations. A mobile recruiting capability should be established to target specific areas, ethnic groups, or tribes to ensure demographic representation within the security forces. Moderate groups and factions within hostile or potentially hostile ethnic groups should be encouraged to join the HN security forces. Most HN governments will likely resist recruiting disaffected ethnic groups into their security forces. However, even moderate success in recruiting from these groups provides enormous payoffs. It builds the security forces' legitimacy and often quiets legitimate fears of such groups regarding their relationship to the HN government. Effectively disarming, demobilizing, and reintegrating former insurgents and their groups must be part of the overall COIN plan. It must be included in the recruiting effort.

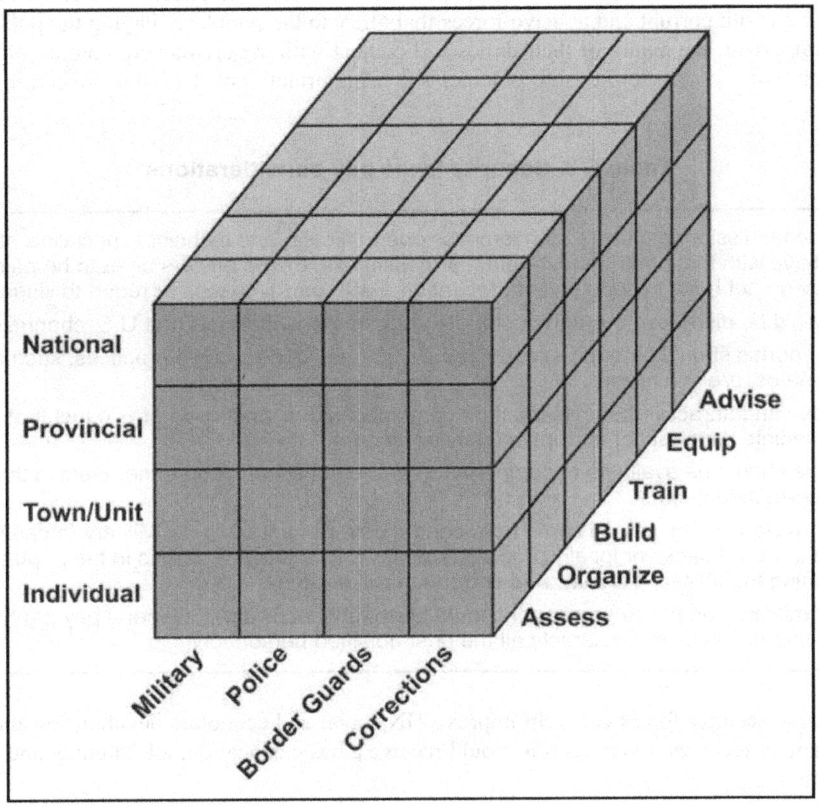

Figure 6-1. Factors affecting security force development

6-46. A proper recruiting program requires a clear set of appropriate mental, physical, and moral standards. Ideally, recruits are centrally screened and inducted. Recruiting centers need to be located in areas safe from insurgent attacks; these centers are attractive targets. All recruits should undergo a basic security check and be vetted against lists of suspected insurgents. As much as possible, HN agencies and personnel should perform this screening. Membership in illegal organizations needs to be carefully monitored. Past membership need not preclude a person from joining the security forces; however, any ongoing relationship with an illegal organization requires constant monitoring. HN personnel need to ensure that no single military or police unit contains too many prior members of an illegal unit, tribal militia, or other militant faction.

Promotion Screening and Selection

6-47. Selection for promotion must stem from proven performance and aptitude for increased responsibility. Objective evaluations ensure promotion is by merit and not through influence or family ties. Two methods for selecting leaders may be worth considering. One method identifies the most competent performers, trains them, and recommends them for promotion. The second method identifies those with social or professional status within the training group, then trains and recommends them for promotion. The first method may lead to more competent leaders but could be resisted for cultural reasons. The second method ensures the new leader will be accepted culturally but may sacrifice competence. The most effective solution comes from combining the two methods.

Pay and Benefits

6-48. Appropriate compensation levels help prevent a culture of corruption in the security forces. It is cheaper to spend the money needed for adequate wages and produce effective security forces than to pay less and end up with corrupt and abusive forces that alienate the populace. Paying the police adequately is especially important; the nature of their duties and contact with the civilian community often expose them to opportunities for corruption. (Table 6-3 lists some important considerations concerning security force pay.)

Table 6-3. Security force pay considerations

- Pay for commissioned officers, noncommissioned officers, and technical specialists should be competitive with that of other host-nation professionals. Police officers need to be paid enough that they do not have to supplement their income with part-time jobs or resort to illegal behavior.
- Pay should be disbursed through host-nation government channels, not U.S. channels.
- Cultural norms should be addressed to ensure that any questionable practices, such as "taxing" subordinates, are minimized.
- Good pay and attractive benefits must be combined with a strict code of conduct that allows for the immediate dismissal of corrupt security personnel.
- Pensions should be available to compensate families of security force members in the event of a service-related death.
- Pay for military forces should come from central government budgets. Military forces must not be paid from kickbacks or locally procured revenue. This practice results in the populace questioning the military's loyalty, and corruption will be likely.
- Pay for military and police personnel should be roughly equivalent. General pay parity helps ensure that no single force attracts all the best qualified personnel.

6-49. Effective security forces can help improve HN social and economic development through the benefits each member receives. Every recruit should receive a basic education, job training, and morals and values inculcation.

Leader Recruiting and Selection

6-50. Officer candidate standards should be high. Candidates should be in good health and pass an academic test with a higher standard than the test for enlisted recruits. Officer candidates should be carefully vetted to ensure that they do not have close ties to any radical or insurgent organization.

6-51. NCOs should be selected from the best enlisted security force members. Objective standards, including proficiency tests, should be established and enforced to ensure that promotion to the NCO ranks comes from merit, not through influence or family ties. Many armies lack a professional NCO corps; establishing one for a host nation may be difficult. In the meantime, adjustments will have to be made, placing more responsibility on commissioned officers.

Personnel Accountability

6-52. HN leaders must carefully track and account for security force personnel. Proper personnel accountability reduces corruption, particularly in countries with manual banking systems where security force personnel are paid in cash. In addition, large numbers of personnel failing to report for duty can indicate possible attacks, low unit morale, or insurgent and militia influences on the security forces.

Demobilization of Security Force Personnel

6-53. Host nations should develop programs to keep a class of impoverished and disgruntled former officers and soldiers from forming. As the security forces mature, commissioned officers and NCOs who perform poorly or fail to meet the new, higher standards of the force will need to be removed. Providing some form of government-provided education grants or low-interest business loans enables discharged personnel to earn a living outside the military. Commissioned officers and NCOs who serve for several years and are then removed should receive a lump-sum payment or small pension to ease their transition to civilian life. These programs should not apply to those guilty of major human rights abuses or corruption. Demobilization planning should start as soon as commanders anticipate the need. (It may not be required in all cases.) Any plan should evolve with HN security force development to ensure its feasibility. Similar programs may be required when demobilizing nongovernment militias.

6-54. As a conflict ends, some security forces may need to be disarmed, demobilized, and reintegrated into civil society. To avoid producing a pool of recruits for the insurgency, the host nation should establish programs to keep large numbers of demobilized security force members from becoming immediately unemployed. Civil service departments should provide a hiring preference to people completing an honorable term of service. Government-financed education programs for demobilized members are another possibility.

BUILD OR REBUILD FACILITIES

6-55. HN security forces need infrastructure support. People need buildings for storage, training, and shelter. Often requirements include barracks, ranges, motor pools, and other military facilities. Construction takes time; the host nation needs to invest early in such facilities if they are to be available when needed. Protection must be considered in any infrastructure design, including headquarters facilities, as infrastructure provides attractive targets for insurgents. (See FM 5-104 for information on hardening measures to increase infrastructure survivability and improve protection.)

6-56. During an insurgency, HN military and police forces often operate from local bases. Building training centers and unit garrisons requires a long-term force-basing plan. If possible, garrisons should include housing for the commissioned officers, NCOs, and families; government-provided medical care for the families; and other benefits that make national service attractive.

6-57. The host nation may need to make large investments in time and resources to restore or create the nationwide infrastructure necessary to effectively command and control HN security forces. Communications facilities are especially important. Besides building local bases and police stations, the host nation will need functional regional and national headquarters and ministries.

TRAIN

6-58. U.S. and multinational training assistance should address shortfalls at every level with the purpose of establishing self-sustaining training systems.

Training the U.S. Trainers

6-59. Soldiers and Marines assigned training missions should receive training on the specific requirements of developing HN forces. The course should emphasize the host nation's cultural background, introduce its language, and provide cultural tips for developing a good rapport with HN personnel. The course should also include protection training for troops working with HN forces. U.S. trainees must become familiar with the HN organization and equipment, especially weapons not in the U.S. inventory. This training must emphasize the following:

- Sustaining training and reinforcing individual and team skills.
- Using the smallest possible student-to-instructor ratio.
- Developing HN trainers.
- Training to standards—not to time.
- Providing immediate feedback; using after-action reviews.

- Respecting the HN culture, but learning to distinguish between cultural practices and excuses.
- Learning the HN language.
- Working with interpreters.

6-60. U.S. forces should show respect for local religions and traditions. Soldiers and Marines should willingly accept many aspects of the local and national culture, including food (if sanitation standards permit). U.S. forces must make clear that they do not intend to undermine or change the local religion or traditions. However, Soldiers and Marines have a mission to reduce the effects of dysfunctional social practices that affect the ability to conduct effective security operations. U.S. trainers and advisors must have enough awareness to identify and stop inappropriate behavior, or at least report it to the multinational and HN chains of command.

Establishing Training Standards

6-61. Insurgent approaches and their corresponding responses from targeted governments vary widely; however, the host nation and trainers can still establish clear measures for evaluating the training of individuals, leaders, and units. COIN operations require many of the same individual and collective skills performed in conventional military operations but also include additional requirements for COIN. Small units execute most COIN operations; therefore, effective COIN forces require strong junior leaders. All levels of training for all components should include values training. Metrics for evaluating units should include subjective measures, such as loyalty to the HN government, as well as competence in military tasks. Soldiers and Marines know how to evaluate military training. However, the acceptance of values, such as ethnic equality or the rejection of corruption, may be a better measure of training effectiveness in some COIN situations. Gauging this acceptance is far more difficult than evaluating task performance.

6-62. Effective training programs require clear, detailed individual, leader, and unit performance standards. These standards take into account cultural factors that directly affect the ability of the individual or unit to operate. For example, training a unit to conduct effective operations requires more time in countries where the average soldier is illiterate. Similarly, staff training proves more difficult in countries with a low educational level. Building a security force from the ground up takes far more time than creating one around a trained cadre of HN personnel. With this in mind, it is usually better to use existing military personnel to form units and cadres for units, rather than creating novice security forces. Vetting may be required to determine loyalties and validate the abilities of existing security forces.

6-63. Poorly trained leaders and units are more prone to committing human rights violations than well-trained, well-led units. Leaders and units unprepared for the pressure of active operations tend to use indiscriminate force, target civilians, and abuse prisoners. These actions can threaten the popular support and government legitimacy essential for COIN success. Badly disciplined and poorly led security forces become effective recruiters and propagandists for insurgents.

6-64. Setting realistic metrics, both objective and subjective, for HN security forces and following through on training plans consume time. The pressure is strong to find training shortcuts, employ "quick fixes," or to train personnel on the job. Trainers should resist such approaches. In the long term, such approaches create more problems than they solve. However, trainers should also avoid the temptation to create long, complex training programs based on unrealistic standards. Effective programs account for the host nation's culture, resources, and short-term security needs. No firm rules exist on how long particular training programs should take, but trainers can use existing and historical U.S. or multinational training programs as starting points for planning. To a certain extent, the insurgent threat dictates how long training can take. As security improves, training programs can expand to facilitate achievement of the long-term end state.

Training Methods

6-65. Training programs should be designed to prepare HN personnel to train themselves. HN trainers are the best trainers and should be used as much as possible. Many training methods have proven successful; some also enhance developing HN training capability. (Table 6-4 [page 6-14] lists several successful training methods.)

Table 6-4. Methods of training host-nation security forces

- **Formal schools** initially run by U.S. forces with selected graduates returning as instructors. This includes entry-level individual training.
- **Mobile training teams** to reinforce individual and collective training as needed.
- **Partnership training** with U.S. forces tasked to train and advise host-nation units with whom they are partnered. U.S. forces support host-nation units. As training progresses, host-nation squads, platoons, and companies may work with their U.S. partners in security or combat operations. In this manner, the whole U.S. unit mentors their partners. Habitual training relationships should be maintained between partners until host-nation units meet standards for full capability.
- **Advisor teams** detailed to assist host-nation units with minimal segregation between U.S. and host-nation personnel. Advisor teams work especially well in training senior ministry personnel.
- **U.S. personnel embedded in key positions in host-nation units**. This may be required where host-nation security forces are needed but leader training is still in its early stages. This approach increases dependency on U.S. forces and should be used only in extreme circumstances. As host-nation capabilities improve, host-nation personnel should replace the embedded U.S. or multinational personnel.
- **Contractors** can also be used to assist with training, though care is required to ensure the training is closely supervised and meets standards.

Training Soldiers

6-66. Security force members must be developed through a systematic training program. The program first builds their basic skills, then teaches them to work together as a team, and finally allows them to function as a unit. Basic military training should focus first on COIN-related skills, such as first aid, marksmanship, and fire discipline. Leaders must be trained in tactics, including patrolling and urban operations. Everyone must master rules of engagement and the law of armed conflict. HN units should train to standard for conducting the major COIN missions they will face. Required skills include the following:

- Manage their security.
- Provide effective personnel management.
- Conduct logistic (planning, maintenance, sustainment, and movement) operations.
- Conduct basic intelligence functions.
- Coordinate indirect fires.
- Provide for effective medical support.

Training Leaders

6-67. The effectiveness of the HN security forces directly relates to the quality of their leadership. Building effective leaders requires a comprehensive program of officer, staff, and specialized training. The ultimate success of any U.S. involvement in a COIN effort depends on creating viable HN leaders able to carry on the fight at all levels and build their nation on their own.

Leader Training Standards

6-68. The leader training methodology must reinforce the different levels of authority within the HN security force. The roles and responsibilities of each commissioned officer and NCO rank must be firmly established so recruits understand what is expected of them. Their subordinate relationship to civilian authorities must also be reinforced to ensure civilian control. In addition, training should establish team dynamics. In some cultures, security forces may need training to understand the vital role of members not in primary leadership positions.

6-69. In addition to tactical skills, commissioned officers should be trained in accountability, decision making, delegating authority, values, and ethics. Special requirements for COIN should be the primary focus of the initial curriculum. These subjects include the following:

- Intelligence collection.
- Day and night patrolling.
- Point security.
- Cordon and search operations.
- Operations with police.
- Treatment of detainees and prisoners.
- Psychological operations.
- Civic action.

As the insurgency declines, the curriculum can be adjusted for a long-term focus.

6-70. In addition, leader training should be conducted in a way that shows—

- How to work as a team.
- How to develop and take advantage of subordinates' skills.
- How to train subordinates.
- How to maintain discipline and assume responsibility for one's own and subordinates' actions.
- How to understand and enforce the rules of engagement.

Basic Commissioned Officer Education

6-71. Various models for basic officer education exist. These include the following:

- One-year military college.
- Four-year military college.
- Officer candidate school (OCS).
- Military training at civilian universities.

6-72. Time available may determine which model to use as the primary commissioning source. If the situation allows, four-year programs at military or civilian institutions may be the best choice. If not, the OCS and one-year college models may be better. Theoretically, having a few high-quality officers may be better than many adequate ones, but the insurgents may not allow this luxury. Citizens under attack would rather have an adequate officer and unit now than a better leader and organization years later.

6-73. The British Army uses the one-year military college model. Prospective officers attend Sandhurst, an intensive course that includes a rigorous program of basic training and a thorough grounding in the British Army's history and culture. It also emphasizes developing each future officer as a leader. At the end of the year, each new officer attends a shorter specialty branch course.

6-74. The four-year military and civilian college models provide the best overall education, while preparing officers for work at the tactical and operational levels. The longer programs also are good for inculcating values. However, they require significant time and resources.

6-75. Under an OCS-style program, outstanding individuals come from the enlisted ranks or society. They receive intensive training in the military skills junior officers require. OCS programs often last less than a year. An OCS-style course should be followed by specialized branch training.

Intermediate and Advanced Commissioned Officer Education

6-76. Military units only become effective when their commanders and staffs can effectively plan, prepare, execute, and assess operations. Initial intermediate-level commissioned officer training should focus on building effective commanders and staffs for small units, then progressively move to higher echelons. Thus, initial intermediate-level officer training focuses on the company and battalion levels (or police station level). Later courses address higher echelons, depending on the size of the overall force to be developed.

6-77. A cadre of carefully selected low- and mid-level commissioned officers can receive an advanced education at existing formal schools in the United States or other partner nations through IMET-like programs. This type of program builds a qualified leadership cadre. This cadre can, upon their return home, assume leadership positions and become the faculty for HN schools. These officers should have increased credibility when they return to their country. Officer students usually make and maintain strong personal connections with their foreign hosts during and after their stay abroad. One of the key goals for developing HN security forces is to professionalize them. The first-hand experience of their officers training at foreign military schools, living abroad, and seeing military professional standards practiced proves invaluable. As with officer commissioning programs, time is a key consideration. IMET-like programs are expensive and time consuming. The best officers—those normally selected for such training—may be needed more in the country's combat forces fighting the insurgency.

Operational Employment of Newly Trained Forces

6-78. Building the morale and confidence of security forces should be a primary strategic objective. Committing poorly trained and badly led forces results in high casualties and invites tactical defeats. While defeat in a small operation may have little strategic consequence in a conventional war, even a small tactical defeat of HN forces can have serious strategic consequences in a COIN. Insurgent warfare is largely about perceptions. Effective insurgent leaders can quickly turn minor wins into major propaganda victories. Defeat of one government force can quickly degrade the morale of others. If a HN force fails, the local populace may begin to lose confidence in the government's ability to protect them. A string of relatively minor insurgent victories can cause widespread loss of morale in the HN forces and encourage the "neutral majority" to side with the insurgents. In short, the HN security forces must be prepared for operations so that they have every possible advantage. The decision to commit units to their first actions and their employment method requires careful consideration. As much as possible, HN forces should begin with simpler missions. As their confidence and competence grows, these forces can assume more complex assignments. Collaborating with U.S. or multinational units can help new HN units to become accustomed to the stresses of combat.

6-79. Newly trained units should enter their first combat operation in support of more experienced HN, U.S., or multinational forces. Operational performance of such inexperienced organizations should be carefully monitored and evaluated so that weaknesses can be quickly corrected. The employment plan for HN security units should allow enough time for additional training after each operation. Introducing units into combat gradually allows identification of poor leaders to the HN leadership for retraining or other action. Competent leaders are also identified and given greater authority and responsibility.

Training Defense Ministry Civilians

6-80. U.S. forces tasked with training HN personnel must also ensure that the military and security forces have capable management in the top ranks. Combatant commanders place experienced U.S. officers and Department of Defense personnel inside the HN defense and interior ministries as trainers and advisors for HN managers and leaders. U.S. forces should also develop a training program for civilian personnel of the ministry of defense. Personnel training should address the following:

- Equipment acquisition.
- Departmental administration.
- Personnel management.
- Financial management.
- Maintenance and inventory controls.
- Strategic (or national) level operations.

Selected ministry of defense personnel may receive specialized training in defense management through U.S. or multinational partners' schools (for example, the National Defense University) or in civilian institutions that specialize in graduate programs for security studies.

EQUIP

6-81. The strategic plan for security force development should outline HN equipment requirements. Equipment should meet the host nation's requirements. Appropriate equipment is affordable and suitable against the threat. The host nation must also be able to train on the equipment. Interoperability may be desired in some cases. A central consideration includes the host nation's long-term ability to support and maintain the equipment.

6-82. The initial development plan should include phases with goals for HN forces to meet over a period of three to four years. Due to the highly adaptive nature of insurgents and the often rapidly changing situation on the ground, commanders must continually asses the direction and progress of developing HN security forces.

6-83. The requirement to provide equipment may be as simple as assisting with maintenance of existing formations or as extensive as providing everything from shoes and clothing to vehicles, communications, and investigation kits. If insurgents use heavy machine guns and rocket-propelled grenades, HN security forces need comparable or better equipment. This especially applies to police forces, which are often lightly armed and vulnerable to well-armed insurgents.

6-84. Primary considerations should include maintainability, ease of operation, and long-term sustainment costs. Few developing nations can support highly complex equipment. In COIN operations, having many versatile vehicles that require simple maintenance is often better than having a few highly capable armored vehicles or combat systems that require extensive maintenance. Developing an effective HN maintenance system often begins with major maintenance performed by contractors. The program then progresses to partnership arrangements with U.S. forces as HN personnel are trained to perform the support mission.

6-85. Sources for HN materiel include U.S. foreign military sales, multinational or third-nation resale of property, HN contracts with internal suppliers, or HN purchases on the international market. (Paragraphs D-27 through D-34 discusses relevant legal considerations.) The HN, U.S., and multinational organizations responsible for equipping HN forces should have the flexibility necessary to obtain equipment that meets the HN force needs for quality, timeliness and cost. As part of their training, HN security forces also need to learn the practices and importance of property accountability to reduce corruption and ensure proper equipment usage. Part of equipping HN forces includes training them in the practices and importance of property accountability. HN forces are expected to provide equipment the same level of control and protection that U.S. forces provide for similar equipment. (See AR 12-1 and DODD 5105.38M.)

ADVISE

6-86. Advisors are the most prominent group of U.S. personnel that serve with HN units. Advisors live, work, and (when authorized) fight with their HN units. Segregation is kept at an absolute minimum. The relationship between advisors and HN forces is vital. U.S. commanders must remember that advisors are not liaison officers, nor do they command HN units.

6-87. Effective advisors are an enormous force enhancer. The importance of the job means that the most capable individuals should fill these positions. Advisors should be Soldiers and Marines known to take the initiative and who set the standards for others. (FM 31-20-3 provides additional information and guidelines for advisors.)

6-88. More than anything else, professional knowledge and competence win the respect of HN troops. Effective advisors develop a healthy rapport with HN personnel but avoid the temptation to adopt HN positions contrary to U.S. or multinational values or policy.

6-89. Advisors who understand the HN military culture understand that local politics have national effects. Effective advisors recognize and use cultural factors that support HN commitment and teamwork. A good advisor uses the culture's positive aspects to get the best performance from each security force member and leader. Table 6-5 (page 6-18) lists important guidelines for advisors.

Table 6-5. Guidelines for advisors

- Try to learn enough of the language for simple conversation.
- Be patient. Be subtle. In guiding host-nation counterparts, explain the benefits of an action and convince them to accept the idea as their own. Respect the rank and positions of host-nation counterparts.
- Be diplomatic in correcting host-nation forces. Praise each success and work to instill pride in the unit.
- Understand that an advisor is not the unit commander but an enabler. The host-nation commander makes decisions and commands the unit. Advisors help with this task.
- Keep host-nation counterparts informed; try not to hide agendas.
- Work to continually train and improve the unit, even in the combat zone. Help the commander develop unit standing operating procedures.
- Be prepared to act as a liaison to multinational assets, especially air support and logistics. Maintain liaison with civil affairs and humanitarian teams in the area of operations.
- Be ready to advise on the maintenance of equipment and supplies.
- Have a thorough knowledge of light infantry tactics and unit security procedures.
- Use “confidence” missions to validate training.
- Stay integrated with the unit. Eat their food. Do not become isolated from them.
- Be aware of the operations in the immediate area to prevent fratricide.
- Insist on host-nation adherence to the recognized human rights standards concerning treatment of civilians, detainees, and captured insurgents. Report any violations to the chain of command.
- Be objective in reports on host-nation unit and leader proficiency. Report gross corruption or incompetence.
- Train host-nation units to standard and fight alongside them. Consider host-nation limitations and adjust. Flexibility is key. It is impossible to plan completely for everything in this type of operation. Therefore, constantly look forward to the next issue and be ready to develop solutions to problems that cannot be answered with a doctrinal solution.
- Remember that most actions have long-term strategic implications.
- Maintain a proper military bearing and professional manner.

Multinational Security Transition Command–Iraq

The experience of Multinational Security Transition Command–Iraq (MNSTC-I) in 2005 demonstrates the challenges of developing and assisting host-nation security forces facing an intensive insurgency. MNSTC-I programs were built on three pillars:

- Training and equipping the Iraqi security forces to standard.
- Using transition teams to guide the development of leaders and staffs.
- Partnerships between U.S. and multinational forces on the ground and developing Iraqi forces.

Initial plans called for only a small army (to deal with external threats) supplemented by conventional police forces (to maintain internal law and order). As the insurgency matured, the decision was made to develop a larger Iraqi Army and to focus it on the internal threat. A more robust police force was also developed. Training programs matured, becoming longer and more focused on counterinsurgency tasks. Decisions by the Iraqi government also necessitated training and organizational changes. Training programs were adjusted, based on the experience of recruits, and eventually lengthened and changed in response to the increasing lethality of the insurgency and lessons learned. Advisors assigned to Iraqi units were termed “transition teams” and instructed to focus on the development of leaders and staffs at battalion level and above.

As security forces grew significantly, it became evident that civilian ministry infrastructure development was not keeping pace. So MNSTC-I assumed the additional mission of creating ministerial-level organizations, finding resources, and changing culture. Another significant challenge involved selecting and training competent host-nation leadership—commissioned and noncommissioned officers. This grew more challenging as the host nation regained sovereignty and asserted its authority, which led to a greater focus on developing staffs by transition teams.

The MNSTC-I experience shows the advantage of doing a prompt initial assessment and then adjusting as conditions change and lessons are learned. Developing HN security forces from scratch in an active insurgency environment is often more about overcoming friction than about perfect planning. MNSTC-I also demanded robust interagency and multinational participation in the training effort and fought for the funding necessary to make it effective. MNSTC-I found that easily measured measures of effectiveness, such as soldiers equipped or battalions fielded, were not as useful as more complicated and more subjective metrics, such as the training level of fielded units and their loyalty to the national government.

POLICE IN COUNTERINSURGENCY

6-90. The primary frontline COIN force is often the police—not the military. The primary COIN objective is to enable local institutions. Therefore, supporting the police is essential. But the police are only a part of the rule of law. Police require support from a law code, judicial courts, and a penal system. Such support provides a coherent and transparent system that imparts justice. Upholding the rule of law also requires other civil institutions and a HN ability to support the legal system. Commanders should ensure that robust coordination mechanisms linking their efforts with the larger developmental process exist. If parts of the rule of law do not work, then commanders must be prepared to meet detention requirements.

6-91. Few military units can match a good police unit in developing an accurate human intelligence picture of their AO. Because of their frequent contact with populace, police often are the best force for countering small insurgent bands supported by the local populace. In COIN operations, special police strike units may move to different AOs, while patrol police remain in the local area on a daily basis and build a detailed intelligence picture of the insurgent strength, organization, and support.

ORGANIZING THE POLICE

6-92. Police often consist of several independent but mutually supporting forces. These may include—

- Criminal and traffic police.
- Border police.
- Transport police for security of rail lines and public transport.
- Specialized paramilitary strike forces.

In addition, a host nation may establish various reserve police units or home guards to provide local security. The force may include paramilitary units. Police might be organized on a national or local basis. Whatever police organization is established, Soldiers and Marines must understand it and help the host nation effectively organize and use it. This often means dealing with several police organizations and developing plans for training and advising each one.

6-93. A formal link or liaison channel must exist between the HN police and military forces. This channel for coordination, deconfliction, and information sharing enables successful COIN operations.

6-94. Military forces might have to perform police duties at the start of an insurgency; however, it is best to establish police forces to assume these duties as soon as possible. U.S., multinational, and HN partners should institute a comprehensive program of police training. Moreover, plans for police training need to envision a several-year program to systematically build institutions and leadership.

6-95. Although roles of the police and military forces in COIN operations may blur, important distinctions between the two forces exist. If security forces treat insurgents as criminals, the police may retain the primary responsibility for their arrest, detention, and prosecution.

6-96. Countering an insurgency requires a police force that is visible day and night. The host nation will not gain legitimacy if the populace believes that insurgents and criminals control the streets. Well-sited and protected police stations can establish a presence in communities as long as the police do not hide in those stations. Police presence provides security to communities and builds support for the HN government. When police have daily contact with the local populace, they can collect information for counterinsurgents.

6-97. Good pay and attractive benefits must be combined with a strict code of conduct that follows the rule of law and allows for the immediate dismissal of police officers for gross corruption. Good planning ensures that police pay, housing, benefits, and work conditions attract a high quality of police recruit as well as discourage petty corruption. Such corruption undermines the populace's confidence in the police and government. An important step in organizing a police force involves setting up an independent review board composed of experts, government officials, or nongovernmental organization members. It should not be under the direct command of the police force. This board should have the authority to investigate charges of police abuse and corruption, oversee the complaints process, and dismiss and fine police found guilty of misconduct.

Training the Police in Counterinsurgency

6-98. Police training is best conducted as an interagency and multinational operation. In a multinational effort, a separate multinational police training and advisory command could work with the military training command. Ideally, leaders for police training are civilian police officers from the Departments of Justice and State along with senior police officers from multinational partners. Civilian police forces have personnel with extensive experience in large city operations. Department of Justice and multinational police organizations have extensive experience operating against organized crime groups. Experience countering organized crime is especially relevant to COIN; most insurgent groups are more similar to organized crime in their organizational structure and relations with the populace than they are to military units. U.S. military police units serve best when operating as a support force for the professional civilian police trainers. However, military forces may be assigned the primary responsibility for police training; they must be prepared to assume that role if required. (See paragraph D-3 for legal considerations associated with this mission.)

6-99. Military police can provide much of the initial police training. They are especially suited to teach the HN police forces the following skills:

- Weapons handling.
- Small-unit tactics.
- Special weapons employment.
- Convoy escort.
- Riot control.
- Traffic control.
- Prisoner and detainee handling and processing.
- Police intelligence.
- Criminal intelligence.
- Criminal handling.
- Stations management.

Higher level police skills—such as civilian criminal investigation procedures, antiorganized crime operations, and police intelligence operations—are best taught by civilian experts.

6-100. Military police or corrections personnel can also provide training for detention and corrections operations. HN personnel should be trained to handle and interrogate detainees and prisoners according to internationally recognized human rights norms. Prisoner and detainee management procedures should provide for the security and the fair and efficient processing of those detained.

6-101. Police forces, just like military forces, need quality support personnel to be effective. This requires training teams to ensure that training in support functions is established. Specially trained personnel required by police forces include the following:

- Armorers.
- Supply specialists.
- Communications personnel.
- Administrative personnel.
- Vehicle mechanics.

6-102. Effective policing also requires an effective justice system that can process arrests, detentions, warrants, and other judicial records. Such a system includes trained judges, prosecutors, defense counsels, prison officials, and court personnel. These people are important to establishing the rule of law.

6-103. Advisors should help the host nation establish and enforce police roles and authority. The authority to detain and interrogate, procedures for detention facilities, and human rights standards are important considerations.

POLICE/MILITARY OPERATIONS

6-104. In COIN operations, police forces from the host nation, United States, and multinational partners often conduct operations together. To work effectively together, the police and military coordinate rules of engagement. These forces also—

- Establish common standing operating procedures.
- Conduct supporting information operations.
- Perform combined planning.
- Ensure command and control interoperability.

6-105. Military forces can support the police in carrying out numerous COIN functions in accordance with U.S. law. This support can include the following:

- Assisting in the arrest of war criminals.
- Supporting police presence and search patrols.
- Providing logistic support.
- Controlling crowds and urban unrest.
- Detaining suspected felons.
- Securing key facilities.
- Providing advisors to the police.

6-106. The Departments of Justice and State normally take the lead in helping the host nation develop a workable judicial system through a ministerial-level advisory program. The military's staff judge advocate and civil affairs personnel may help develop the HN judicial system. As this system is developed and reformed, the military commander's legal and political advisors should ensure that the military's concerns are addressed.

Developing a Police Force in Malaya

In 1948, the Malayan Communist Party, whose members were primarily ethnic Chinese, began an insurgency against the British colonial government. The British first responded by dramatically expanding the Malayan security forces. The police, not the army, served as the lead counterinsurgency force. Between 1948 and 1950, the number of Malayan police expanded fivefold to 50,000, while the British army garrison expanded to 40,000. However, there was only time to provide a few weeks of rudimentary training to the new police officers before throwing them into operations.

Police with little training and little competent leadership were ineffective in conducting operations. They also abused the civilian population and fell into corrupt practices. The population largely regarded the police as hostile; they were reluctant to give them information on the insurgents.

By 1952, the insurgency had reached a stalemate. The British then established a new strategy. The strategy included reforming and retraining the entire Malaya Police Force. First, 10,000 corrupt or incompetent police officers were removed from the force. Then, police officers who had proven the most competent in operations were made instructors in new police schools. During 1952 and 1953, every police officer attended a four-month basic training course. Police commissioned and noncommissioned officers were sent to three- to four-month advanced courses. All senior Malayan police officers were required to attend the police intelligence school. There they learned the latest criminal investigation techniques. Teams of Britain's top police officers taught them intelligence collection and analysis methods as well. Dozens of the most promising Malayan officers attended the full yearlong course in advanced police operations in Britain.

To win the ethnic Chinese away from the insurgents, the British worked closely with ethnic Chinese organizations to recruit Chinese for the Malaya Police Force. In 1952, the number of ethnic Chinese in the force more than doubled. Although the percentage of ethnic Chinese in the police force did not equal their percentage in the population, the ethnic Chinese saw this reaching out as a sign that the government was addressing their interests. At the same time, some Chinese and Malay political groups were building a coalition to establish an independent Malaya in which all the major ethnic groups would participate. The two efforts complemented each other.

Better trained police officers and soldiers led by fully trained commissioned and noncommissioned officers dramatically improved the Malayan security forces' discipline. Better relations between the population and security forces resulted, and the people began to provide information on the insurgents. Thanks to their intelligence training, the security forces could develop intelligence from that information and act on it. They begin to break the insurgent organization. In 1953, the government gained the initiative. After that, the insurgent forces and support structure declined rapidly. In late 1953, the British began withdrawing forces. They progressively turned the war over to the Malayans, who were fully prepared to conduct counterinsurgency operations without a drop in efficiency.

The Malaya insurgency provides lessons applicable to combating any insurgency. Manpower is not enough; well-trained and well-disciplined forces are required. The Malayan example also illustrates the central role that police play in counterinsurgency operations. British leaders concentrated on training the Malayan leadership. The British insisted that chosen personnel receive the full British Army and police officer courses. These actions built the Malayan security forces on a sound foundation. By taking a comprehensive approach to security force training and reform, the British commanders transformed a demoralized organization into a winning force. This transformation required only 15 months.

SUMMARY

6-107. A successful COIN effort establishes HN institutions that can sustain government legitimacy. Developing effective HN security forces—including military, police, and paramilitary forces—is one of the highest priority COIN tasks. Soldiers and Marines can make vital contributions to this mission by training and advising the HN security forces. Effective commanders understand the importance of this mission and select the right personnel as trainers and advisors. Developing all necessary HN security forces require a considerable interagency effort and normally includes a significant multinational involvement as well.

Chapter 7
Leadership and Ethics for Counterinsurgency

Leaders must have a strong sense of the great responsibility of their office; the resources they will expend in war are human lives.

Marine Corps Doctrinal Publication 1, 1997

There are leadership and ethical imperatives that are prominent and, in some cases, unique to counterinsurgency. The dynamic and ambiguous environment of modern counterinsurgency places a premium on leadership at every level, from sergeant to general. Combat in counterinsurgency is frequently a small-unit leader's fight; however, commanders' actions at brigade and division levels can be more significant. Senior leaders set the conditions and the tone for all actions by subordinates. Today's Soldiers and Marines are required to be competent in a broad array of tasks. They must also rapidly adapt cognitively and emotionally to the perplexing challenges of counterinsurgency and master new competencies as well as new contexts. Those in leadership positions must provide the moral compass for their subordinates as they navigate this complex environment. Underscoring these imperatives is the fact that exercising leadership in the midst of ambiguity requires intense, discriminating professional judgment.

LEADERSHIP IN COUNTERINSURGENCY

7-1. Army and Marine Corps leaders are expected to act ethically and in accordance with shared national values and Constitutional principles, which are reflected in the law and military oaths of service. These leaders have the unique professional responsibility of exercising military judgment on behalf of the American people they serve. They continually reconcile mission effectiveness, ethical standards, and thoughtful stewardship of the Nation's precious resources—human and material—in the pursuit of national aims.

7-2. Army and Marine Corps leaders work proactively to establish and maintain the proper ethical climate of their organizations. They serve as visible examples for every subordinate, demonstrating cherished values and military virtues in their decisions and actions. Leaders must ensure that the trying counterinsurgency (COIN) environment does not undermine the values of their Soldiers and Marines. Under all conditions, they must remain faithful to basic American, Army, and Marine Corps standards of proper behavior and respect for the sanctity of life.

7-3. Leaders educate and train their subordinates. They create standing operating procedures and other internal systems to prevent violations of legal and ethical rules. They check routinely on what Soldiers and Marines are doing. Effective leaders respond quickly and aggressively to signs of illegal or unethical behavior. The Nation's and the profession's values are not negotiable. Violations of them are not just mistakes; they are failures in meeting the fundamental standards of the profession of arms.

LARGE- AND SMALL-UNIT LEADERSHIP TENETS

7-4. There are basic leadership tenets that apply to all levels of command and leadership in COIN, though their application and importance may vary.

7-5. Effective leaders ensure that Soldiers and Marines are properly trained and educated. Such training includes cultural preparation for the operational environment. In a COIN environment, it is often counterproductive to use troops that are poorly trained or unfamiliar with operating close to the local populace. COIN forces aim to mobilize the good will of the people against the insurgents. Therefore, the populace must feel protected, not threatened, by COIN forces' actions and operations.

7-6. Proper training addresses many possible scenarios of the COIN environment. Education should prepare Soldiers and Marines to deal with the unexpected and unknown. Senior commanders should, at a minimum, ensure that their small-unit leaders are inculcated with tactical cunning and mature judgment. Tactical cunning is the art of employing fundamental skills of the profession in shrewd and crafty ways to out-think and out-adapt enemies. Developing mature judgment and cunning requires a rigorous regimen of preparation that begins before deployment and continues throughout. Junior leaders especially need these skills in a COIN environment because of the decentralized nature of operations.

7-7. Senior leaders must determine the purpose of their operations. This entails, as discussed in chapter 4, a design process that focuses on learning about the nature of unfamiliar problems. Effective commanders know the people, topography, economy, history, and culture of their area of operations (AO). They know every village, road, field, population group, tribal leader, and ancient grievance within it. The COIN environment changes continually; good leaders appreciate that state of flux and constantly assess their situation.

7-8. Another part of analyzing a COIN mission involves assuming responsibility for everyone in the AO. This means that leaders feel the pulse of the local populace, understand their motivations, and care about what they want and need. Genuine compassion and empathy for the populace provide an effective weapon against insurgents.

7-9. Senior leaders exercise a leadership role throughout their AO. Leaders directly influence those in the chain of command while indirectly leading everyone else within their AO. Elements engaged in COIN efforts often look to the military for leadership. Therefore, military actions and words must be beyond reproach. The greatest challenge for leaders may be in setting an example for the local populace. Effective senior and junior leaders embrace this role and understand its significance. It involves more than just killing insurgents; it includes the responsibility to serve as a moral compass that extends beyond the COIN force and into the community. It is that moral compass that distinguishes Soldiers and Marines from the insurgents.

7-10. Senior commanders must maintain the "moral high ground" in all their units' deeds and words. Information operations complement and reinforce actions, and actions reinforce the operational narrative. All COIN force activity is wrapped in a blanket of truth. Maintaining credibility requires commanders to immediately investigate all allegations of immoral or unethical behavior and provide a prudent degree of transparency.

7-11. Army and Marine Corps leaders emphasize that on the battlefield the principles of honor and morality are inextricably linked. Leaders do not allow subordinates to fall victim to the enormous pressures associated with prolonged combat against elusive, unethical, and indiscriminate foes. The environment that fosters insurgency is characterized by violence, immorality, distrust, and deceit; nonetheless, Army and Marine Corps leaders continue to demand and embrace honor, courage, and commitment to the highest standards. They know when to inspire and embolden their Soldiers and Marines and when to enforce restraint and discipline. Effective leaders at all levels get out and around their units, and out among the populace. Such leaders get a true sense of the complex situation in their AO by seeing what subordinates are actually doing, exchanging information with military and interagency leaders, and—most importantly—listening.

7-12. Leaders at every level establish an ethical tone and climate that guards against the moral complacency and frustrations that build up in protracted COIN operations. Leaders remain aware of the emotional toll that constant combat takes on their subordinates and the potential for injuries resulting from combat stress. Such injuries can result from cumulative stress over a prolonged period, witnessing the death of a comrade, or killing other human beings. Caring leaders recognize these pressures and provide emotional "shock absorbers" for their subordinates. Soldiers and Marines must have outlets to share their feelings and reach closure on traumatic experiences. These psychological burdens may be carried for a long time. Leaders watch for signs of possible combat stress within individuals and units. These signs include—

- Physical and mental fatigue.
- Lack of respect for human life.
- Loss of appetite, trouble with sleep, and no interest in physical hygiene.
- Lack of unit cohesion and discipline.
- Depression and fatalism.

(See FM 6-22.5/MCRP 6-11C for techniques first-line leaders can use to prevent, identify, and treat combat stress reactions.)

7-13. Combat requires commanders to be prepared to take some risk, especially at the tactical level. Though this tenet is true for the entire spectrum of conflict, it is particularly important during COIN operations, where insurgents seek to hide among the local populace. Risk takes many forms. Sometimes accepting it is necessary to generate overwhelming force. However, in COIN operations, commanders may need to accept substantial risk to de-escalate a dangerous situation. The following vignette illustrates such a case.

Defusing a Confrontation

[On 3 April 2005, a] small unit of American soldiers was walking along a street in Najaf [en route to a meeting with a religious leader] when hundreds of Iraqis poured out of the buildings on either side. Fists waving, throats taut, they pressed in on the Americans, who glanced at one another in terror. ...The Iraqis were shrieking, frantic with rage.... [It appeared that a shot would] come from somewhere, the Americans [would] open fire, and the world [would] witness the My Lai massacre of the Iraq war. At that moment, an American officer stepped through the crowd holding his rifle high over his head with the barrel pointed to the ground. Against the backdrop of the seething crowd, it was a striking gesture.... "Take a knee," the officer said.... The Soldiers looked at him as if he were crazy. Then, one after another, swaying in their bulky body armor and gear, they knelt before the boiling crowd and pointed their guns at the ground. The Iraqis fell silent, and their anger subsided. The officer ordered his men to withdraw [and continue on their patrol].

7-14. Leaders prepare to indirectly inflict suffering on their Soldiers and Marines by sending them into harm's way to accomplish the mission. At the same time, leaders attempt to avoid, at great length, injury and death to innocents. This requirement gets to the very essence of what some describe as "the burden of command." The fortitude to see Soldiers and Marines closing with the enemy and sustaining casualties day in and day out requires resolve and mental toughness in commanders and units. Leaders must develop these characteristics in peacetime through study and hard training. They must maintain them in combat.

7-15. Success in COIN operations requires small-unit leaders agile enough to transition among many types of missions and able to adapt to change. They must be able to shift through a number of activities from nation building to combat and back again in days, or even hours. Alert junior leaders recognize the dynamic context of a tactical situation and can apply informed judgment to achieve the commander's intent in a stressful and ambiguous environment. COIN operations are characterized by rapid changes in tactical and operational environments. The presence of the local populace within which insurgents may disappear creates a high degree of ambiguity. Adaptable leaders observe the rapidly changing situation, identify its key characteristics, ascertain what has to be done in consultation with subordinates, and determine the best method to accomplish the mission.

7-16. Cultural awareness has become an increasingly important competency for small-unit leaders. Perceptive junior leaders learn how cultures affect military operations. They study major world cultures and put a priority on learning the details of the new operational environment when deployed. Different solutions are required in different cultural contexts. Effective small-unit leaders adapt to new situations, realizing their words and actions may be interpreted differently in different cultures. Like all other competencies, cultural awareness requires self-awareness, self-directed learning, and adaptability.

7-17. Self-aware leaders understand the need to assess their capabilities and limitations continually. They are humble, self-confident, and brave enough to admit their faults and shortcomings. More important, self-aware leaders work to improve and grow. After-action reviews, exchanging information with subordinate and interagency leaders, and open discussions throughout a COIN force are essential to achieve understanding and improvement. Soldiers and Marines can become better, stronger leaders through a similar habit of self-examination, awareness, and focused corrective effort.

7-18. Commanders exercise initiative as leaders and fighters. Learning and adapting, with appropriate decision-making authority, are critical to gaining an advantage over insurgents. Effective senior leaders establish a climate that promotes decentralized modes of command and control—what the Army calls mission command and the Marine Corps calls mission command and control. Under mission command, commanders create the conditions for subordinates' success. These leaders provide general guidance and the commander's intent and assign small-unit leaders authority commensurate with their responsibilities. Commanders establish control measures to monitor subordinates' actions and keep them within the bounds established by commander's intent without micromanaging. At the same time, Soldiers and Marines must feel the commander's presence throughout the AO, especially at decisive points. The operation's purpose and commander's intent must be clearly understood throughout the force.

7-19. The practice of leaders sharing hardship and danger with subordinates builds confidence and esprit. Soldiers and Marines are more confident in their chances of success when they know that their leaders are involved. They understand their leaders are committing them to courses of action based on firsthand knowledge. However, this concept of leaders being fighters does not absolve leaders from remembering their position and avoiding needless risk.

7-20. COIN operations require leaders to exhibit patience, persistence, and presence. While leading Soldiers and Marines, commanders cooperate with, and leverage the capabilities of, multinational partners, U.S. Government agencies, and nongovernmental organizations. Commanders also gain the confidence of the local populace while defeating and discrediting the insurgents.

Patience, Presence, and Courage

For the first two months of 2006, the Marine platoon of the 22d Marine Expeditionary Unit had walked the streets in Iraq on foot without serious incident. Their patrols had moved fearlessly around lines of cars and through packed markets. For the most part, their house calls began with knocks, not kicks. It was their aim to win the respect of the city's Sunni Arab population.

Suddenly things changed. An armored HMMWV on night patrol hit an improvised explosive device. The bomb destroyed the vehicle. Five Marines were wounded and two died shortly thereafter. A third Marine, a popular noncommissioned officer, later died of his wounds as well.

The platoon was stunned. Some of the more veteran noncommissioned officers shrugged it off, but the younger Marines were keyed up and wanted to make the elusive enemy pay a price. A squad leader stood up in the squad bay asserted that there would be a pile of dead Arabs on the street when the platoon went out the next day.

Just then, the company commander walked in. He was widely respected and generally short on words. He quickly sensed the unit's mood and recognized the potential danger in their dark attitude. Speaking directly to his Marines, the commander urged them to remember why they were there. He reminded them that a very small percentage of the populace was out to create problems. It was that minority that benefited from creating chaos. The enemy would love to see an overreaction to the attack, and they would benefit from any actions that detracted from the Marines' honor or purpose. The commander urged his Marines not to get caught up in the anger of the moment and do something they all would regret for a long time. Rather, they needed to focus on what the force was trying to accomplish and keep their minds on the mission. They had taken some hits and lost some good men, the commander said, but escalating the violence would not help them win. It would fall for the insurgents' strategy instead of sticking to the Marines' game plan of winning the respect of the populace.

The commander knew his Marines and understood the operational environment. He assessed the situation and acted aggressively to counter a dangerous situation that threatened mission accomplishment. By his actions, the commander demonstrated patience, presence, and courage.

ETHICS

7-21. Article VI of the U.S. Constitution and the Army Values, Soldier's Creed, and Core Values of U.S. Marines all require obedience to the law of armed conflict. They hold Soldiers and Marines to the highest standards of moral and ethical conduct. Conflict brings to bear enormous moral challenges, as well as the burden of life-and-death decisions with profound ethical considerations. Combat, including counterinsurgency and other forms of unconventional warfare, often obligates Soldiers and Marines to accept some risk to minimize harm to noncombatants. This risk taking is an essential part of the Warrior Ethos. In conventional conflicts, balancing competing responsibilities of mission accomplishment with protection of noncombatants is difficult enough. Complex COIN operations place the toughest of ethical demands on Soldiers, Marines, and their leaders.

7-22. Even in conventional combat operations, Soldiers and Marines are not permitted to use force disproportionately or indiscriminately. Typically, more force reduces risk in the short term. But American military values obligate Soldiers and Marines to accomplish their missions while taking measures to limit the destruction caused during military operations, particularly in terms of collateral harm to noncombatants. It is wrong to harm innocents, regardless of their citizenship.

7-23. Limiting the misery caused by war requires combatants to consider certain rules, principles, and consequences that restrain the amount of force they may apply. At the same time, combatants are not required to take so much risk that they fail in their mission or forfeit their lives. As long as their use of force is proportional to the gain to be achieved and discriminates in distinguishing between combatants and noncombatants. Soldiers and Marines may take actions where they knowingly risk, but do not intend, harm to noncombatants.

7-24. Ethically speaking, COIN environments can be much more complex than conventional ones. Insurgency is more than combat between armed groups; it is a political struggle with a high level of violence. Insurgents try to use this violence to destabilize and ultimately overthrow a government. Counterinsurgents that use excessive force to limit short-term risk alienate the local populace. They deprive themselves of the support or tolerance of the people. This situation is what insurgents want. It increases the threat they pose. Sometimes lethal responses are counterproductive. At other times, they are essential. The art of command includes knowing the difference and directing the appropriate action.

7-25. A key part of any insurgent's strategy is to attack the will of the domestic and international opposition. One of the insurgents' most effective ways to undermine and erode political will is to portray their opposition as untrustworthy or illegitimate. These attacks work especially well when insurgents can portray their opposition as unethical by the opposition's own standards. To combat these efforts, Soldiers and Marines treat noncombatants and detainees humanely, according to American values and internationally recognized human rights standards. In COIN, preserving noncombatant lives and dignity is central to mission accomplishment. This imperative creates a complex ethical environment.

WARFIGHTING VERSUS POLICING

7-26. In counterinsurgencies, warfighting and policing are dynamically linked. The moral purpose of combat operations is to secure peace. The moral purpose of policing is to maintain the peace. In COIN operations, military forces defeat enemies to establish civil security; then, having done so, these same forces preserve it until host-nation (HN) police forces can assume responsibility for maintaining the civil order.

When combatants conduct stability operations in a way that undermines civil security, they undermine the moral and practical purposes they serve. There is a clear difference between warfighting and policing. COIN operations require that every unit be adept at both and capable of moving rapidly between one and the other.

7-27. The COIN environment frequently and rapidly shifts from warfighting to policing and back again. There are many examples from Iraq and Afghanistan where U.S. forces drove insurgents out of urban areas only to have the insurgents later return and reestablish operations. Insurgents were able to return because U.S. forces had difficulty maintaining civil security. U.S. forces then had to deal with insurgents as an organized combatant force all over again. To prevent such situations, counterinsurgents that establish civil security need to be prepared to maintain it. Maintaining civil security entails very different ethical obligations than establishing it.

7-28. Civil security holds when institutions, civil law, courts, prisons, and effective police are in place and can protect the recognized rights of individuals. Typically this requires that—

- The enemy is defeated or transformed into a threat not capable of challenging a government's sovereignty.
- Institutions necessary for law enforcement—including police, courts, and prisons—are functioning.
- These institutions are credible, and people trust them to resolve disputes.

7-29. Where a functioning civil authority does not exist, COIN forces must work to establish it. Where U.S. forces are trying to build a HN government, the interim government should transition to HN authority as soon as possible. Counterinsurgents must work within the framework of the institutions established to maintain order and security. In these conditions, COIN operations more closely resemble police work than combat operations.

PROPORTIONALITY AND DISCRIMINATION

7-30. The principle of proportionality requires that the anticipated loss of life and damage to property incidental to attacks must not be excessive in relation to the concrete and direct military advantage expected to be gained. Proportionality and discrimination require combatants not only to minimize the harm to noncombatants but also to make positive commitments to—

- Preserve noncombatant lives by limiting the damage they do.
- Assume additional risk to minimize potential harm.

7-31. Proportionality requires that the advantage gained by a military operation not be exceeded by the collateral harm. The law of war principle of proportionality requires collateral damage to civilians and civilian property not be excessive in relation to the military advantage expected to be gained by executing the operation. Soldiers and Marines must take all feasible precautions when choosing means and methods of attack to avoid and minimize loss of civilian life, injury to civilians, and damage to civilian objects.

7-32. In conventional operations, proportionality is usually calculated in simple utilitarian terms: civilian lives and property lost versus enemy destroyed and military advantage gained. But in COIN operations, advantage is best calculated not in terms of how many insurgents are killed or detained, but rather which enemies are killed or detained. If certain key insurgent leaders are essential to the insurgents' ability to conduct operations, then military leaders need to consider their relative importance when determining how best to pursue them. In COIN environments, the number of civilian lives lost and property destroyed needs to be measured against how much harm the targeted insurgent could do if allowed to escape. If the target in question is relatively inconsequential, then proportionality requires combatants to forego severe action, or seek noncombative means of engagement.

7-33. When conditions of civil security exist, Soldiers and Marines may not take any actions that might knowingly harm noncombatants. This does not mean they cannot take risks that might put the populace in danger. But those risks are subject to the same rules of proportionality. The benefit anticipated must outweigh the risk taken.

7-34. Discrimination requires combatants to differentiate between enemy combatants, who represent a threat, and noncombatants, who do not. In conventional operations, this restriction means that combatants cannot intend to harm noncombatants, though proportionality permits them to act, knowing some noncombatants may be harmed.

7-35. In COIN operations, it is difficult to distinguish insurgents from noncombatants. It is also difficult to determine whether the situation permits harm to noncombatants. Two levels of discrimination are necessary:

- Deciding between targets.
- Determining an acceptable risk to noncombatants and bystanders.

7-36. Discrimination applies to the means by which combatants engage the enemy. The COIN environment requires counterinsurgents to not only determine the kinds of weapons to use and how to employ them but also establish whether lethal means are desired—or even permitted. (FM 27-10 discusses forbidden means of waging war.) Soldiers and Marines require an innate understanding of the effects of their actions and weapons on all aspects of the operational environment. Leaders must consider not only the first-order, desired effects of a munition or action but also possible second- and third-order effects—including undesired ones. For example, bombs delivered by fixed-wing close air support may effectively destroy the source of small arms fire from a building in an urban area; however, direct-fire weapons may be more appropriate due to the risk of collateral damage to nearby buildings and noncombatants. The leader at the scene assesses the risks and makes the decision. Achieving the desired effects requires employing tactics and weapons appropriate to the situation. In some cases, this means avoiding the use of area munitions to minimize the potential harm inflicted on noncombatants located nearby. In situations where civil security exists, even tenuously, Soldiers and Marines should pursue nonlethal means first, using lethal force only when necessary.

7-37. The principles of discrimination in the use of force and proportionality in actions are important to counterinsurgents for practical reasons as well as for their ethical or moral implications. Fires that cause unnecessary harm or death to noncombatants may create more resistance and increase the insurgency's appeal—especially if the populace perceives a lack of discrimination in their use. The use of discriminating, proportionate force as a mindset goes beyond the adherence to the rules of engagement. Proportionality and discrimination applied in COIN require leaders to ensure that their units employ the right tools correctly with mature discernment, good judgment and moral resolve.

DETENTION AND INTERROGATION

7-38. Detentions and interrogations are critical components to any military operation. The nature of COIN operations sometimes makes it difficult to separate potential detainees from innocent bystanders, since insurgents lack distinctive uniforms and deliberately mingle with the local populace. Interrogators are often under extreme pressure to get information that can lead to follow-on operations or save the lives of noncombatants, Soldiers, or Marines. While enemy prisoners in conventional war are considered moral and legal equals, the moral and legal status of insurgents is ambiguous and often contested. What is not ambiguous is the legal obligation of Soldiers and Marines to treat all prisoners and detainees according to the law. All captured or detained personnel, regardless of status, shall be treated humanely, and in accordance with the Detainee Treatment Act of 2005 and DODD 2310.01E. No person in the custody or under the control of DOD, regardless of nationality or physical location, shall be subject to torture or cruel, inhuman, or degrading treatment or punishment, in accordance with, and as defined in, U.S. law. (Appendix D provides more guidance on the legal issues concerning detention and interrogation.)

LIMITS ON DETENTION

7-39. Mistreatment of noncombatants, including prisoners and detainees is illegal and immoral. It will not be condoned. The Detainee Treatment Act of 2005 makes the standard clear:

> *No person in the custody or under the effective control of the Department of Defense or under detention in a Department of Defense facility shall be subject to any treatment or technique of interrogation not authorized by and listed in the United States Army Field Manual on Intelligence Interrogation* [FM 2-22.3].

> *No individual in the custody or under the physical control of the United States Government, regardless of nationality or physical location, shall be subject to cruel, inhuman, or degrading treatment or punishment.*

7-40. In COIN environments, distinguishing an insurgent from a civilian is difficult and often impossible. Treating a civilian like an insurgent, however, is a sure recipe for failure. Individuals suspected of insurgent or terrorist activity may be detained for two reasons:

- To prevent them from conducting further attacks.
- To gather information to prevent other insurgents and terrorists from conducting attacks.

These reasons allow for two classes of persons to be detained and interrogated:

- Persons who have engaged in, or assisted those who engage in, terrorist or insurgent activities.
- Persons who have incidentally obtained knowledge regarding insurgent and terrorist activity, but who are not guilty of associating with such groups.

People engaging in insurgent activities may be detained as enemies. Persons not guilty of associating with insurgent or terrorist groups may be detained and questioned for specific information. However, since these people have not—by virtue of their activities—represented a threat, they may be detained only long enough to obtain the relevant information. Since persons in the second category have not engaged in criminal or insurgent activities, they must be released, even if they refuse to provide information.

7-41. At no time can Soldiers and Marines detain family members or close associates to compel suspected insurgents to surrender or provide information. This kind of hostage taking is both unethical and illegal.

Limits on Interrogation

7-42. Abuse of detained persons is immoral, illegal, and unprofessional. Those who engage in cruel or inhuman treatment of prisoners betray the standards of the profession of arms and U.S. laws. They are subject to punishment under the Uniform Code of Military Justice. The Geneva Conventions, as well as the Convention against Torture and Other Cruel, Inhuman or Degrading Treatment or Punishment, agree on unacceptable interrogating techniques. Torture and cruel, inhuman, and degrading treatment is never a morally permissible option, even if lives depend on gaining information. No exceptional circumstances permit the use of torture and other cruel, inhuman, or degrading treatment. Only personnel trained and certified to interrogate can conduct interrogations. They use legal, approved methods of convincing enemy prisoners of war and detainees to give their cooperation. Interrogation sources are detainees, including enemy prisoners of war. (FM 2-22.3 provides the authoritative doctrine and policy for interrogation. Chapter 3 and appendix D of this manual also address this subject.)

7-43. The ethical challenges posed in COIN operations require commanders' attention and action. Proactive commanders establish procedures and checks to ensure proper handling of detainees. Commanders verify that subordinate leaders do not allow apparent urgent requirements to result in violations of these procedures. Prohibitions against mistreatment may sometimes clash with leaders' moral imperative to accomplish their mission with minimum losses. Such situations place leaders in difficult situations, where they must choose between obedience to the law and the lives of their Soldiers and Marines. U.S. law and professional values compel commanders to forbid mistreatment of noncombatants, including captured enemies. Senior commanders clearly define the limits of acceptable behavior to their subordinates and take positive measures to ensure their standards are met.

7-44. To the extent that the work of interrogators is indispensable to fulfilling the state's obligation to secure its citizens' lives and liberties, conducting interrogations is a moral obligation. The methods used, however, must reflect the Nation's commitment to human dignity and international humanitarian law. A commander's need for information remains valid and can be met while observing relevant regulations and ethical standards. Acting morally does not necessarily mean that leaders give up obtaining critical information. Acting morally does mean that leaders must relinquish certain methods of obtaining information, even if that decision requires Soldiers and Marines to take greater risk.

Lose Moral Legitimacy, Lose the War

During the Algerian war of independence between 1954 and 1962, French leaders decided to permit torture against suspected insurgents. Though they were aware that it was against the law and morality of war, they argued that—

- This was a new form of war and these rules did not apply.
- The threat the enemy represented, communism, was a great evil that justified extraordinary means.
- The application of torture against insurgents was measured and nongratuitous.

This official condoning of torture on the part of French Army leadership had several negative consequences. It empowered the moral legitimacy of the opposition, undermined the French moral legitimacy, and caused internal fragmentation among serving officers that led to an unsuccessful coup attempt in 1962. In the end, failure to comply with moral and legal restrictions against torture severely undermined French efforts and contributed to their loss despite several significant military victories. Illegal and immoral activities made the counterinsurgents extremely vulnerable to enemy propaganda inside Algeria among the Muslim population, as well as in the United Nations and the French media. These actions also degraded the ethical climate throughout the French Army. France eventually recognized Algerian independence in July 1963.

THE LEARNING IMPERATIVE

7-45. Today's operational environment requires military organizations at all echelons to prepare for a broader range of missions than ever before. The Services are preparing for stability operations and post-conflict reconstruction tasks with the same degree of professionalism and study given to the conduct of combat operations. Similarly, COIN operations are receiving the attention and study merited by their frequency and potential impact. This broader mission set has significant leader development, education, and training implications, especially for land forces.

7-46. Army and Marine Corps leaders need to visualize the operational and informational impact of many tactical actions and relate their operations to larger strategic purposes. Effectively blending traditional military operations with other forms of influence is necessary. Effective leaders place a stronger emphasis on organizational change, develop subordinates, and empower them to execute critical tasks in consonance with broad guidance. Commanders must influence directly and indirectly the behavior of others outside their chain of command. Leaders are increasingly responsible for creating environments in which individuals and organizations learn from their experiences and for establishing climates that tap the full ingenuity of subordinates. Open channels of discussion and debate are needed to encourage growth of a learning environment in which experience is rapidly shared and lessons adapted for new challenges. The speed with which leaders adapt the organization must outpace insurgents' efforts to identify and exploit weaknesses or develop countermeasures.

7-47. Effective individual professional development programs develop and reward initiative and adaptability in junior leaders. Self-development, life-long learning, and reflection on experience should be encouraged and rewarded. Cultural sensitivity, development of nonauthoritarian interpersonal skills, and foreign language ability must be encouraged. Institutional professional development programs must develop leaders' judgment to help them recognize when situations change from combat to policing. Effective leaders are as skilled at limiting lethal force as they are in concentrating it. Indeed, they must learn that nonlethal solutions may often be preferable.

SUMMARY

7-48. Senior leaders must model and transmit to their subordinates the appropriate respect for professional standards of self-discipline and adherence to ethical values. Effective leaders create command climates that reward professional conduct and punish unethical behavior. They also are comfortable delegating authority. However, as always, accountability for the overall behavior and performance of a command cannot be delegated. Commanders remain accountable for the attainment of objectives and the manner in which they are attained.

Chapter 8
Sustainment

In my experience in previous wars, the logistic soldier was generally regarded as a rear area soldier.... Over here in Vietnam, that is completely changed....There is no rear area soldier, as such. Because of this, more than ever before, the man in logistics has to be first, a soldier, in the full sense of the word, and yet at the same time he has to know his MOS so that he can do his logistics job.

Major General James M. Heiser, USA, 1969

This chapter begins with a general discussion and analysis of how logistics in counterinsurgency (COIN) operations differ from logistics in conventional operations. This is followed by a survey of COIN-specific factors that affect how commanders can leverage available logistic assets and assign logisticians to meet special requirements needed to support different COIN logical lines of operations. Discussions that follow acknowledge that COIN operations may be entered into from various military conditions ranging from unstable peace to general war. The chapter concludes with a discussion of contracting support to COIN operations.

LOGISTIC CONSIDERATIONS IN COUNTERINSURGENCY

8-1. In counterinsurgency (COIN), the support provided by sustainment units often extends beyond sustaining operations; support provided to the population may become an important shaping operation or even the decisive operation. Logistic providers are often no longer the tail but the nose of a COIN force. Some of the most valuable services that military logisticians can provide to COIN operations include the means and knowledge for setting up or restarting self-perpetuating sustainment designs. The development of effective sustainment designs gives the populace a stake in stability and hope for the future. One COIN paradox is that many of the logisticians' best weapons for countering an insurgency do not shoot. Logistic units provide some of the most versatile and effective nonlethal resources available to Soldiers and Marines. Logisticians prepare to provide support across all logical lines of operations (LLOs) visualized and articulated by the commander. Often, logisticians already supporting COIN combat operations may be the only available source of prompt, essential knowledge, capabilities, and materials. This chapter focuses on capabilities and responsibilities of logistic units and logisticians. Commanders of all types of units at all levels must also be aware of the characteristics of COIN support.

WHAT IS DIFFERENT ABOUT LOGISTICS IN COUNTERINSURGENCY

8-2. In COIN operations, logistic units and other logistic providers perform all the functions they do in conventional operations as well as some different ones. Conventional operations usually involve two recognizable military organizations engaging each other in contiguous areas of operations. In COIN operations, the usual logistic functions—as well as COIN-specific activities—are performed in a frequently disorienting environment complicated by important social, political, and economic implications. Security conditions in these environments can rapidly change from moment to moment and every few hundred yards over various terrain conditions. (Table 8-1 [page 8-2] lists differences between the characteristics of logistic support to conventional operations and to COIN operations.)

Table 8-1. Conventional and counterinsurgency operations contrasted

	Conventional operations	*Counterinsurgency operations*
Mission	• Support combat unit missions. • Sustain and build combat power. • Support a mobile force with clear organization and structure. • Typically in direct support. • Logistic units and assets conduct only sustaining operations (focused on the force).	• Same as conventional operations plus support of logical lines of operations specific to counterinsurgency. • Support both a static force and mobile force. • Increased requirements for area support operations. • Logistic units and assets can be assigned as decisive and shaping operations (focused on the environment).
Enemy	• Enemy forces have supply trains and support echelons. • Friendly operational surprise (masking possible). • Difficult for enemy to perform pattern analysis. • Targeting logistic units is the enemy's shaping effort and considered a second front.	• Insurgents use nonstandard, covert supply methods that are difficult to template. • Limited operational surprise. • Easy for enemy to observe patterns in friendly logistic operations. • Insurgents place a high value on attacking logistic units and other less formidable, soft, high-payoff targets.
Terrain	• Fought in a definable area of operations. • Focus on destruction of enemy combat forces. • Few constraints. • Echeloned formations and discernable, hierarchical logistic organizations supporting well-defined, contiguous areas of operations. • Relatively secure lines of communications facilitate distribution operations from theater to corps to division to brigade.	• Operational environment poorly defined with multiple dimensions. • Support of the host nation population is the key objective. • Constrained time to achieve results, yet many counterinsurgency tasks are inherently time consuming. • Noncontiguous areas of operations and wide dispersion of units. • No front; everything is potentially close, yet far. • Need to maximize multiple lines of communications capacity/greater complexity. • Potentially decreased throughput capabilities. • Increased area support requirements. • Lines of communications vulnerable.
Troops and support available	• Uniformed personnel always suitable. • Contractor personnel suitable for secure areas only.	• Uniformed personnel usually suitable. • Suitability of contractor personnel judged case by case. • Task and location dependent; must be part of economic pluralism promotion plan.
Time available	• Tempo quicker. • Geared toward decisive major combat.	• Long-duration operations. • Continuity/logistics hand-off planning often required.
Civil considerations	• Secondary to considerations of how to defeat the enemy.	• May be the primary determinant of victory. • May figure prominently in logistic planning.

8-3. Differences in COIN logistics fit into the following major considerations:

- Logistic units are an essential part of COIN operations.
- Logistic units are perceived by insurgents as high-payoff targets and potential sources of supplies; thus, lines of communications (LOCs) are a main battle area for insurgents.

What is Different: Insurgent Perceptions of Military Logistics

Insurgents have a long history of exploiting their enemies' lines of communications as sources of supply. During the Revolutionary War, American forces significantly provisioned themselves from the British Army's overindulgent and carelessly defended logistic tail. In the 1930s, Mao Zedong codified a doctrine for insurgency logistics during the fight against the Japanese occupation of China. Without exaggerating, Mao stated, "We have a claim on the output of the arsenals of [our enemies],...and, what is more, it is delivered to us by the enemy's transport corps. This is the sober truth, it is not a jest." For Mao's forces, his enemy's supply trains provided a valuable source of supply. Mao believed the enemy's rear was the guerrillas' front; the guerrillas' advantage was that they had no discernable logistic rear.

This relative lack of logistic capacity was not an insurmountable problem for Mao or one of his logistic theorists, Ming Fan. According to Ming, "Weapons are not difficult to obtain. They can be purchased from the people's 'self-preservation corps.' Almost every home has some sort of weapon that can be put to use.... Ammunition can be obtained in the following ways: (1) From supplies given by friendly troops and headquarters on higher echelons. (2) Purchased or appropriated from the people. (3) Captured by ambushing enemy supply columns. (4) Purchased undercover from the enemy army. (5) From salvage in combat areas. (6) From the field of battle. (7) Self-made. (8) Manufactured by guerrilla organizations. (Such items as hand grenades, ammunition, etc.)" Beyond these specifics, this doctrine prescribes a mindset of actively seeking parasitic logistic relationships with not only the conventional enemy forces that the insurgents seek to co-opt and defeat but also active linkages to local black market activities and the cultivation of host-nation sympathizers.

For these reasons, forces conducting counterinsurgency operations must protect all potential supplies. Forces must also vigorously protect their lines of communications, scrupulously collect and positively control dud munitions and access to other convertible materiel, and actively seek ways to separate insurgents from black market activities.

In one moment in time, our service members will be feeding and clothing displaced refugees, providing humanitarian assistance. In the next moment, they will be holding two warring tribes apart—conducting peacekeeping operations—and, finally, they will be fighting a highly lethal mid-intensity battle—all in the same day...all within three city blocks. It will be what we call the "three block war."

General Charles C. Krulak, USMC, 1997

8-4. In a COIN environment, units providing logistic support potentially can be involved in more complicated tasks than even those of the three block war metaphor. COIN operations significantly differ in that logistic units must prepare to provide conventional logistic support to highly lethal, mid-intensity combat operations while supporting humanitarian operations. Logistic units may be required to maintain this support until conditions stabilize locally and civilian organizations can assume those duties.

8-5. The COIN environment requires logisticians to seek distribution efficiencies wherever possible. Logisticians must strive to eliminate backtracking and unnecessary distribution traffic. Ideally, logisticians maximize throughput methods that bypass—either on the ground or by air—population centers and heavily used civilian transportation nets. These practices are especially valuable in COIN operations. They improve logistic security, speed delivery, and minimize adverse effects and stress on the local populace.

8-6. Because of the diverse requirements, logisticians stay involved from the beginning of the planning process. They begin planning in detail as early as possible. Because of the complex logistic requirements and conditions under which COIN operations are pursued, commanders must ensure a careful logistic preparation of the area of operations (AO).

LOGISTIC PREPARATION OF THE COUNTERINSURGENCY AREA OF OPERATIONS

8-7. Logistic preparation of the AO relates to and can be treated as a COIN-specific logistic preparation of the theater. (FM 4-0, paragraphs 5-34 through 5-57, discusses logistics preparation of the theater.) In COIN operations, detailed analysis of civil logistic and economic assets takes on great importance. These assets can potentially support insurgents as well as the development and sustainment of host-nation (HN) security forces and the restoration of other essential services. Some examples of essential information for COIN logistic planning include the following:

- Analysis of the HN conventional force's existing logistic resources as a source of supply for developing HN security forces as well as the potential for insurgent black market activity.
- Effects of requirements generated by combat operations and collateral damage.
- Effects of multinational distribution requirements on HN lines of commerce.
- HN economic base (such as industry, manufacturing, and agriculture).
- HN lines of commerce (such as main supply routes, industrial cities, technical cities, pipelines, rail lines, and air and maritime ports).
- HN public works, utilities, and health, transportation, legal, and justice systems.
- Potential or existing dislocated civilian requirements.

ANALYSIS OF INSURGENT LOGISTIC CAPABILITIES

8-8. In COIN operations, analysis of the insurgents' logistic capabilities and shortfalls is especially significant. Logisticians and intelligence personnel perform what was formerly known as reverse-BOS (battlefield operating systems) analysis. This analysis does not just target enemy logistic capabilities and LOCs; it also assesses the suitability of supply sources for developing and sustaining insurgent forces. Effective analysis includes assessment of black market material, including salvage goods that insurgents might use to improvise equipment.

LOGISTIC SUPPORT TO LOGICAL LINES OF OPERATIONS

8-9. Although logisticians support all LLOs, logistic support during COIN focuses on the following LLOs:

- Conduct combat operations/civil security operations.
- Train and employ HN security forces.
- Establish or restore essential services.
- Support development of better governance.
- Support economic development.

SUPPORT TO COMBAT OPERATIONS/CIVIL SECURITY OPERATIONS

8-10. Most logisticians and nonlogisticians are familiar with the combat operations/civil security operations LLO. The paramount role of logistic units remains to support Soldiers and Marines in accomplishing the mission and meeting other Title 10 responsibilities. Using logistic units to augment civil programs supporting other LLOs must not detract from the logistic system's capability to support combined arms forces engaged in combat operations.

Support to and from Operating Bases

8-11. Logistic support to COIN combat operations is often accomplished from bases (see FM 3-90, paragraph E-19 through E-29) or forward operating bases (see FM 3-05). Operating bases provide combined arms units with relatively secure locations from which to plan and prepare for operations. As a result, these bases take on new significance in operations executed in a noncontiguous COIN environment.

Considerations for Situating Operating Bases

8-12. In COIN operations, base site selection becomes extremely important for more reasons than providing optimal support to combat operations. Under certain geographic conditions, such as in rugged mountains with few passes or desolate desert terrain, placing secure operating bases astride or near the insurgents' LOCs can improve counterinsurgents' interdiction and disruption capabilities. In urban areas and jungles, insurgents may negate advantages of such a position by rerouting their LOCs around the base. This happened when U.S. forces tried to interdict insurgent supplies on the Ho Chi Minh Trail in Vietnam.

8-13. Other reasons for carefully considering base placements involve sensitivities and concerns of the local populace. The potential for ill-considered bases to substantially disrupt the local populace's daily lives and produce other unintentionally negative effects is significant, even if counterinsurgents arrive with positive intentions. Bases must be set up so that they do not project an image of undue permanency or a posture suggesting a long-term foreign occupation. Similarly, logistic postures that project an image of unduly luxurious living by foreign forces while HN civilians suffer in poverty should be avoided. Such postures undermine the COIN message and mission. Insurgent propaganda can twist such images into evidence of bad intentions by counterinsurgents. While these considerations take on special significance in COIN operations, none of them override the primary concern that operating bases be securable and defendable.

8-14. Selecting and developing operating base sites requires the additional consideration and balance of other factors. In COIN operations, logisticians must provide support through a careful mix of supply-based or supply-point ("just-in-case") practices with distribution-based or unit distribution ("just-in-time") logistic methods. Situations can swiftly develop that require equally rapid logistic responses to prevent further deterioration of security conditions. Under these COIN-specific circumstances, just-in-time practices may still not be quick enough; using just-in-case capabilities may be more appropriate, effective, and timely, while conserving resources. A fire-fighting analogy best illustrates this dilemma of COIN logistics. A small fire confined to a pan on the stove can be put out easily with the five-pound extinguisher. But when this extinguisher is not immediately available to put out the fire, half the house may burn before fire fighters arrive. Then extinguishing the fire requires trucks, hoses, and thousands of gallons of water. The house also needs construction materials and time to be restored to its former state. Commanders and logisticians supporting COIN operations must correctly identify which materials equate to "five-pound fire extinguishers" for counterinsurgents. Logisticians must then ensure that items are available at the most appropriate location.

8-15. This carefully considered balance between distribution- and supply-based methods supports the goal of minimizing the size of operating bases. When required, Soldiers and Marines can relocate smaller bases more easily. Such bases are also less intrusive and antagonizing to the local populace.

8-16. Planners must consider an operating base's purpose when selecting its location. If planners anticipate extensive logistic throughput, they pay close attention to entry and exit points. Ideally, more than one entry and exit point should exist. (FM 5-104 discusses the construction of entry control points and facilities.) Where possible, at least one control point should not require convoys to travel though a populated area. In addition, at least one entry point requires a staging area for convoys and should be located to avoid having to transit the base to form up.

8-17. Due to the noncontiguous nature of COIN operations, logisticians develop weblike LOCs and main supply routes between operating bases and logistic bases. Weblike links between bases have two advantages. By dispersing logistic operations, weblike LOCs minimize intrusive effects of these operations on the populace. They also provide redundancy in distribution capabilities, making the system more robust and limiting the effects of any one LOC's interdiction. In addition, more ground LOC routes provide more opportunities to observe the populace and gather information from them. Wherever possible in COIN operations, planners should identify multiple LOCs between bases.

Protection

8-18. Protection of logistic activities takes on greater significance during COIN operations. Historically, insurgents have deliberately sought out and engaged logistic units, particularly poorly defended, easy targets. COIN operations have intensive manpower requirements and a dispersed nature. Logistic units cannot assume that combined arms units will be available to assist them with protection. For this reason, logistic units play a larger role in defending bases and LOCs. These units must assume responsibility to protect civilian logistic augmenters, whether Department of Defense (DOD) civilians or contractors, working in their AOs.

Vietnam: Meeting the Enemy and Convoy Security

The year 1968 proved a turning point for units of the 48th Transportation Group stationed in the Republic of Vietnam. The group was assigned to transport supplies to 25th Infantry Division units operating in Vietnam's Cu Chi province. In August, North Vietnamese Army and Vietcong units ambushed a supply convoy from the group. The maneuver brigade responsible for clearing that part of the main supply route had been recently assigned other missions, and its resources were spread thin. The ambushers chose their moment to attack well. Monsoon conditions prevailed, and the site was outside the range of supporting indirect fire. In addition, the supplies were destined for the unit tasked with responding to such attacks. Under dangerous weather conditions, two UH-1C "Huey" gun ships arrived first to assist the beleaguered convoy. From the air, aviators witnessed enemy soldiers unloading supplies from U.S. vehicles onto trucks hidden in the tree line off the road. Almost three hours later, the first relief force arrived on the ground. This force barely had enough capability to continue a minimal defense of the remaining convoy assets and surviving personnel. Finally, seven hours later, a U.S. armored cavalry force arrived and forced the attackers to withdraw.

Thirty Soldiers were killed, 45 were wounded, and 2 were taken prisoner. This event forced the 48th Group and 25th Infantry Division to rethink their convoy tactics. The two units started to hold detailed convoy planning meetings and renewed their enforcement of Soldier discipline. They placed security guards on every vehicle, hardened cabs of supply trucks with steel plates, and mounted M-60 machine guns on every vehicle possible. The greatest improvement was the clarification of command and support relationships and responsibilities between the 48th Group and 25th Infantry Division. This included publishing common convoy standing operating procedures. With these new practices in place, convoy ambushes soon had different endings. A change in thinking about a logistic problem converted convoy operations from unglamorous defensive activities into valuable opportunities to engage insurgents offensively.

Combat Logistic Convoys

8-19. During COIN operations, every logistic package or resupply operation becomes a mounted combat operation, or combat logistic convoy. Insurgents see attacks on resupply operations as a potential source of dramatic propaganda as well as a source of supplies and materiel. For this reason, combat logistic convoys should project a resolute ("hard and prickly") image that suggests that they will not be an easy ("soft and chewy") target. Combat logistic convoys project their available combat power to the maximum extent possible, as would any other combat convoy or patrol. Under these conditions, logistic units—or anyone else involved in resupply operations—perform a detailed intelligence preparation of the battlefield and prepare a fire support plan. These units also identify usable intelligence, surveillance, and reconnaissance assets. Additionally, combat logistic convoys should gather information, report on road statuses, and contribute to intelligence collection plans. Logisticians must remember that while the materiel is in transit, it is not only unusable but also vulnerable to insurgent attacks. In a COIN environment, distribution-based practices may actually provide insurgents with more opportunities to target resupply activities. These opportunities stem from the large blocks of time the materiel is in ground transit. Likewise, logisticians must remember that insurgents constantly and deliberately seek out adaptive countermeasures to logistic activities. For example, the development and proliferation of improvised explosive devices were a natural counter to U.S. distribution-based doctrine. As a result, engineer assets and other combined arms elements should prepare to execute periodic route clearance operations. Logisticians must carefully analyze conditions and perform thorough combat preparations before launching combat logistic convoys.

Unit Equipment for Counterinsurgency

8-20. COIN operations shift quite rapidly. Logistic units and supported units often deploy with some equipment unsuitable for prevailing operational and tactical trends when they arrive. This dynamic often forces logisticians to seek equipment modifications and new items. For this reason, COIN operations especially benefit from new procurement programs, such as rapid fielding initiatives and the purchase of commercial off-the-shelf (COTS) items. These approaches make sense for counterinsurgents when insurgents' capabilities also come from their creative exploitation of commercially available technologies and materials as well as their lack of bureaucratic encumbrance.

8-21. Counterinsurgents benefit from using more streamlined materiel procurement procedures. They receive what they need when they need it. Specific and localized environmental and cultural conditions create unseen needs. Streamlined procedures often meet these needs more quickly. Examples of COIN requirements that can be met by COTS procurement are—

- Public address systems.
- Language translation devices.
- Nonlethal weapons.
- Backpack drinking water systems.
- Cargo all-terrain vehicles ("Gators").

8-22. Examples of rapid fielding initiatives are—

- Up-armoring kits for light wheeled vehicles.
- Body armoring improvements.
- Improved explosive detectors.
- Improvised explosive device signal jammers.

8-23. A potential drawback when adopting COTS equipment concerns maintenance support packages. Repair parts may be inadequate or difficult to obtain in theater. Many commercial manufacturers lack the experience or infrastructure needed to support their equipment under military conditions and in quantity. Often, they are unaccustomed to operating in hostile austere theaters, far from their regular markets and customer base. It may take time to get needed parts into normal supply channels and trained personnel in theater to fix COTS equipment. Logisticians must consider other measures to assure the continuous operation of this type of new, vital equipment. They should consider establishing pools of low-density COTS items that can provide exchanges through procedures similar to those used for operational readiness floats. These procedures provide time to evacuate exchanged equipment to locations where it can be maintained and repaired and where the required commercial parts are readily available.

8-24. During COIN operations, units may temporarily have to draw additional or specialized equipment in theater. For long-term COIN operations, leaders may have to establish theater property books. These procedures help maintain and account for rotationally issued additions to standard equipment as well as specialized or specially modified equipment. In-theater special issues and fieldings may include materiel and equipment procured through military channels, rapid fielding initiatives, or COTS sources. Units might draw supplementary or modified equipment to a greater or lesser degree depending on the unit's non-COIN primary function. For example, an artillery unit normally equipped with self-propelled howitzers may draw numerous hardened HMMWVs to conduct security missions. This unit would probably leave many of its howitzers at home station. Conversely, a military police unit might already be well equipped to conduct security activities and might only have to draw a few pieces of the latest specialized equipment. Examples of other items that counterinsurgents might need include—

- Up-armored vehicles.
- Cargo trucks.
- All-terrain cargo vehicles ("Gators").
- Improvised explosive device jammers.
- Body armor.
- Specialized mine-clearing equipment ("Buffaloes").

8-25. Units conducting COIN operations can expect somewhat different maintenance requirements than in conventional operations. Units that put high mileage on their wheeled vehicles need more frequent servicing. Armor packages may wear out shock absorbers and springs much faster; these would require replacement sooner than in conventional operations. COIN missions and the remoteness of many operating bases may compel maintenance sections to perform higher echelons of maintenance than normal and encourage greater organic capability.

Unit Basic Loads and Operational Reach

8-26. In places like Somalia and Sadr City during Operation Iraqi Freedom, Soldiers and Marines conducting COIN operations risked being cut off from bases and forward operating bases due to weather changes, enemy action, or civil protests. Such rapidly developing situations can deny units access to resupply for extended periods. Units operating away from supporting bases carry the maximum amount of basic supplies—water, food, ammunition, first aid, and equipment batteries—on their vehicles. Additionally, some COIN operations can consume surprisingly high quantities of ammunition (specifically small arms) because of combined defensive and offensive actions. Logisticians supporting these operations adjust stockage levels for unit basic loads and other sustainment commodities. In turn, logisticians and their supported units should rethink how to best configure their supporting vehicles as supply platforms to meet these COIN-specific needs. Competent authorities should validate successful solutions. This process ensures standardization across formations to ensure safety and to support planning for effective employment.

8-27. When developing their operating base requirements, commanders and logisticians must plan for areas to store ammunition and explosives. Ideally, units are issued ammunition and explosives, anticipating that it may be some time before resupply is available. Units normally carry only their basic load. The rest should be staged appropriately.

Aerial Distribution

8-28. During COIN operations, logisticians should maximize intratheater aerial resupply. This practice reduces the vulnerability of resupply activities to ground-based attacks by insurgents. Its also minimizes negative effects of COIN logistic activities on public roadways and reduces the potential for alienating the populace. Site selection for bases and forward operating bases includes assessing aircraft support capabilities. Site selection also considers maximizing the possible options for aerial delivery by rotary- and fixed-wing aircraft, airdrops, and landings. Additionally bases with medical capability (level II or higher) require a helicopter pad near the medical area. (FM 4-02, paragraph 2-4, and NAVMED P-117, article 19-24, discuss levels of medical care.)

Air Delivery in Iraq: Maximizing Counterinsurgency Potential

For almost five months in 2004, two Marine battalions with attached units operating in remote areas of Iraq were resupplied by airdrops from rotary- and fixed-wing aircraft. Until then, cargo airdrops from helicopters had been suspended since the mid-1990s due to technical difficulties that posed unacceptable peacetime risks. The high tempo in Iraq, coupled with severe challenges in using ground supply methods to reach remote locations, made circumstances opportune to reexamine helicopter airdrop possibilities.

A careful analysis of earlier challenges resulted in a clarification of flying procedures during drops. Furthermore, technological advances made it possible to drop bundles by parachute with some directional adjustment after release. This new ability greatly increased the accuracy and utility of helicopter airdrops. These improved procedures replaced dangerous nine-hour convoys with quicker and more secure flights covering three-quarters the ground distance. During this period, Marines received more than 103 tons of supplies from airdrops. A logistic method that previously had been set aside as unworkable found new utility when operational conditions changed and planners addressed technological shortcomings.

With the need to deliver many supplies to diverse locations quickly, counterinsurgents discovered that adopting various air delivery procedures significantly improved their logistic posture under challenging conditions. Such procedures also minimized the risk of negative encounters with the populace and insurgents.

SUPPORT TO TRAINING AND EMPLOYING HOST-NATION SECURITY FORCES

8-29. One of the most important LLOs for U.S. forces engaged in COIN operations is training and employing HN security forces. Various support and training activities contribute to security sector reform. Each activity can substantially involve military logisticians. Usually, developing HN police forces and sustaining their training falls under the auspices of non-DOD agencies. These agencies include the Department of State, the Department of Justice, or United Nations mandated missions. The development and support of HN military forces is a COIN mission that military logisticians must prepare to support, from planning at the strategic level to practical implementation on the ground. (Chapter 6 covers the support of HN security forces in more detail.)

8-30. Some tasks required to establish HN security forces might initially fall to military logistic units until other government agencies' programs start, logistic support can be contracted, or HN logistic organizations begin to function. Examples of HN security force support tasks include the following:

- Providing operating base space or establishing other supportable, secure locations to recruit, receive, and train HN security forces.
- Providing initial logistic support to forming HN security forces, possibly to include equipping, arming, feeding, billeting, fueling, and providing medical support.
- Providing logistic training to newly formed HN security force logistic organizations.

Equipping and Sustaining Host-Nation Security Forces

8-31. Logisticians can expect to help develop plans and programs for sustaining HN security forces. They must ensure that equipment selected is suitable to and sustainable through the host nation's capabilities. Equipment and support programs must fall within the host nation's resources, including budget and technological capabilities. In many cases "good enough to meet standards" equipment that is indigenously sustainable is preferable to "high-technology, best available" equipment that requires substantial foreign assistance for long-term maintenance. Foreign high-technology equipment can provide the insurgent movement with a valuable propaganda point that could negate any potential technological advantages.

8-32. One acknowledged difficulty in establishing HN security forces is identifying sources of suitable materiel and equipment. Often, multinational partners develop plans to equip the host nation from multiple donor nations and agency sources. Logisticians may need to become familiar with these agencies' capabilities and donor nations' supply and maintenance systems, even though support packages may not be included with the donation.

Host-Nation Security Forces Logistics

8-33. Logisticians involved in training logistic personnel to support HN security forces need to be aware of several special challenges. Part of the problem with previously dysfunctional military cultures in many unstable countries is a pervasive climate of corruption and graft that can cripple attempts to develop effective support services. Logisticians conducting such training should expect to stress repeatedly the long-term benefits of supply discipline and materiel accountability. Logisticians must also emphasize how these practices affect HN security and development. For this reason, emphasis should be placed on inventory procedures. Logisticians should monitor the black market to check for pilfered military equipment and to evaluate the effectiveness of logistic procedures and accountability training. Of all capabilities, logistic functions may take the longest for HN forces to develop. These functions are rife with inherent complexity and potential cultural challenges. Hence, HN forces may take a long time to operate independently of U.S. or multinational logistic support. (AR 12-1 and DODD 5105-38M discuss control and accountability requirements for property transferred to a host nation [end-use monitoring].)

8-34. HN-produced materiel should be procured and used to support HN security forces whenever it can meet requirements and is reasonably and reliably available. This practice helps stimulate the HN economic base and promotes an attitude of self-sufficiency in HN forces. It reinforces the important political message that HN security forces are of the people, not agents of foreign powers. When promoting these practices, logisticians may find themselves outside the normal scope of their duties when they assess the suitability of locally available materials and advise how to make such materials suitable for self-sustainment. The most valuable lesson logisticians may give to HN security forces and those supporting them is not "what to do" but "how to think about the problem of sustainment" and its link to security effectiveness.

Building a Military: Sustainment Failure

By 1969, pressure was on for U.S. forces in Vietnam to turn the war over to the host nation in a process now known as Vietnamization. While assisting South Vietnamese military forces, the United States armed and equipped them with modern small arms, communications, and transportation equipment—all items produced by and sustained from the U.S. industrial base. This modern equipment required an equally sophisticated maintenance and supply system to sustain it. Sustaining this equipment challenged the South Vietnamese economically and culturally, despite the training of several thousand South Vietnamese in American supply and maintenance practices. In short, the American way of war was not indigenously sustainable and was incompatible with the Vietnamese material culture and economic capabilities. South Vietnam's predominately agrarian-based economy could not sustain the high-technology equipment and computer-based systems established by U.S. forces and contractors. Consequently, the South Vietnamese military transformation was artificial and superficial. Many South Vietnamese involved in running the sustainment systems had little faith in them. Such attitudes encouraged poor administration and rampant corruption. After U.S. forces left and most U.S. support ended, the logistic shortcomings of the supposedly modern South Vietnamese military contributed to its rapid disintegration when the North Vietnamese advanced in 1975.

Support to Establishing or Restoring Essential Services

8-35. According to existing U.S. military logistic doctrine, no provision exists for U.S. forces to become decisively or exclusively engaged in providing essential services to the populace. However, this doctrinal position does not prohibit units from applying skills and expertise to help assess HN essential services needs. Along with these assessments, logistic and other units may be used to meet immediate needs, where possible and in the commander's interest. These units can also assist in the handoff of essential service functions to appropriate U.S. agencies, HN agencies, and other civilian organizations.

Assessing Essential Services Requirements

8-36. Logistic preparation of the theater and detailed assessments of COIN-specific issues should give logisticians good insights into HN capabilities, requirements, and shortfalls. (See paragraph 8-7.) Logistic assessments should be combined with information from civil affairs area assessments. Logisticians working closely with other branches can contribute to these area assessments. (Table 8-2 [page 8-12] identifies the areas that concern civil affairs. Logisticians can help assess these areas.)

Table 8-2. Civil affairs capabilities used by logisticians

Service	*Branch*	*Capability*
Police and Fire	Military police	• Assessment of police capabilities (law enforcement, crowd control, traffic control, crime trends, criminal investigations) and detention and correction operations.
	Military firefighter	• Assessment of firefighting capabilities and equipment.
Water	Quartermaster	• Water purification.
	Medical service	• Preventive medicine and sanitation specialists.
	Engineer	• Water pipelines, facilities, and drilling support.
	Ordnance	• Pumps and mechanicals repair.
Electricity	Engineer	• Power generation and facility repair and construction specialists.
Schools	Engineer	• Vertical construction specialists.
	All	• Training and education.
Transportation network	Engineer	• Assessment of rail, bus, and port facilities; roadway and bridge assessments; and airfield capabilities and upgrades.
	Transportation	• Assessment of rail, bus, and ferry capacities.
	Ordnance	• Assessment of mechanical maintenance of rail, bus, truck, and ferry operating equipment.
	Military police	• Roadway and traffic flow assessments.
Medical	Medical service and medical corps	• Assessment of local health care capabilities and needs.
Sanitation (trash and sewage)	Medical service	• Preventive medicine and sanitation specialists.
	Quartermaster	• Water specialists.
	Engineer	• Earthmoving specialists, plumbing construction, soil analysis, concrete casting, and sanitary landfill management.
	Ordnance	• Heavy wheeled vehicle, pump, and mechanical repair.
Food supply	Veterinary corps	• Food source and quality; vector control.
	Quartermaster	• Food packaging and distribution.
Fuel	Quartermaster	• Fuel specialists; testing of locally procured fuel supplies.
Financial	Finance	• Identify requirements to reestablish accountability and security of host-nation and captured funds. • Assess financial support requirements and banking and currency needs.

Time as a Logistic Commodity

8-37. The speed with which COIN operations are executed may determine their success and whether the populace supports them. This is especially true for operations that involve restoring essential services. Planners must strive to have the smallest possible gap of time between when they assess essential services and when U.S. forces begin remediation efforts. To keep this time gap as small as possible and manage the development of popular expectations, logistic units may need to initiate remedial services until HN authorities and agencies can assume these functions. For example, think of the populace as a patient destabilized by the trauma of insurgency. In COIN operations, logistic and other units may need to function much like the first-responder medic on the scene. The medic conducts initial assessments of a patient's needs, provides lifesaving first aid, and lets the hospital know what more specific higher level care the patient requires. Medical and COIN first responders work best when they can assess and initiate life support treatment immediately. In COIN operations, "treatment" is harder to determine without obvious calibrated vital signs—such as blood pressure, pulse, and temperature—and "immediately" may last weeks or months. For this reason, COIN logistic units may have to take whatever measures they can for immediately stabilizing essential services and preventing deteriorating conditions. (Table 8-3 [page 8-37] lists some examples of how military logistic assets and capabilities can be used to bridge the essential services gap and meet immediate and essential service needs.)

Handoff of Essential Services

8-38. Frequently, logisticians who have provided stopgap essential services may be the only ones with accurate knowledge of essential services needs and priorities. Logisticians providing these services may stay actively involved in the handoff to other government agencies and their designated civil organizations. Their involvement should continue until HN agencies and activities can function and meet essential services needs. A poor handoff can provide insurgents with propaganda opportunities and evidence of the "insincerity" of COIN efforts.

8-39. When U.S. forces restore and transition essential services to the HN government, they remove one of the principal causes insurgents exploit. This action greatly assists the HN government in its struggle for legitimacy. Competent leaders can expect insurgents to conduct attacks against restored services. During this handoff, multinational logistic assets may need to maintain a logistic quick reaction force. This force ensures the continuity of services and marginalizes counteractions and messages by insurgents.

Public Transportation, Population Movement, and Life Support to Internally Displaced Persons and Refugees

8-40. Under conditions of national crisis or insurgency, public transportation systems often fall into disarray. Counterinsurgents may have to recover stolen or misappropriated buses, trucks, cars, and other government vehicles (including former military equipment) and restore them to public service. This action helps alleviate urgent requirements for public transportation. It also sends an unmistakable message of resumption of governmental authority and can substantially reduce the amount of replacement equipment counterinsurgents must procure. Logistic units and personnel can expect to assist in this process—from reestablishing accountability procedures to assessing repair and maintenance needs—until competent HN public or government authorities can resume these duties.

8-41. An insurgency often creates many groups of internally displaced persons and refugees on short notice. Attending to internally displaced person and refugee needs can quickly become an urgent logistic requirement. Planners draw on all essential services to provide secure emergency shelter, camps for internally displaced persons and refugees, and life support (food, water, and medical care). Nongovernmental organizations and other civilian agencies normally furnish this support to internally displaced persons and refugees. However, conditions may prevent these agencies from providing these services quickly. Furthermore, in COIN operations, internally displaced person and refugee security may take on heightened military importance. Traumatized and dislocated persons may become vulnerable to insurgent threats and recruitment. The restoration and maintenance of public transportation services can help internally displaced persons and refugees. Figure 8-1 (page 8-15) shows that as essential services projects take root, they start to provide tangible benefits for the populace. Progress in these individual endeavors may experience individual setbacks as planners calibrate programs and projects to localized needs. Before implanting agency transfers, planners should measure overall progress. As essential services become more effective, insurgent activities lose influence and popular support. As a general rule, it is best to provide essential services to people in their native areas and thereby discourage their displacement.

Table 8-3. Logistic units and their capabilities

Logistics activities	*Capabilities*
Contingency contracting officers	• Procure commercial public utility equipment with Title 22 funds. • Employ theater contractors and external theater subject matter experts and trainers to maintain assets.
Reverse osmosis water purification units	• Provide immediate sources of potable water until water pumps at purification plants are restored.
Distribution companies and supply support activities	• Provide temporary storage and distribution of foreign humanitarian assistance.
Explosive ordnance disposal units	• Dispose of munitions and other unexploded explosive ordnance in populated areas. • May require transportation or engineer support.
Combat logistic convoys	• Provide security for nongovernmental organization (NGO) transportation of critical humanitarian assistance. (Not all NGOs agree to this.)
Medical units	• Provide medical civic action program teams to conduct visiting clinics at small or remote villages. • May augment NGOs, such as Doctors Without Borders.
Brigade surgeons and brigade engineers	• Work with contracting officer representatives to restore clinic or hospital services.
Medical personnel and units	• Assist with upgrading and restoring host-nation medical training programs to meet civil health care providers' critical shortfalls.
Senior power generation technicians	• Provide advice concerning and troubleshoot municipal power sources.
Class I (rations) sections	• Account for, preserve, and distribute humanitarian daily rations (a Department of State-controlled item).
Medium truck companies	• Move internally displaced persons.
Logistic units	• Provide life support to internally displaced person (or refugee) camps—billeting, food service, personnel (biometric), accountability, and work placement.
Preventive medicine teams	• In conjunction with veterinary support, conduct vector and parasite analysis on farm livestock (host-nation food source).
Engineers	• Repair critical highways, renovate bridges, or build buildings (such as clinics or schools).

Support to Developing Better Governance

8-42. COIN operations strive to restore order, the rule of law, and civil procedures to the authority of the HN government. All counterinsurgent actions must be those of agents of a legitimate and law-abiding HN government. Multinational and U.S. forces brought in to support this objective must remember that the populace will scrutinize their actions. People will watch to see if Soldiers and Marines stay consistent with this avowed purpose. Inconsistent actions furnish insurgents with valuable issues for manipulation in propaganda.

Legal Support to Operations

8-43. Legal support to COIN operations can cover many areas. (See FM 4-0, chapter 12; MCWP 4-12, page A-4; and appendix D of this manual.) It includes continuously monitoring and evaluating rules of engagement. This legal status may affect the conduct of contractors and their requirements for protection. Legal support also includes status of forces agreements. These agreements need to be negotiated and revised as the HN government becomes able to responsibly assume and exercise sovereignty. Status of forces agreements affect how legal disputes between U.S. forces and local nationals are handled, including those disputes emerging from contracting and other commercial activities. Contracts and claims require sensitive and fair construction and execution so perceptions of exploitation and favoritism do not undermine overall COIN initiatives. COIN operations often depend on many funding sources. Leaders get judge advocate legal advice on fiscal law to ensure compliance with domestic statutes governing the funding of military and nonmilitary activities. Leaders may ask judge advocates to advise the HN government at all political levels about how to establish and administer appropriate legal safeguards.

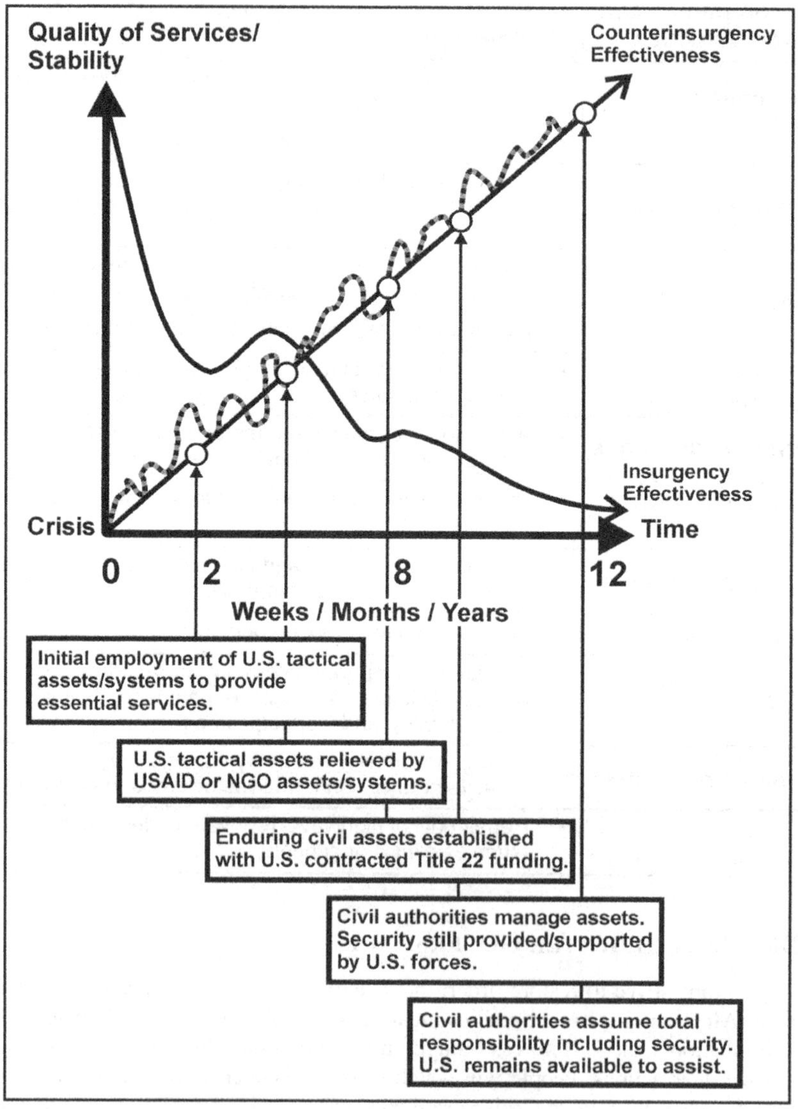

Figure 8-1. Comparison of essential services availability to insurgency effectiveness

Legal Aspects of Contracting and Claims

8-44. In COIN operations, two circumstances may require extensive civil law support. The first situation is when counterinsurgents engage in commercial contracts with people for provision of goods and or services. The second occasion is when people seek compensation for damages, injuries, or deaths that they or their relatives claim to have suffered due to actions by counterinsurgents.

8-45. Legally reviewing COIN contracts negotiated with HN contractors establishes several important conditions. First, the process makes clear to HN contractors that established procedures rooted in law govern such transactions. It sends the message that favoritism and partisanship are not part of the process in a legitimate government. Second, this review potentially forestalls contracts going to individuals who may be part of the insurgency and may already be named or identified as subjects of other ongoing investigations or legal actions.

8-46. In the case of claims for damages allegedly caused by counterinsurgents, legal reviews show genuinely wronged people that the HN government takes their grievances seriously. (See paragraphs D-35 through D-37.) When insurgents or opportunists misrepresent terms or conditions under which "damages" occur, legal reviews effectively assess the validity or falsehood of such claims and thereby prevent counterinsurgents from squandering resources or inadvertently supporting insurgents.

Restoration of Civil Judicial Functions

8-47. In periods of extreme unrest and insurgency, HN legal structures—courts, prosecutors, defense assistance, prisons—may fail to exist or function at any level. Under these conditions, to establish legal procedures and precedents for dealing with captured insurgents and common criminals, U.S. forces may make provisions to establish special tribunals under the auspices of either a provisional authority or a United Nations mandate. While legal actions fall under these provisions, counterinsurgents can expect to provide sustainment and security support as well as legal support and advice on these functions.

8-48. Even when HN authorities have restored judicial functions, counterinsurgents may have to provide logistic and security support to judicial activities. If insurgents continue to disrupt activities supporting the rule of law, the support may be prolonged. With restoration of legislative and judicial functions to the HN government, counterinsurgents must recognize and acknowledge that not all laws will look familiar. Laws passed by the legislature of the HN government may differ from those experienced by multinational forces in their home countries. Under such conditions, counterinsurgents need to consult their legal advisors, commanders, and diplomatic representatives for guidance on dealing with sensitive matters.

SUPPORT TO ECONOMIC DEVELOPMENT

8-49. Many commanders are unfamiliar with the tools and resources required for promoting economic pluralism. In COIN operations, economic development is probably the LLO with the greatest logistic significance. Military commanders using resource managers (comptrollers) and contingency contracting officers are usually involved with this LLO. Staff officers will need to deliver financial resources that—

- Maximally benefit the HN population.
- Support achieving objectives along other LLOs.
- Ensure the funds stay out of insurgent hands.

Achieving these objectives depends on logisticians keeping a thorough and accurate logistic preparation of the theater as well as commanders and contracting officers obtaining goods and services consistent with its assessments. Such purchasing must also promote vendors and businesses whose practices support widespread job stimulation and local investment. In addition to the logistic preparation of the AO issues discussed in paragraph 8-7, other areas for assessment and analysis include the following:

- HN economic capabilities and shortfalls suitable for filling by external means.
- Methods of determining land and other real property ownership, means of transfers, and dispute resolution.
- Methods for promoting and protecting property and asset rights as well as open access to trade goods and services.

- Prevailing wage rate standards and correlation to occupational category (unskilled, skilled, and professional labor).
- Historic market demographics.
- Identification of potential vendors with local sources of supply in the AO.

8-50. Various funding sources usually support COIN operations. (See paragraphs D-27 through D-31.) U.S. forces most commonly operate under two types of funds:

- Title 10 funds, which are strictly for the supply, support, and sustainment of DOD service members and employees.
- Title 22 funds, which are appropriated for foreign relations purposes and used solely for the benefit and support of the HN government and population.

8-51. Other sources of funding in COIN operations may include those provided by other government agencies. Such agencies can include the U.S. Agency for International Development, the Department of State, other donor nations and agencies, the United Nations, or even the host nation. In some cases, counterinsurgents may seize or capture misappropriated or illicit funds. Confiscated funds may be redistributed to fund COIN activities. Under these complex fiscal circumstances, resource managers and staff judge advocates provide the best guidance on the legal use of different types of funds. (Appendix D covers this subject in more detail.)

CONTRACTED LOGISTIC SUPPORT

8-52. In COIN operations, logistic contractors might support U.S. forces. Contractor activities fall into three different categories:

- Theater support contractors.
- External support contractors.
- System contractors.

(See FM 4-0, paragraphs 5-92 through 5-95; MCWP 4-12, page 4-8.)

8-53. Theater support contractors can make significant contributions to promoting economic pluralism because they rely the most on HN employees and vendors. External support contractors and the Logistic Civilian Augmentation Program (LOGCAP) provide logistic services, usually through large-scale prearranged contracts with major contractors, who may in turn subcontract various components of their large contracts to smaller theater-based providers. Systems contracts are designed by systems program managers to support special or complex equipment; they generally have little influence on promoting economic pluralism.

THEATER SUPPORT CONTRACTORS

8-54. Theater support contractors can be obtained either under prearranged contracts or by contracting officers serving under the direct authority of the theater principle assistant responsible for contracting. Theater support contractors usually obtain most of their materials, goods, and labor from the local manufacturing and vendor base. Some examples of goods and services that theater support contractors provide include—

- Construction, delivery, and installation of concrete security barriers for the defense of counterinsurgent bases and HN public buildings.
- Construction of security fencing.
- Public building construction and renovations (such as site preparation, structure construction, electrical and plumbing installation, and roofing).
- Sanitation services.
- Maintenance augmentation in motor pools.
- Road construction and repair.
- Trucking and cartage.

- Manual labor details (such as grounds maintenance and sandbag filling).
- Housekeeping (such as warehouses).

Counterinsurgency Contractor Considerations

8-55. In a COIN environment, commanders carefully consider when to use theater support contractors and local hires. Commanders also supervise contracted personnel to ensure they do not undermine achieving COIN objectives. Due to the subversive nature of many insurgent activities, all contractors and their employees require vetting through the intelligence section. All contractors and their employees require tamperproof, photograph, biometric-tagged identification. This identification needs to be coded to indicate access areas, security level, and supervision required. In the case of HN employees, "badging" can also be an accountability tool if U.S. forces issue and receive badges at entry control points daily. Contractor security breaches are one concern; another is the security and safety of the contractor's employees. Though insurgents may target logistic contractors and their employees, the employees are not combatants. They are classified as "civilians accompanying the force." This status must not be jeopardized and the military units with which they work must keep them secure in the workplace. Units employing HN contractors and employees must watch for signs of exploitive or corrupt business practices that may alienate segments of the local populace and inadvertently undermine COIN objectives. Treated fairly and respectfully, HN employees can provide good insights into the local language, culture, and perceptions of COIN activities as well as other issues affecting communities in the AO.

Host-Nation Contracting: A Potential Double-Edged Sword

Early in Operation Iraqi Freedom, a brigade from the 101st Airborne Division was assigned a large area of operations near Tal Afar, in northern Iraq. The terrain the unit was required to cover and support exceeded the distribution capabilities of its ground transportation assets. Logistic officers supporting the brigade sought out and found a local business leader with a family-owned transportation company. He was positive towards U.S. aims for improving Iraq and willing to work with U.S. forces by providing various truck and bus services.

After two months of ad hoc daily arrangements for services at the U.S. forces' compound entry point, the unit established a six-month contract to make this transportation support more regular. As the working relationship became more solid, the contractor and his employees also furnished insights into the effectiveness of U.S. information operations as well as information on the presence and activities of suspicious persons possibly affiliated with the insurgency. The arrangement worked exceptionally well, effectively supported counterinsurgency activities, and maintained peace and security—as long as the original unit that established the services was stationed in the area.

Eventually, a smaller task force replaced the first unit that established the contract, and the security situation in the area began to deteriorate. Upon detecting this change in security posture, insurgents quickly found the contractor and killed him. No doubt their intent was to degrade the U.S. forces' logistic posture and to send the message to other local vendors that doing business with the Americans was costly. Eventually the contactor's brother took over operations, but understandably support deteriorated.

When setting up logistic contracting arrangements with HN contractors in a COIN environment, U.S. logisticians and contracting officers must remember the grave risks people take by accepting these jobs. Insurgents are exceptionally adept at finding ways to attack logistics. When insurgents attack people branded as traitors, there is an added terror or political message benefit. Inadequate or shifting U.S. security arrangements can provide openings for insurgents to more easily attack host-nation contractors and logistic providers.

External Support Contractors

8-56. Many of the same considerations that apply to theater contractors apply to external support contractors and their subcontracted employees, particularly if they are local or in-theater hires who are not U.S. or multinational partner citizens.

System Contractors

8-57. System contractors generally work on technologically complicated military systems, such as vehicles, weapons systems, aircraft, and information systems. This support is provided under contracts negotiated by program executive officers and program managers. These contractors provide systems expertise. Most are U.S. citizens, and many of them are former U.S. military members. These contractors generally meet deployment and security requirements similar to DOD civilian requirements.

Contingency Contracting Officers and Other Agents

8-58. In COIN operations, the timely and well-placed distribution of funds at the local level can serve as an invaluable force multiplier. Many challenges to accomplishing payments and purchases in the COIN environment exist. These challenges can include—

- Problems with the security of financial institutions, agents, and instruments.
- Potential for sudden volatility in the HN economy.
- Reliability issues with local supplier and vendors.
- Peculiarities of local business cultures.

8-59. Contracting officers and other agents authorized to make payments to support COIN activities often find it difficult to obtain reliable information upon which to make decisions and conduct negotiations. Military means to accomplish this type of purchasing are found at two levels:

- Contingency contracting officer, who acts on unit-generated purchasing requests and committals.
- Ordering officer for smaller purchases.

8-60. Legal requirements keep U.S. funds for different purposes separate and distinct. (See paragraphs D-27 through D-31.) COIN units must maintain two types of purchasing officer teams:

- Field ordering officer teams for Title 10 funds.
- Project ordering officer teams for Title 22 funds.

(Figure 8-2 [page 8-20] portrays the relationships and roles among different contracting and ordering officers, and the types of funds that they manage.)

Contingency Contracting

8-61. Because contingency contracting officers can set contracts for larger sums than ordering officers can, they normally place purchasing request and committal requirements out for bid to local vendors. During COIN operations, contacting officers must spread contracts across different vendors to forestall any appearance of partiality.

Field Ordering and Project Ordering Officer Teams

8-62. Both field ordering and project ordering officer teams consist of the respective contracting officer agent, a paying agent, and a security detail. Both teams operate under similar regulatory constraints. These officers' duties differ not only with respect to the type of funds they disburse but also in the increment caps applied. Field ordering officers with Title 10 funds are limited to individual payments not to exceed $2,500. Project ordering officers can make individual contract payments of up to $20,000, since projects are associated with higher costs and scales. In both cases, these teams are an invaluable asset for reaching into HN communities and promoting economic pluralism while assessing the economic effects of purchasing activities and economic stability initiatives.

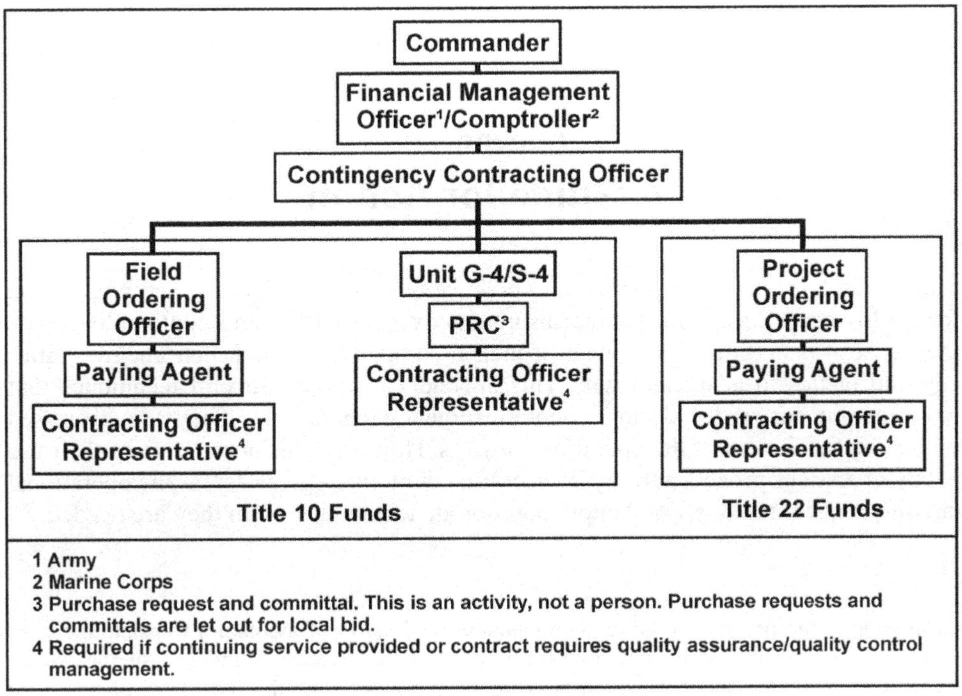

Figure 8-2. Tactical financial management organizations

8-63. Due to their activities throughout HN communities and their cultivation of local business connections, field ordering and project ordering officer teams should be viewed as potential means of gathering and distributing information. Vendor observations and actions may also reveal much about the real status of the COIN effort. Vendors may notice commonplace events that, taken together, are significant. For example, vendors may notice outsiders moving into the area, contractors failing to deliver goods, and HN employees asking to leave early. Together these events might indicate an impending attack. By doing business with counterinsurgents, local contractors and vendors may put themselves at great risk. Protection of their activities can pose a great challenge to U.S. forces and must be seriously considered when doing business with them.

SUMMARY

8-64. Logistic activities are an integral part of COIN operations. These activities take the traditional form of support to combat and security forces as well as the unconventional form of providing mixes of essential and timely support to many HN security and stability-enhancing activities that may seem purely civil in character. Initially, uniformed military logistic providers may have to provide this support. However, COIN logistic objectives should include encouraging and promoting HN providers as soon as security conditions make this feasible. This transition is a delicate one. Logistic providers must constantly determine whether their practices are contributing to achieving the end state and adjust their methods if necessary. If there is a final paradox in counterinsurgency, it is that logistic postures and practices are a major part of the effort and may well determine the operation's success.

Appendix A

A Guide for Action

Translating lessons of this manual into practice begins with planning and preparation for deployment. Successful counterinsurgents execute wisely and continually assess their area of operations, the impact of their own operations, and their enemy's strategy and tactics to adapt and win. This appendix discusses several techniques that have proven successful during counterinsurgency operations. They are discussed within the framework of the operations process. However, this does not limit their use to any operations process activity. Successful counterinsurgents assess the operational environment continuously and apply appropriate techniques when they are needed.

PLAN

A-1. *Planning* is the process by which commanders (and staffs, if available) translate the commander's visualization into a specific course of action for preparation and execution, focusing on the expected results (FMI 5-0.1). Planning for counterinsurgency (COIN) operations is no different from planning for conventional operations. However, effective COIN planning requires paying at least as much attention to aspects of the environment as to the enemy force.

Assess During Planning: Perform Mission Analysis

A-2. Learn about the people, topography, economy, history, religion, and culture of the area of operations (AOs). Know every village, road, field, population group, tribal leader, and ancient grievance. Become the expert on these topics. If the precise destination is unknown, study the general area. Focus on the precise destination when it is determined. Ensure leaders and staffs use the Secret Internet Protocol Router Network (SIPRNET) to immerse themselves virtually in the AO into which the unit is deploying. Understand factors in adjacent AOs and the information environment that can influence AOs. These can be many, particularly when insurgents draw on global grievances.

A-3. Read the map like a book. Study it every night before sleep and redraw it from memory every morning. Do this until its patterns become second nature. Develop a mental model of the AO. Use it as a frame into which to fit every new piece of knowledge.

A-4. Study handover notes from predecessors. Better still, get in touch with personnel from the unit in theater and pick their brains. In an ideal world, intelligence officers and area experts provide briefings. This may not occur. Even if it does, there is no substitute for personal mastery.

A-5. Require each subordinate leader, including noncommissioned officers, to specialize on some aspect of the AO and brief the others.

Analyze the Problem

A-6. Mastery of the AO provides a foundation for analyzing the problem. Who are the insurgents? What drives them? What makes local leaders tick? An insurgency is basically a competition among many groups, each seeking to mobilize the populace in support of its agenda. Thus, COIN operations are always more than two sided.

A-7. Understand what motivates the people and how to mobilize them. Knowing why and how the insurgents are getting followers is essential. This requires knowing the real enemy, not a cardboard cutout. Insurgents are adaptive, resourceful, and probably from the area. The local populace has known them since they were young. U.S. forces are the outsiders. The worst opponents are not the psychopathic terrorists of the movies; rather, they are charismatic warriors who would excel in any armed force. Insurgents are not necessarily misled or naive. Much of their success may stem from bad government policies or security forces that alienate the local populace.

A-8. Work the problem collectively with subordinate leaders. Discuss ideas and explore possible solutions. Once leaders understand the situation, seek a consensus on how to address it. If this sounds unmilitary, get over it. Such discussions help subordinates understand the commander's intent. Once in theater, situations requiring immediate action will arise too quickly for orders. Subordinates will need to exercise subordinates' initiative and act based on the commander's intent informed by whatever knowledge they have developed. Corporals and privates will have to make quick decisions that may result in actions with strategic implications. Such circumstances require a shared situational understanding. They also require a command climate that encourages subordinates to assess the situation, act on it, and accept responsibility for their actions. Employing mission command is essential in this environment. (Mission command, subordinates' initiative and commander's intent are defined in the glossary. See FM 6-0, paragraphs 1-67 through 1-80, 2-83 through 2-92, and 4-26 through 4-31 for discussions of the principles involved.)

PREPARE

A-9. *Preparation* consists of activities by the unit before execution to improve its ability to conduct the operation, including, but not limited to, the following: plan refinement, rehearsals, reconnaissance, coordination, inspection, and movement (FM 3-0). Compared with conventional operations, preparing for COIN operations requires greater emphasis on organizing for intelligence and for working with nonmilitary organizations. These operations also require more emphasis on preparing small-unit leaders for increased responsibility and maintaining flexibility.

ORGANIZE FOR INTELLIGENCE

A-10. Intelligence and operations are always complementary, especially in COIN operations. COIN operations are intelligence driven, and units often develop much of their own intelligence. Commanders must organize their assets to do that.

A-11. Each company may require an intelligence section, including analysts and an individual designated as the "S-2." Platoon leaders may also have to designate individuals to perform intelligence and operations functions. A reconnaissance and surveillance element is also essential. Augmentation for these positions is normally not available, but companies still must perform the tasks. Put the smartest Soldiers and Marines in the intelligence section and the reconnaissance and surveillance element. This placement results in one less rifle squad, but an intelligence section pays for itself in lives and effort saved.

A-12. There are never enough linguists. Commanders consider with care where best to use them. Linguists are a battle-winning asset, but like any other scarce resource, commanders must allocate them carefully. During predeployment, the best use of linguists may be to train Soldiers and Marines in basic language skills.

ORGANIZE FOR INTERAGENCY OPERATIONS

A-13. Almost everything in COIN is interagency. Everything from policing to intelligence to civil-military operations (CMO) to trash collection involves working with interagency and host-nation (HN) partners. These agencies are not under military control, but their success is essential to accomplishing the mission. Train Soldiers and Marines in conducting interagency operations. Get a briefing from the Department of State, aid agencies, and the local police or fire departments. Designate interagency subject matter experts in each subordinate element and train them. Look at the situation through the eyes of a civilian who knows nothing about the military. Many civilians find rifles, helmets, and body armor intimidating. Learn how not to scare them. Seek advice from those who come from that nation or culture. Most importantly, know that military operations create temporary breathing space. But to prevail, civilian agencies need long-term development and stabilization.

Travel Light and Harden Your Sustainment Assets

A-14. A normal combat load for Soldiers and Marines includes body armor, rations, extra ammunition, communications gear, and many other things—all of which are heavy. Insurgents may carry a rifle or rocket-propelled grenade, a headscarf, and a water bottle. Without the extra weight, insurgents can run and maneuver easily. U.S. forces must lighten their combat loads and enforce a habit of speed and mobility. Otherwise, insurgents consistently outrun and outmaneuver them. However, make sure Soldiers and Marines can always reach back for fires or other support.

A-15. Remember to harden sustainment bases. Insurgents often consider them weak points and attack there. Most attacks on coalition forces in Iraq in 2004 and 2005, other than combat actions, were against sustainment installations and convoys. Ensure sustainment assets are hardened and have communications. Make sure to prepare Soldiers and Marines whose primary task is providing logistic support to fight ground combat operations. While executing their sustaining operations, they may do more fighting than some rifle squads.

Find a Political and Cultural Advisor

A-16. A force optimized for COIN operations would have political and cultural advisors at company level. The current force structure gives corps and division commanders a political advisor. Lower echelon commanders must improvise. They select a political and cultural advisor from among their troops. This person may be a commissioned officer, but may not. The position requires someone with "people skills" and a feel for the environment. Commanders should not try to be their own cultural advisor. They must be fully aware of the political and cultural dimension, but this is a different role. In addition, this position is not suitable for intelligence professionals. They can help, but their task is to understand the environment. The political advisor's job is to help shape the environment.

Train the Squad Leaders—Then Trust Them

A-17. Squads and platoons execute mostly COIN operations. Small-unit actions in a COIN environment often have more impact than similar actions during major combat operations. Engagements are often won or lost in moments; whoever can bring combat power to bear in seconds wins. The on-scene leader controls the fight. This situation requires mission command and subordinates' initiative. Train leaders at the lowest echelons to act intelligently and independently.

A-18. Training should focus on basic skills: marksmanship, patrolling, security on the move and at the halt, and basic drills. When in doubt, spend less time on company and platoon training and more time on squads. Ruthlessly replace ineffective leaders. Once trained, give Soldiers and Marines a clear commander's intent and trust them to exercise subordinates' initiative within it. This allows subordinates to execute COIN operations at the level at which they are won.

Identify and Use Talent

A-19. Not everyone is good at counterinsurgency. Many leaders do not understand it, and some who do cannot execute it. COIN operations are difficult and anyone can learn the basics. However, people able to intuitively grasp, master, and execute COIN techniques are rare. Learn how to spot these people and put them into positions where they can make a difference. Rank may not indicate the required talent. In COIN operations, a few good Soldiers and Marines under a smart junior noncommissioned officer doing the right things can succeed, while a large force doing the wrong things will fail.

Continue to Assess and Plan During Preparation: Be Flexible

A-20. *Commander's visualization* is the mental process of developing situational understanding, determining a desired end state, and envisioning how the force will achieve that end state (FMI 5-0.1). It begins with mission receipt and continues throughout any operation. The commander's visualization forms the basis for conducting (planning, preparing for, executing, and assessing) an operation.

A-21. Commanders continually refine their visualization based on their assessment of the operational environment. They describe and direct any changes they want made as the changes are needed. They do not wait for a set point in any process. This flexibility is essential during preparation for COIN operations. Some are tempted to try to finalize a plan too early. They then prepare to execute the plan rather than what changes in the operational environment require. However, as commanders gain knowledge, their situational understanding improves. They get a better idea of what to do and of their own limitations. This lets them refine their visualization and direct changes to the plan and their preparations. Even with this, any plan will change once operations begin. If there is a major shift in the environment, commanders may need to scrap the plan. However, a plan is still needed. Developing it gives leaders a simple robust idea of what to achieve, even if the methods change. Directing changes to it based on continuous assessment is one aspect of the art of command.

A-22. One planning approach is to identify phases of the operation in terms of major objectives to achieve such as establishing dominance, building local networks, and marginalizing the enemy. Make sure forces can easily transition between phases, both forward to exploit successes and backward to recover from setbacks. Insurgents can adapt their activity to friendly tactics. The plan must be simple enough to survive setbacks without collapsing. This plan is the solution that began with the shared analysis and consensus that began preparation. It must be simple and known to everyone.

EXECUTE

A-23. *Execute* means to put a plan into action by applying combat power to accomplish the mission and using situational understanding to assess progress and make execution and adjustment decisions (FM 6-0). The execution of COIN operations demands all the skills required to execute conventional operations. In addition, it also requires mastery of building alliances and personal relationships, attention to the local and global media, and additional skills that are not as heavily tasked in conventional operations.

ESTABLISH AND MAINTAIN A PRESENCE

A-24. The first rule of COIN operations is to establish the force's presence in the AO. If Soldiers and Marines are not present when an incident happens, they usually cannot do much about it. The force cannot be everywhere at once. The more time Soldiers and Marines spend in the AO, the more likely they are where the action is. If the force is not large enough to establish a presence throughout the AO, then determine the most important places and focus on them. This requires living in the AO close to the populace. Raiding from remote, secure bases does not work. Movement on foot, sleeping in villages, and night patrolling all seem more dangerous than they are—and they are what ground forces are trained to do. Being on the ground establishes links with the local people. They begin to see Soldiers and Marines as real people they can trust and do business with, rather than as aliens who descended from armored boxes. Driving around in an armored convoy actually degrades situational awareness. It makes Soldiers and Marines targets and is ultimately more dangerous than moving on foot and remaining close to the populace.

ASSESS DURING EXECUTION: AVOID HASTY ACTIONS

A-25. Do not act rashly; get the facts first. Continuous assessment, important during all operations, is vital during COIN operations. Violence can indicate several things. It may be part of the insurgent strategy, interest groups fighting among themselves, or individuals settling vendettas. Or, it may just be daily life. Take the time to learn what normalcy looks like. Insurgents may try to goad Soldiers and Marines into lashing out at the local populace or making a similar mistake. Unless leaders are on the spot when an incident occurs, they receive only second-hand reports and may misunderstand the local context or interpretation. This means that first impressions are often highly misleading, particularly in urban areas. Of course, leaders cannot avoid making judgments. When there is time, ask an older hand or trusted local people for their opinions. If possible, keep one or two officers from your predecessor unit for the first part of the tour. Avoid rushing to judgment.

Build Trusted Networks

A-26. Once the unit settles into the AO, its next task is to build trusted networks. This is the true meaning of the phrase "hearts and minds," which comprises two separate components. "Hearts" means persuading people that their best interests are served by COIN success. "Minds" means convincing them that the force can protect them and that resisting it is pointless. Note that neither concerns whether people like Soldiers and Marines. Calculated self-interest, not emotion, is what counts. Over time, successful trusted networks grow like roots into the populace. They displace enemy networks, which forces enemies into the open, letting military forces seize the initiative and destroy the insurgents.

A-27. Trusted networks are diverse. They include local allies, community leaders, and local security forces. Networks should also include nongovernmental organizations (NGOs), other friendly or neutral nonstate actors in the AO, and the media.

A-28. Building trusted networks begins with conducting village and neighborhood surveys to identify community needs. Then follow through to meet them, build common interests, and mobilize popular support. This is the true main effort; everything else is secondary. Actions that help build trusted networks support the COIN effort. Actions that undermine trust or disrupt these networks—even those that provide a short-term military advantage—help the enemy.

Go with the Grain and Seek Early Victories

A-29. Do not try to crack the hardest nut first. Do not go straight for the main insurgent stronghold or try to take on villages that support insurgents. Instead, start from secure areas and work gradually outwards. Extend influence through the local people's networks. Go with, not against, the grain of the local populace. First, win the confidence of a few villages, and then work with those with whom they trade, intermarry, or do business. This tactic develops local allies, a mobilized populace, and trusted networks.

A-30. Seek a victory early in the operation to demonstrate dominance of the AO. This may not be a combat victory. Early combat without an accurate situational understanding may create unnecessary collateral damage and ill will. Instead, victories may involve resolving a long-standing issue or co-opting a key local leader. Achieving even a small early victory can set the tone for the tour and help commanders seize the initiative.

Practice Deterrent Patrolling

A-31. Establish patrolling tactics that deter enemy attacks. An approach using combat patrols to provoke, then defeat, enemy attacks is counterproductive. It leads to a raiding mindset, or worse, a bunker mentality. Deterrent patrolling is a better approach. It keeps the enemy off balance and the local populace reassured. Constant, unpredictable activity over time deters attacks and creates a more secure environment. Accomplishing this requires one- to two-thirds of the force to be on patrol at any time, day or night.

Be Prepared for Setbacks

A-32. Setbacks are normal in counterinsurgencies, as in all operations. Leaders make mistakes and lose people. Soldiers and Marines occasionally kill or detain the wrong person. It may not be possible to build or expand trusted networks. If this happens, drop back to the previous phase of the plan, recover, and resume operations. It is normal in company-level COIN operations for some platoons to do well while others do badly. This situation is not necessarily evidence of failure. Give subordinate leaders the freedom to adjust their posture to local conditions. This creates flexibility that helps survive setbacks.

Remember the Global Audience

A-33. The omnipresence and global reach of today's news media affects the conduct of military operations more than ever before. Satellite receivers are common, even in developing countries. Bloggers and print, radio, and television reporters monitor and comment on everything military forces do. Insurgents use terrorist tactics to produce graphics that they hope will influence public opinion—both locally and globally.

A-34. Train Soldiers and Marines to consider how the global audience might perceive their actions. Soldiers and Marines should assume that the media will publicize everything they say or do. Also, treat the media as an ally. Help reporters get their story. That helps them portray military actions favorably. Trade information with media representatives. Good relationships with nonembedded media, especially HN media, can dramatically increase situational awareness.

Engage the Women; Be Cautious Around the Children

A-35. Most insurgent fighters are men. However, in traditional societies, women are hugely influential in forming the social networks that insurgents use for support. When women support COIN efforts, families support COIN efforts. Getting the support of families is a big step toward mobilizing the local populace against the insurgency. Co-opting neutral or friendly women through targeted social and economic programs builds networks of enlightened self-interest that eventually undermine insurgents. Female counterinsurgents, including interagency people, are required to do this effectively.

A-36. Conversely, be cautious about allowing Soldiers and Marines to fraternize with local children. Homesick troops want to drop their guard with kids. But insurgents are watching. They notice any friendships between troops and children. They may either harm the children as punishment or use them as agents. It requires discipline to keep the children at arm's length while maintaining the empathy needed to win local support.

Assess During Execution

A-37. Develop measures of effectiveness early and continuously refine them as the operation progresses. These measures should cover a range of social, informational, military, and economic issues. Use them to develop an in-depth operational picture. See how the operation is changing, not just that it is starting or ending. Typical measures of effectiveness include the following:

- Percentage of engagements initiated by friendly forces versus those initiated by insurgents.
- Longevity of friendly local leaders in positions of authority.
- Number and quality of tips on insurgent activity that originate spontaneously.
- Economic activity at markets and shops.

These mean virtually nothing as a snapshot; trends over time indicate the true progress.

A-38. Avoid using body counts as a measure of effectiveness. They actually measure very little and may provide misleading numbers. Using body counts to measure effectiveness accurately requires answers to the following questions:

- How many insurgents were there at the start?
- How many insurgents have moved into the area?
- How many insurgents have transferred from supporter to combatant status?
- How many new fighters has the conflict created?

Accurate information of this sort is usually not available.

Maintain Mission Focus Throughout

A-39. Once a unit is established in its AO, Soldiers and Marines settle into a routine. A routine is good as long as the mission is being accomplished. However, leaders should be alert for the complacency that often accompanies routines.

A-40. It often takes Soldiers and Marines at least one-third of the tour to become effective. Toward the tour's end, leaders struggle against the "short-timer" mentality. Thus, the middle part of the tour is often the most productive. However, leaders must work to keep Soldiers and Marines focused on the mission and attentive to the environment.

Exploit a Single Narrative

A-41. Since counterinsurgency is a competition to mobilize popular support, it pays to know how people are mobilized. Most societies include opinion-makers—local leaders, religious figures, media personalities, and others who set trends and influence public perceptions. This influence often follows a single narrative—a simple, unifying, easily expressed story or explanation that organizes people's experience—and provides a framework for understanding events. Nationalist and ethnic historical myths and sectarian creeds are examples of such narratives. Insurgents often try to use the local narrative to support their cause. Undercutting their influence requires exploiting an alternative narrative. An even better approach is tapping into an existing narrative that excludes insurgents.

A-42. Higher headquarters usually establishes the COIN narrative. However, only leaders, Soldiers, and Marines at the lowest levels know the details needed to tailor it to local conditions and generate leverage from it. For example, a nationalist narrative can be used to marginalize foreign fighters. A narrative of national redemption can undermine former regime elements seeking to regain power. Company-level leaders apply the narrative gradually. They get to know local opinion makers, win their trust, and learn what motivates them. Then they build on this knowledge to find a single narrative that emphasizes the inevitability and rightness of the COIN operation's success. This is art, not science.

Have Local Forces Mirror the Enemy, Not U.S. Forces

A-43. By mid-tour, U.S. forces should be working closely with local forces, training or supporting them and building an indigenous security capability. The natural tendency is to create forces in a U.S. image. This is a mistake. Instead, local HN forces need to mirror the enemy's capabilities and seek to supplant the insurgent's role. This does not mean they should be irregular in the sense of being brutal or outside proper control. Rather, they should move, equip, and organize like insurgents but have access to U.S. support and be under the firm control of their parent societies. Combined with a mobilized populace and trusted networks, these characteristics allow HN forces to separate the insurgents from the population.

A-44. U.S. forces should support HN forces. At the company level, this means raising, training, and employing local HN auxiliary forces (police and military). These tasks require high-level clearance, but if permission is given, companies should each establish a training cell. Platoons should aim to train one local squad and then use that squad as a nucleus for a partner platoon. The company headquarters should train an HN leadership team. This process mirrors the development of trusted networks. It tends to emerge naturally with the emergence of local allies willing to take up arms to defend themselves.

Conduct Civil-Military Operations

A-45. COIN operations can be characterized as armed social work. It includes attempts to redress basic social and political problems while being shot at. This makes CMO a central COIN activity, not an afterthought. Civil-military operations are one means of restructuring the environment to displace the enemy from it. They must focus on meeting basic needs first. A series of village or neighborhood surveys, regularly updated, are invaluable to understanding what the populace needs and tracking progress in meeting them.

A-46. Effective CMO require close cooperation with national, international, and local interagency partners. These partners are not under military control. Many NGOs, for example, do not want to be too closely associated with military forces because they need to preserve their perceived neutrality. Interagency cooperation may involve a shared analysis of the problem, building a consensus that allows synchronization of military and interagency efforts. The military's role is to provide protection, identify needs, facilitate CMO, and use improvements in social conditions as leverage to build networks and mobilize the populace.

A-47. There is no such thing as impartial humanitarian assistance or CMO in COIN. Whenever someone is helped, someone else is hurt, not least the insurgents. So civil and humanitarian assistance personnel often become targets. Protecting them is a matter not only of providing a close-in defense, but also of creating a secure environment by co-opting local beneficiaries of aid and their leaders.

Remember Small is Beautiful

A-48. Another tendency is to attempt large-scale, mass programs. In particular, Soldiers and Marines tend to apply ideas that succeed in one area to another area. They also try to take successful small programs and replicate them on a larger scale. This usually does not work. Often small-scale programs succeed because of local conditions or because their size kept them below the enemy's notice and helped them flourish unharmed. Company-level programs that succeed in one AO often succeed in another; however, small-scale projects rarely proceed smoothly into large programs. Keep programs small. This makes them cheap, sustainable, low-key, and (importantly) recoverable if they fail. Leaders can add new programs—also small, cheap, and tailored to local conditions—as the situation allows.

Fight the Enemy's Strategy

A-49. When COIN efforts succeed, insurgents often transition to the offensive. COIN successes create a situation that threatens to separate insurgents from the populace. Insurgents attack military forces and the local populace to reassert their presence and continue the insurgency. This activity does not necessarily indicate an error in COIN tactics (though it may, depending on whether insurgents successfully mobilized the population). It is normal, even in the most successful operations, to have spikes of offensive insurgent activity.

A-50. The obvious military response is a counteroffensive to destroy enemy's forces. This is rarely the best choice at company level. Only attack insurgents when they get in the way. Try not to be distracted or forced into a series of reactive moves by a desire to kill or capture them. Provoking combat usually plays into the enemy's hands by undermining the population's confidence. Instead, attack the enemy's strategy. If insurgents are seeking to recapture a community's allegiance, co-opt that group against them. If they are trying to provoke a sectarian conflict, transition to peace enforcement operations. The possible situations are endless, but the same principle governs the response: fight the enemy's strategy, not enemy forces.

Assess During Execution: Recognize and Exploit Success

A-51. Implement the plan developed early in the campaign and refined through interaction with local partners. Focus on the environment, not the enemy. Aim at dominating the whole district and implementing solutions to its systemic problems. Continuously assess results and adjust as needed.

A-52. Achieving success means that, particularly late in the campaign, it may be necessary to negotiate with the enemy. Local people supporting the COIN operation know the enemy's leaders. They even may have grown up together. Valid negotiating partners sometimes emerge as the campaign progresses. Again, use close interagency relationships to exploit opportunities to co-opt segments of the enemy. This helps wind down the insurgency without alienating potential local allies who have relatives or friends among insurgents. As an insurgency ends, a defection is better than a surrender, a surrender better than a capture, and a capture better than a kill.

Prepare During Execution: Get Ready for Handover from Day One

A-53. It is unlikely the insurgency will end during a troop's tour. There will be a relief in place, and the relieving unit will need as much knowledge as can be passed to them. Start handover folders in every platoon and specialist squad immediately upon arrival, if they are not available from the unit being relieved. The folders should include lessons learned, details about the populace, village and patrol reports, updated maps, and photographs—anything that will help newcomers master the environment. Computerized databases are fine. Keep good back-ups and ensure a hard copy of key artifacts and documents exists. Developing and keeping this information current is boring, tedious work. But it is essential to both short- and long-term success. The corporate memory this develops gives Soldiers and Marines the knowledge they need to stay alive. Passing it on to the relieving unit does the same for them. It also reduces the loss of momentum that occurs during any handover.

ENDING THE TOUR

A-54. As the end of the tour approaches, the key leadership challenge becomes keeping the Soldiers and Marines focused. They must not drop their guard. They must continue to monitor and execute the many programs, projects, and operations.

A-55. The tactics discussed above remain applicable as the end-of-tour transition approaches. However, there is an important new one: keep the transition plan secret. The temptation to talk about home becomes almost unbearable toward the end of a tour. The local people know that Soldiers and Marines are leaving and probably have a good idea of the generic transition plan. They have seen units come and go. But details of the transition plan must be protected; otherwise, the enemy might use the handover to undermine any progress made during the tour. Insurgents may stage a high-profile attack. They may try to recapture the populace's allegiance by scare tactics. Insurgents will try to convince the local populace that military forces will not protect them after the transition. Insurgents may try to persuade the local populace that the successor unit will be oppressive or incompetent. Set the follow-on unit up for success. Keep the transition plan details secret within a tightly controlled compartment in the headquarters. Tell the Soldiers and Marines to resist the temptation to say goodbye to local allies. They can always send a postcard from home.

THREE "WHAT IFS"

A-56. The discussion above describes what should happen, but things do go wrong. Here are some "what ifs" to consider.

WHAT IF YOU GET MOVED TO A DIFFERENT AREA?

A-57. Efforts made preparing for operations in one AO are not wasted if a unit is moved to another area. In mastering the first area, Soldiers and Marines learned techniques applicable to the new one. For example, they know how to analyze an AO and decide what matters in the local society. The experience provides a mental structure for analyzing the new AO. Soldiers and Marines can focus on what is different, making the process easier and faster. They need to apply this same skill when they are moved within battalion or brigade AOs.

WHAT IF YOU HAVE NO RESOURCES?

A-58. Things can be things done in a low-priority AO. However, commanders need to focus on self-reliance, keeping things small and sustainable and ruthlessly prioritizing efforts. Local leaders can help. They know what matters to them. Commanders should be honest with them, discuss possible projects and options, and ask them to recommend priorities. Often commanders can find translators, building supplies, or expertise. They may only expect support and protection in making their projects work. Negotiation and consultation can help mobilize their support and strengthen social cohesion. Setting achievable goals is key to making the situation work.

WHAT IF THE THEATER SITUATION SHIFTS?

A-59. Sometimes everything goes well at the tactical level, but the theater situation changes and invalidates those efforts. When that happens, drop back a stage, consolidate, regain balance, and prepare to expand again when the situation allows. A flexible, adaptive plan helps in such situations. Friendly forces may have to cede the initiative for a time; however, they must regain it as soon as the situation allows.

SUMMARY

A-60. This appendix has summarized one set of tactics for conducting COIN operations. Like all tactics, they need interpretation. Constant study of the AO is needed to apply them to the specific circumstances a unit faces. Observations and experience helps Soldiers and Marines apply them better. Whatever else is done, the focus must remain on gaining and maintaining the support of the population. With their support, victory is assured; without it, COIN efforts cannot succeed.

This page intentionally left blank.

Appendix B

Social Network Analysis and Other Analytical Tools

Situational understanding involves determining the relationships among the factors of METT-TC. This appendix discusses several tools used to describe the effects of the operational environment and evaluate the threat. One of the most important of these is social network analysis, a powerful threat evaluation tool. Commanders and staffs use these tools to help them understand the operational environment. This understanding facilitates making decisions and developing plans and orders.

DESCRIBE THE EFFECTS OF THE OPERATIONAL ENVIRONMENT

B-1. Describing the effects of the operational environment requires an analysis of the terrain, weather, and civil considerations. This discussion addresses terrain and civil considerations at length because of their importance in a counterinsurgency (COIN) environment. Terrain and civil considerations have distinct aspects in COIN that analysts must understand to effectively describe the operational environment. Imagery, geospatial analysis tools, overlays, and graphics can help depict these aspects.

DESCRIBE TERRAIN EFFECTS

B-2. As in conventional operations, terrain analysis in COIN includes examining the terrain's effects on the movement of military units and enemy personnel. However, because COIN focuses on people, terrain analysis usually centers on populated areas and the effects of terrain on the local populace. During COIN operations, Soldiers and Marines spend a lot of time in suburban and urban areas interacting with the populace. This battlefield is three dimensional. Multistory buildings and underground lines of communications, such as tunnels and sewers, can be very important. Insurgents also use complex natural terrain to their advantage as well. Mountains, caves, jungles, forests, swamps, and other complex terrain are potential bases of operations for insurgents. (See FM 34-130 for additional information on terrain analysis. See FMI 2-91.4 for terrain analysis in urban operations.)

B-3. An important terrain consideration in COIN is urban and suburban land navigation. This can be difficult in areas without an address system and in cities where 10-digit grids may not be accurate enough to locate a specific apartment. Knowledge of how local people find one another's houses and what type of address system they use are beneficial. Recent, accurate maps that use overhead imagery are also helpful. In addition, tourist maps and locally produced maps facilitate understanding the names local people use to describe places.

Military Aspects of Terrain for Counterinsurgency

B-4. At the tactical level, Soldiers and Marines consider different details of the military aspects of terrain to describe the operational environment.

- **Observation and fields of fire**. In COIN operations, Soldiers and Marines look for areas with good fields of fire that may serve as ambush points. In addition, they also consider different ways insurgents might observe them. These ways include surveillance, the use of spies and infiltrators, and locations with line of sight on counterinsurgent positions.
- **Avenues of approach**. Insurgents use any means possible to get into counterinsurgent installations. Possible entry points include sewers, rooftops, roads, and sidewalks. Insurgents exploit their ability to blend with the populace. They may try to infiltrate by posing as contractors working for counterinsurgents or the host-nation (HN) government. Along border regions, insurgents

may use smuggling routes to move people and materiel in and out of the country. An additional avenue of approach to consider is how insurgents influence public opinion.

- **Key terrain**. Tactically, key terrain may be important structures, economically and politically important areas, areas of religious or cultural significance, access control points, and lines of communications.
- **Obstacles**. In addition to terrain obstacles, obstacles in a COIN environment include anything that hinders insurgent freedom of operation or counterinsurgent operations. Traffic control points, electronic security systems, and guard plans are examples of obstacles to insurgents. Use of places protected under rules of engagement, translators, the ability to communicate with the populace, culture, and politics may all be obstacles for U.S. and HN government forces.
- **Cover and concealment**. In COIN, cover and concealment includes the means by which insurgents hide themselves and their activities. These include using disguises and false identification badges, and hiding supplies underground or in buildings.

Geospatial Intelligence

B-5. Geospatial intelligence (GEOINT) is the intelligence derived from the exploitation, analysis, and fusion of imagery with geospatial information to describe, assess, and visually depict physical features and geographically referenced activities in an area of operations (AO). GEOINT consists of imagery, imagery intelligence, and geospatial information. Geospatial information and services remains a core mission of the engineer branch and provides the foundation for GEOINT. Imagery intelligence remains a core mission of the military intelligence branch and provides the intelligence layers and analytic fusion for GEOINT. The result is digitally integrated intelligence products that support all-source analysis, planning, decision making and support to current operations.

Geospatial Tools

B-6. Geospatial products (tools) that can be provided by the geospatial information and services team include the following:

- Terrain databases.
- Special terrain studies and products prepared by U.S. or HN agencies, and special maps, charts, and geodetic studies.
- Current photography.
- Real-time terrain reconnaissance.
- Terrain factor matrices.

Imagery

B-7. Imagery products include both aerial photography and satellite imagery. In many cases, aerial reconnaissance platforms, such as unmanned aircraft systems, respond directly to commanders. This practice aids timely, focused data collection. Each collection system has its own capabilities. The situation determines whether black and white or infrared imagery offers the better view of a target. (Figure B-1 shows an example of an imagery product.)

The objective is a cinder block, three-story apartment building located in the town of Braggville. Braggville is a close-order, block-pattern village of approximately 800-1,200 people. The town is located in a medium wooded area with numerous roads (dirt and paved), trails, and firebreaks through it.

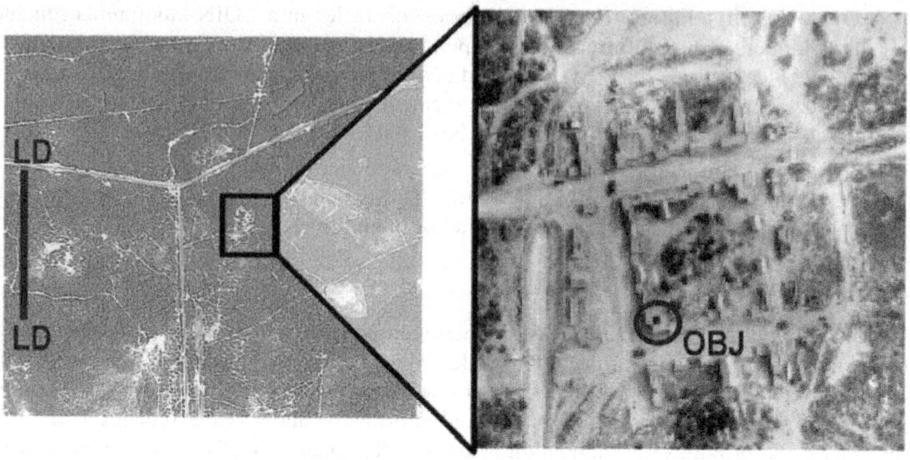

Figure B-1. Example imagery photograph

B-8. A key element in future operations may be the imagery downlink capabilities of space-based intelligence collection platforms. Space-based systems use state-of-the-art spectral, infrared, electro-optical, and synthetic aperture radar imaging. They can provide important information. Data collected from such sources is transferred in a digital format that can be manipulated to address specific requirements. Intelligence staffs remain aware of the capabilities and limitations of these systems and the procedures for requesting this support.

B-9. Advanced GEOINT products are produced using any combination of imaging platforms—visible, infrared, radar, or spectral—depending on requestor needs. These products have many applications. Presenting imagery in an oblique perspective by combining it with digital terrain elevation data provides a perspective view. Spectral imagery uses heat distribution patterns and changes in a scene imaged at various times to discover and distinguish manmade from indigenous activity. Other uses include facility analysis, structural analysis, target detection, soil analysis, and damage assessment.

DESCRIBE CIVIL CONSIDERATIONS (ASCOPE)

B-10. *Civil considerations* concern the manmade infrastructure, civilian institutions, and attitudes and activities of the civilian leaders, populations, and organizations within an area of operations influence the conduct of military operations (FM 6-0). Because the purpose of COIN is to support a HN government in gaining legitimacy and the support of the populace, civil considerations are often the most important factors to consider during mission analysis.

B-11. Civil considerations generally focus on the immediate impact of civilians on operations in progress. However, at higher levels, they also include larger, long-term diplomatic, informational, and economic issues. At the tactical level, civil considerations directly relate to key civilian areas, structures, capabilities, organizations, people, and events within the AO. These characteristics are represented by the memory aid ASCOPE. Socio-cultural factors analysis, discussed in paragraphs 3-19 through 3-73, provides a more in-depth evaluation of civil considerations.

Areas

B-12. Key civilian areas are localities or aspects of the terrain within an AO that have significance to the local populace. This characteristic approaches terrain analysis from a civilian perspective. Commanders analyze key civilian areas in terms of how they affect the missions of individual units as well as how military operations affect these areas. (Table B-1 lists examples of key civilian areas.)

Table B-1. Examples of key civilian areas

- Areas defined by political boundaries, such as—
  - Districts or neighborhoods within a city.
  - Municipalities within a region.
  - Provinces within a country.
- Areas of high economic value, such as industrial centers, farming regions, and mines.
- Centers of government and politics.
- Culturally important areas.
- Social, ethnic, tribal, political, religious, criminal, or other important enclaves.
- Trade routes and smuggling routes.
- Possible sites for the temporary settlement of dislocated civilians or other civil functions.

Structures

B-13. Analyzing a structure involves determining how its location, functions, and capabilities can support operations. Commanders also consider the consequences of using it. Using a structure for military purposes often competes with civilian requirements. Commanders carefully weigh the expected military benefits against costs to the community that will have to be addressed in the future. (Table B-2 lists examples of important structures in an AO.)

Capabilities

B-14. Capabilities can refer to the ability of local authorities—those of the host nation or some other body—to provide a populace with key functions or services. Commanders and staffs analyze capabilities from different perspectives but generally put priority on understanding the capability of the HN government to support the mission. The most essential capabilities are those required to save, sustain, or enhance life, in that order. Some of the more important capabilities are—

- Public administration—effectiveness of bureaucracy, courts, and other parts of the HN government.
- Public safety—provided by the security forces and military, police, and intelligence organizations.
- Emergency services—such agencies as fire departments and ambulance services.
- Public health—clinics and hospitals.
- Food.
- Water.
- Sanitation.

Organizations

B-15. Organizations are nonmilitary groups or institutions in the AO. They influence and interact with the populace, counterinsurgents, and each other. They generally have a hierarchical structure, defined goals, established operations, fixed facilities or meeting places, and a means of financial or logistic support. Some organizations may be indigenous to the area. These may include—

- Religious organizations.
- Political parties.
- Patriotic or service organizations.

- Labor unions.
- Criminal organizations.
- Community organizations.

B-16. Other organizations may come from outside the AO. Examples of these include—

- Multinational corporations.
- Intergovernmental organizations (IGOs), such as United Nations agencies.
- Nongovernmental organizations (NGOs), such as the International Red Cross.

B-17. Operations often require commanders to coordinate with IGOs and NGOs. Information required for evaluation includes these groups' activities, capabilities, and limitations. Situational understanding includes knowing how the activities of different organizations may affect military operations and how military operations may affect these organizations' activities. From this analysis, commanders can determine how organizations and military forces can work together toward common goals.

Table B-2. Examples of important structures

- **Government centers**—necessary for the government to function.
- **Headquarters and bases for security forces**—necessary for security forces to function.
- **Police stations, courthouses, and jails**—necessary for countering crime and beneficial for counterinsurgency operations.
- **Communications and media infrastructure**—important to information flow and the opinions of the populace. These include the following:
  - **Radio towers**.
  - **Television stations**.
  - **Cellular towers**.
  - **Newspaper offices**.
  - **Printing presses**.
- **Roads and bridges**—allow for movement of people, goods, insurgents, and counterinsurgents.
- **Ports of entry, such as airports and seaports**—allow for movement of people, goods, insurgents, and counterinsurgents.
- **Dams**—provide electric power, drinking water, and flood control.
- **Electrical power stations and substations**—enable the economy's functioning and often important for day-to-day life of the populace.
- **Refineries and other sources of fuel**—enable the economy's functioning and often important for day-to-day life of the populace.
- **Sources of potable water**—important for public health.
- **Sewage systems**—important for public health.
- **Clinics and hospitals**—important for the health of the populace; these often are protected sites.
- **Schools and universities**—affect the opinions of the populace; these often are protected sites.
- **Places of religious worship**—affect opinions of the populace; often of great cultural importance; these often are protected sites.

B-18. In almost every case, military forces have more resources than civilian organizations. However, some civilian organizations possess specialized capabilities that they may be willing to share. Commanders do not command civilian organizations in their AOs. However some operations require achieving unity of effort with these groups. These situations require commanders to influence the leaders of these organizations through persuasion, relying on the force of argument and the example of actions.

People

B-19. *People* refers to nonmilitary personnel encountered by military forces. The term includes all civilians within an AO (the populace) as well as those outside the AO whose actions, opinions, or political influence can affect the mission. To display different aspects of the populace, analysts can use population support overlays and religion, race, and ethnicity overlays. (FMI 2-91.4 contains information about these overlays.) Perception is another significant people factor in COIN. The perception assessment matrix is a tool that compares the intent of friendly operations to the populace's perception of those operations.

Population Support Overlay

B-20. The population support overlay can graphically depict the sectors of the populace that are progovernment, antigovernment, proinsurgent, anti-insurgent, uncommitted, and neutral. (See figure B-2.) These overlays are important because they help analysts determine whether the local populace is likely support the HN government or support the insurgency.

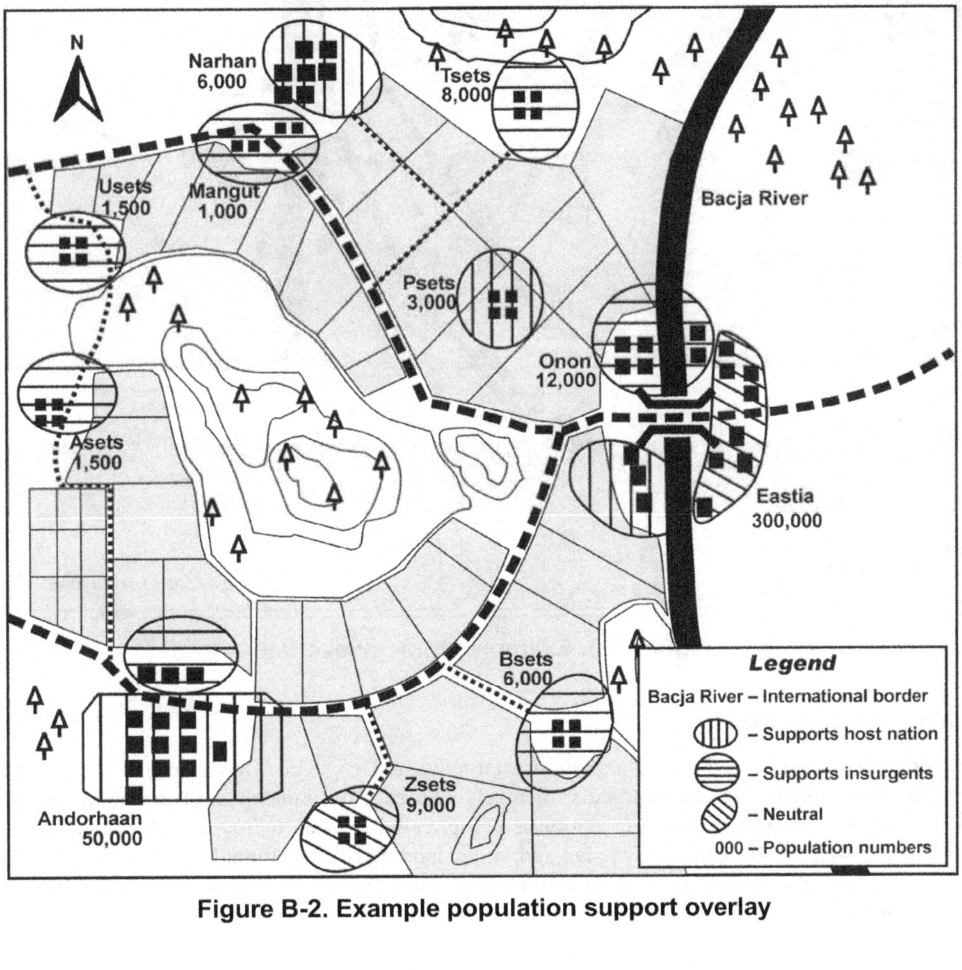

Figure B-2. Example population support overlay

Religion, Race, and Ethnicity Overlay

B-21. Religion, race, and ethnicity issues often contribute to conflicts. Religious, race, and ethnicity overlays depict the current ethnic and religious make-up of an AO. These overlays can also display any specific religious-, racial-, or ethnicity-specific areas and any zones of separation agreed upon by peace accords. These three overlays may be separate or combined. (Figure B-3 shows an example of an ethnicity overlay.)

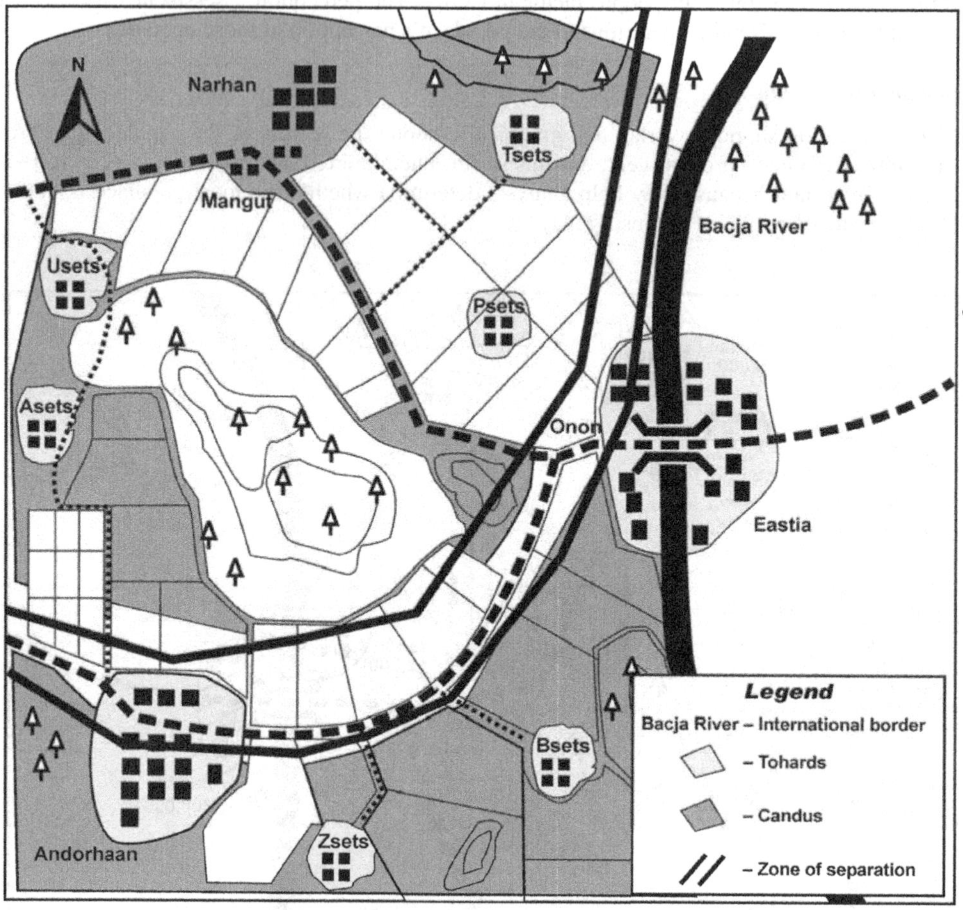

Figure B-3. Example ethnicity overlay

Perception Assessment Matrix

B-22. Perceptions influence how insurgents are targeted and engaged. Important considerations include how insurgents perceive counterinsurgents, themselves, their environment, the nature and reasons for the conflict, and their success criteria. Perception is complicated but key to successfully targeting, engaging, and evaluating success. In-depth knowledge and understanding of the national, regional, and local cultures, norms, moralities, and taboos are needed to understand the operational environment and reactions of the insurgents and populace.

B-23. Perception assessment matrices are often used by psychological operations personnel and other staff elements and can be a valuable tool for intelligence analysts. (See figure B-4 [page B-9].) Counterinsurgent activities intended to be benign or benevolent might have negative results if the populace's perceptions are not considered, and then evaluated or measured. This is true because perceptions—more than reality—drive a commander's decision making and can influence the populace's reactions. A perception assessment matrix displays how well counterinsurgents are able to achieve an effect during an operation. In this sense, the matrix can be used to directly display the effectiveness of the unit's civil affairs, public affairs, and psychological operations efforts.

The Importance of Perceptions in Operation Uphold Democracy

One proposed psychological operations action developed for Operation Uphold Democracy in Haiti in 1994 illustrates why perception assessment is necessary. Before deployment, leaflets were prepared informing the Haitian populace of U.S. intentions. The original leaflet was printed in Dutch, the language of the Haitian elite. However, the one actually used was published in Creole, the official language of Haiti, because an astute team member realized the need to publish to the wider audience.

If a flier in Dutch had been dropped, it could have undermined the American mission to the country in several ways. The majority of the population would have been unable to read the flier. The subsequent deployment of U.S. forces into the country, therefore, could have been perceived as hostile. The mission, which was intended in part to restore equity within Haiti's social structure, could have backfired if the Haitians viewed the Dutch flier as an indication of U.S. favoritism toward the Haitian elite.

B-24. Perception can work against operational objectives. Perceptions should therefore be assessed before and during an operation. It is not possible to read the minds of the local populace; however, there are several means of measuring its perceptions. These include the following:

- Demographic analysis and cultural intelligence are key components of perception analysis.
- Understanding a population's history can help predict expectations and reactions.
- Human intelligence can provide information on perceptions.
- Reactions and key activities can be observed to determine whether people act based on real or perceived conditions.
- Editorial and opinion pieces of relevant newspapers can be monitored for changes in tone or opinion shifts that can steer, or may be reacting to, the opinions of a population group.

B-25. Perception assessment matrices aim to measure the disparities between friendly force actions and what population groups perceive. In addition to assessing the perceptions of the population groups within an AO, commanders may also want to assess the perceptions that their Soldiers and Marines have of unit activities. Assessing counterinsurgents' perceptions can begin to answer the following questions:

- Are counterinsurgents exhibiting Western or American values that the populace does not appreciate?
- Are embedded American beliefs preventing the unit from understanding the HN population or its multinational partners?
- Is what the intelligence and command staff perceives really what is happening?
- Does the populace believe what the unit believes?
- Is there something that is part of the populace's (or a subgroup's) perception that can be detrimental to the unit?

Condition	Cultural norm	Alternative proposed by COIN force	Population's perception	Acceptable difference in perception?	Root of difference	Possible to change perception?	Proposed solution	Possible consequences of unchanged perception
Food shortages	Rice	Meat and potatoes	Inadequate/ inconsiderate	No	Culturally accepted norms/ standards; no known physically detrimental effects	No; logistically restricted	Offer potatoes; seek exchange for rice	Starvation; rioting
Use of guns	Criminal elements carry weapons openly	Confiscate all weapons	Unfair/ population not protected by traditional means	No	Culture/ criminal element provides a measure of security for the local populace	No; population and friendly forces at risk	Psychological operations campaign; weapons turn-in program	Civil unrest; armed backlash
Government structure	Tribal	Hierarchical	Tolerable as long as needs are fulfilled by group in charge	Yes	History	No	Bargain	Unknown
Language	Dual languages: Creole and Dutch	Respect all languages	Unfair/show of favoritism	Yes	History/ national language	Yes	Communicate in all languages when possible	Backlash against elite and friendly forces

Figure B-4. Example perception assessment matrix

Events

B-26. Events are routine, cyclical, planned, or spontaneous activities that significantly affect organizations, people, and military operations. They are often symbols, as described in paragraph 3-51. Examples include the following:

- National and religious holidays.
- Agricultural crop, livestock, and market cycles.
- Elections.
- Civil disturbances.
- Celebrations.

B-27. Other events include disasters from natural, manmade, or technological sources. These create hardships and require emergency responses. Examples of events precipitated by military forces include combat operations, deployments, redeployments, and paydays. Once significant events are determined, it is important to template the events and analyze them for their political, economic, psychological, environmental, and legal implications.

EVALUATE THE THREAT

B-28. Evaluating the threat involves analyzing insurgent organizations, capabilities, and tactics to identify vulnerabilities to exploit. Tools like social network analysis, link diagrams, and association matrices help analysts do this. Other tools such as historical time lines and pattern analysis tools assist in developing event and doctrinal templates to depict enemy tactics.

SOCIAL NETWORK ANALYSIS

B-29. Social network analysis (SNA) is a tool for understanding the organizational dynamics of an insurgency and how best to attack or exploit it. It allows analysts to identify and portray the details of a network structure. Its shows how an insurgency's networked organization behaves and how that connectivity affects its behavior. SNA allows analysts to assess the network's design, how its member may or may not act autonomously, where the leadership resides or how it is distributed among members, and how hierarchical dynamics may mix or not mix with network dynamics.

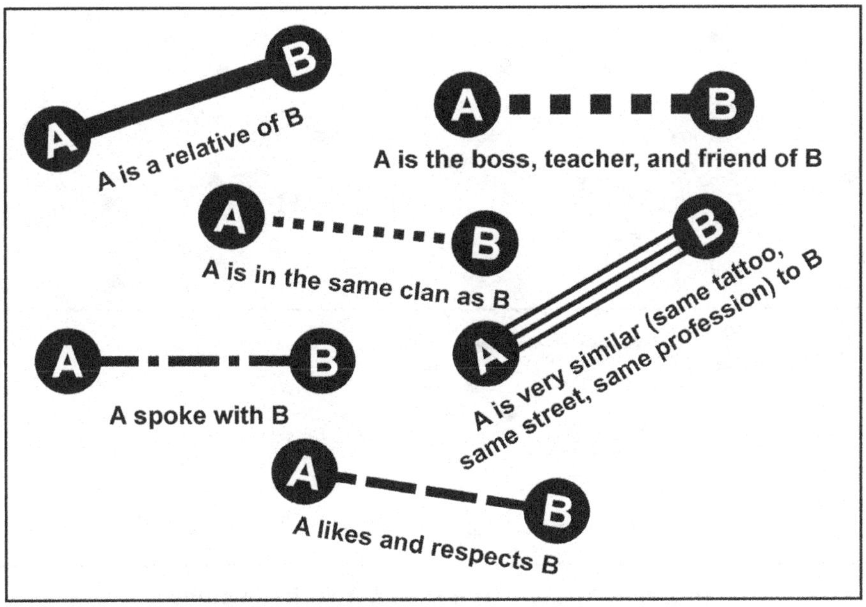

Figure B-5. Examples of dyads

B-30. SNA supports a commander's requirement to describe, estimate, and predict the dynamic structure of an insurgent organization. It also provides commanders a useful tool to gauge their operations' effectiveness. SNA allows analysts assess the insurgency's adaptation to the operational environment and friendly operations.

Social Network Analysis—Terms and Concepts

B-31. The social network graph is the building block of social network analysis. A social network graph consists of individuals and connections between them. Individuals in a network are called *actors* or *nodes*. (Actor and node are often used interchangeably.) The contacts between nodes are called *links*. The basic element of a social network graph is the *dyad*. A dyad consists of two nodes and a single link. In the simplest form of a network, the two nodes represent people and the link represents a relationship between them. (See figure B-5.)

B-32. *Social network measures* allow units to analyze and describe networks. They fall into two categories: organizational-level and individual-level.

Organizational-Level Analysis

B-33. *Organizational-level analysis* provides insight about the insurgent organization's form, efficiency, and cohesion. A regional insurgency may consist of large numbers of disconnected subinsurgencies. As a result, each group should be analyzed based on its capacities as compared to the other groups. Organizational-level capacities can be described in terms of network density, cohesion, efficiency, and core-periphery. Each measure describes a characteristic of a networked organization's structure. Different network structures can support or hinder an organization's capabilities. Therefore, each organizational measure supports the analyst's assessment of subgroup capabilities.

B-34. *Network density* is a general indicator of how connected people are in the network. Network or global-level density is the proportion of ties in a network relative to the total number possible. Comparing network densities between insurgent subgroups provides commanders with an indication of which group is most capable of a coordinated attack and which group is the most difficult to disrupt. (Figure B-6 shows three networks with different densities.)

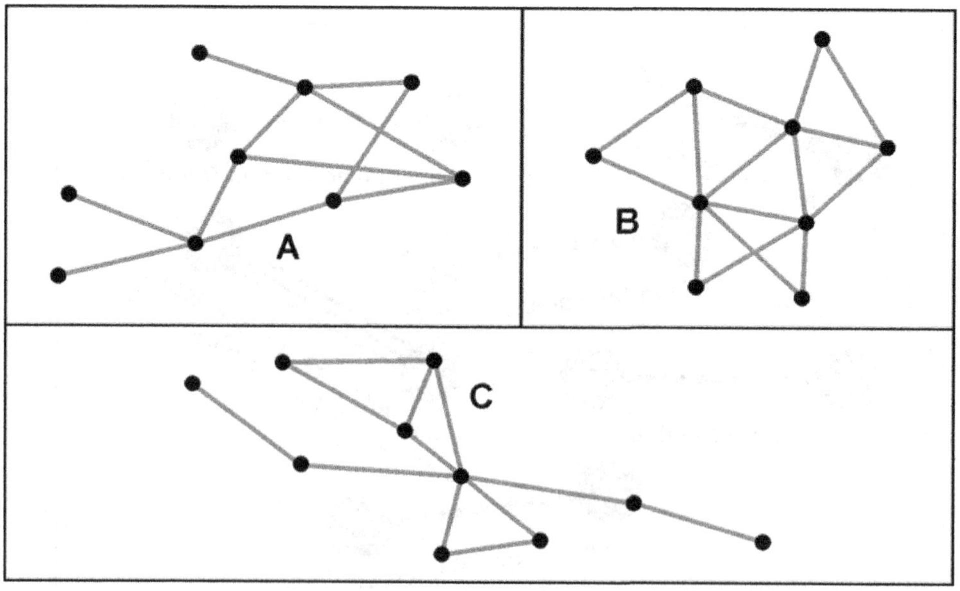

Figure B-6. Comparison of network densities

B-35. Most network measures, including network density, can be mapped out to evaluate performance over time. Based on changes in network density over time, a commander can—

- Monitor enemy capabilities.
- Monitor the effects of recent operations.
- Develop tactics to further fragment the insurgency.

B-36. An increase in network density indicates the likelihood that the insurgent group can conduct coordinated attacks. A decrease in network density means the group is reduced to fragmented or individual-level attacks. (Figure B-7 illustrates an example of how tactics and activities can change based on network density.) A well-executed COIN eventually faces only low-network-density subgroups. This is because high-network-density subgroups require only the capture of one highly connected insurgent to lead counterinsurgents to the rest of the group. So while high-network-density groups are the most dangerous, they are also the easiest to defeat and disrupt.

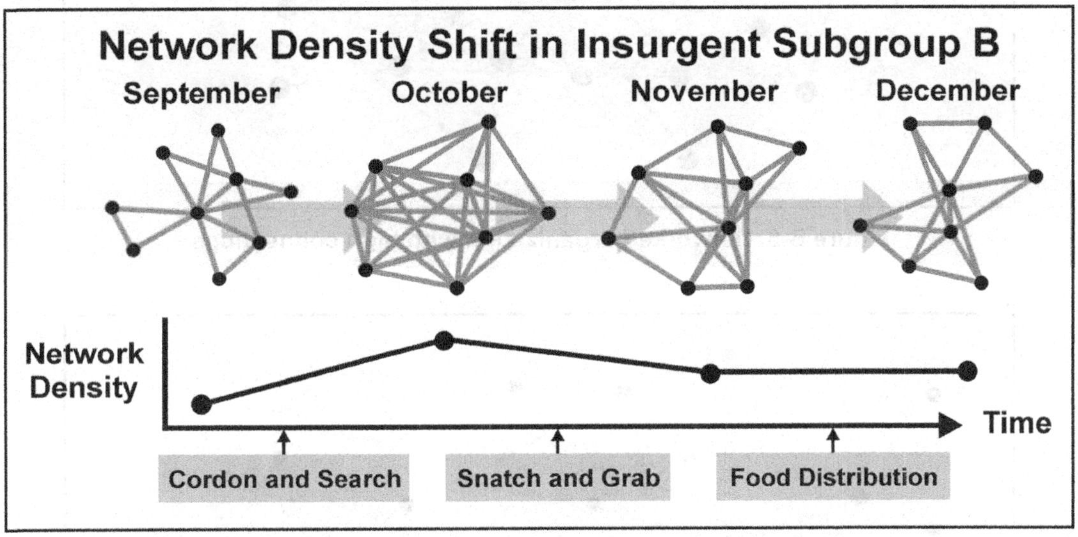

Figure B-7. Example of changes to tactics based on density shift

B-37. Network density does not consider how distributed the connections are between the nodes in a network. Better metrics of group and organizational performance would be network centrality, core-periphery, and diameter. A few nodes with a high number of connections can push up the group network density, even though the majority of the people nodes are only marginally linked to the group. In the case of a highly centralized network dominated by one or a few very connected nodes, these nodes can be removed or damaged to fragment the group further into subnetworks.

B-38. A fully connected network like the one figure B-8 portrays is an unlikely description of the enemy insurgent order of battle. A regional insurgency can be fragmented within itself.

B-39. Sometimes a region may actually contain multiple subinsurgencies that are either unaware of, or even competing with, other subinsurgent groups. In this case, the insurgency resembles a fragmented network. (See figure B-9.)

Individual-Level Analysis

B-40. *Individual-level analysis* characterizes every member of an organization and identifies its key members. Effective SNA allows analysts to identify key individuals from a large mass of data. SNA describes individuals based on their network position in relation to the network position of every other individual in the network. Descriptions are in terms of the following individual-level measures: degree centrality, betweenness centrality, and diameter. Individual network centralities provide insight into an individual's location in the network. The relationship between the centralities of all nodes can reveal much about the overall network structure.

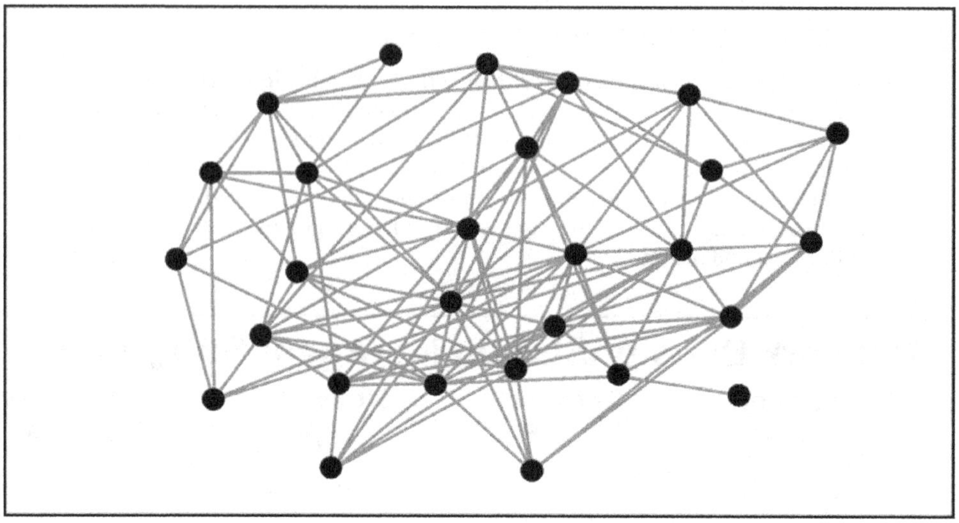

Figure B-8. Networked organization with high connections

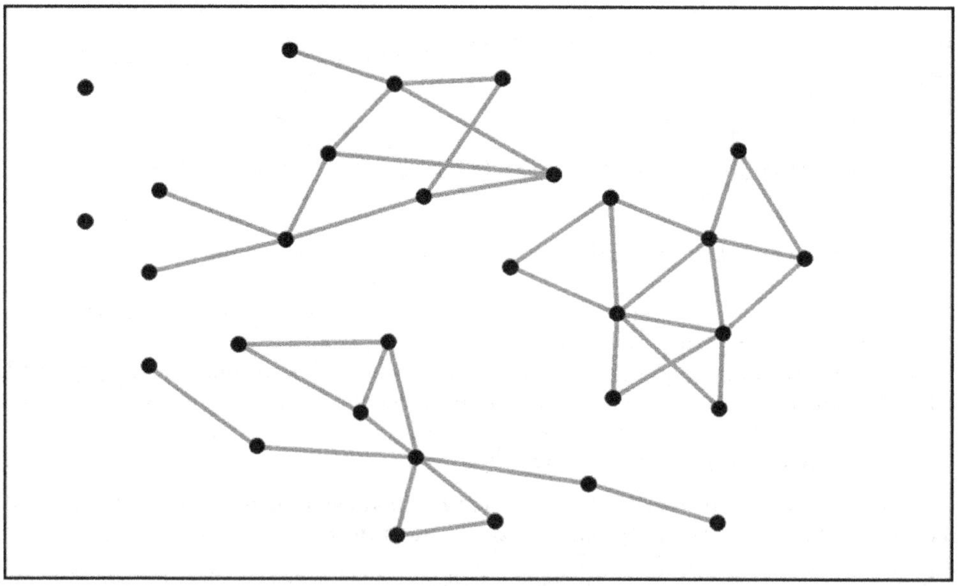

Figure B-9. Fragmented network

B-41. One node or a very few central nodes dominate a very centralized network. If these nodes are removed or damaged, the network may quickly fragment into unconnected subnetworks. *Hubs* are nodes with a very high degree of centrality. A network centralized around a well-connected hub can fail abruptly if that hub is disabled or removed.

B-42. A less centralized network has no single points of failure. It is resilient in the face of many intentional attacks or random failures. Many nodes or links can fail while allowing the remaining nodes to still reach each other over other, redundant network paths.

B-43. *Degree centrality* describes how active an individual is in the network. Network activity for a node is measured using the concept of degrees—the number of direct connections a node has. Nodes with the most direct connections are the most active in their networks. Common wisdom in organizations is "the more connections, the better." This is not always so. What really matters is where those connections lead and how they connect the otherwise unconnected. If a node has many ties, it is often said to be either prominent or influential.

B-44. *Betweenness centrality* indicates the extent to which an individual lies between other individuals in the network, serving as an intermediary, liaison, or bridge. A node with high "betweenness" has great influence over what flows in the network. Depending on position, a person with high betweenness plays a "broker" role in the network. A major opportunity exists for counterinsurgents if, as in group C of figure B-6 (page B-11), the high betweenness centrality person is also a single point of failure which, if removed, would fragment the organization.

B-45. Nodes on the periphery receive very low centrality scores. However, peripheral nodes are often connected to networks that are not currently mapped. The outer nodes may be resource gatherers or individuals with their own network outside their insurgent group. These characteristics make them very important resources for fresh information not available inside their insurgent group.

The Capture of Saddam Hussein

The capture of Saddam Hussein in December 2003 was the result of hard work along with continuous intelligence gathering and analysis. Each day another piece of the puzzle fell into place. Each led to coalition forces identifying and locating more of the key players in the insurgent network—both highly visible ones like Saddam Hussein and the lesser ones who sustained and supported the insurgency. This process produced detailed diagrams that showed the structure of Hussein's personal security apparatus and the relationships among the persons identified.

The intelligence analysts and commanders in the 4th Infantry Division spent the summer of 2003 building link diagrams showing everyone related to Hussein by blood or tribe. Those family diagrams led counterinsurgents to the lower level, but nonetheless highly trusted, relatives and clan members harboring Hussein and helping him move around the countryside. The circle of bodyguards and mid-level military officers, drivers, and gardeners protecting Hussein was described as a "Mafia organization," where access to Hussein controlled relative power within the network.

Over days and months, coalition forces tracked how the enemy operated. Analysts traced trends and patterns, examined enemy tactics, and related enemy tendencies to the names and groups on the tracking charts. This process involved making continual adjustments to the network template and constantly determining which critical data points were missing.

Late in the year, a series of operations produced an abundance of new intelligence about the insurgency and Hussein's whereabouts. Commanders then designed a series of raids to capture key individuals and leaders of the former regime who could lead counterinsurgents to him. Each mission gained additional information, which shaped the next raid. This cycle continued as a number of mid-level leaders of the former regime were caught, eventually leading coalition forces into Hussein's most trusted inner circle and finally to Hussein's capture.

Social Network Analysis and Counterinsurgency

B-46. Figure B-10 shows a simple, social network of key individuals and relationships. The nodes in this data set are from a modified, subnetwork of the link diagram representing Saddam Hussein and his connections to various family members, former regime members, friends, and associates. The original diagram contained hundreds of names and took shape on a large 36-by-36-inch board. Each "box" in the network contained personal information on a particular individual. This information included roles and positions of certain people within the network—for example, chief of staff, chief of operations, and personal secretary. These were not necessarily positions the individuals occupied before the fall of Hussein; rather they were based on an understanding of the role they were filling in the insurgency or Saddam's underground operations. Analysts assigned these roles based on an assessment of various personalities and recent reports. Such a process helped coalition forces focus their efforts in determining those who were closest to Hussein and their importance.

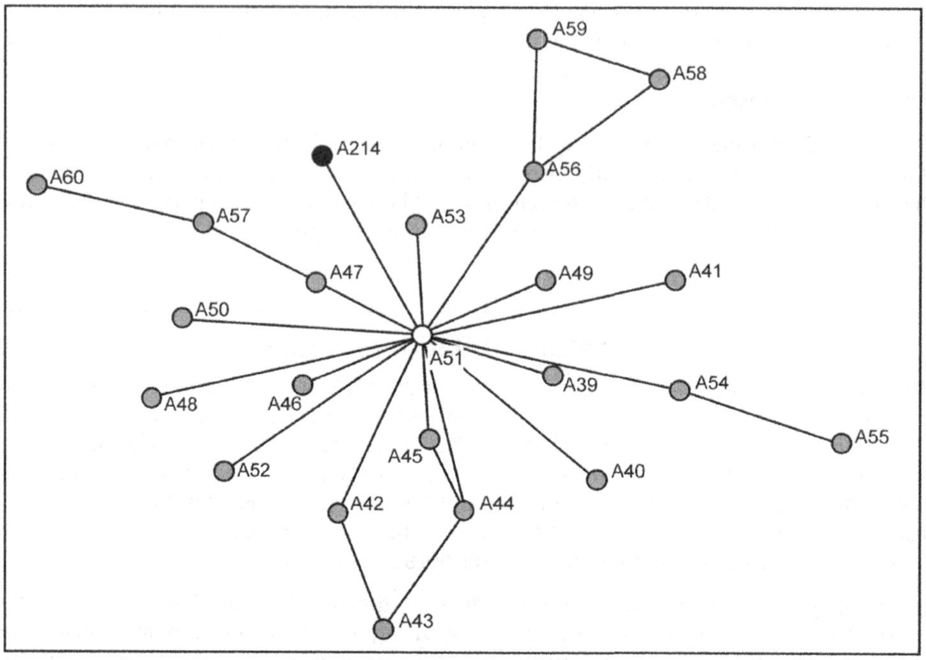

Figure B-10. Simple network

B-47. For an insurgency, a social network is not just a description of who is in the insurgent organization; it is a picture of the population, how it is put together and how members interact with one another. A tribal society already has affiliated social, economic, and military networks easily adapted to warfighting. The ways in which insurgents exploit a tribal network does not represent an evolved form of insurgency but the expression of inherent cultural and social customs. The social dynamic that sustains ongoing fighting is best understood when considered in tribal terms—in particular, from the perspective of a traditionally networked society. It is the traditional tribal network that offers rebels and insurgents a ready-made insurrectionary infrastructure on which to draw.

B-48. The full functioning of a network depends on how well, and in what ways, its members are personally known and connected to one another. This is the classic level of SNA, where strong personal ties, often ones that rest on friendship and bonding experiences, ensure high degrees of trust and loyalty. To function well, networks may require higher degrees of interpersonal trust than do other approaches to organization, like hierarchies. Kinship ties, be they of blood or brotherhood, are a fundamental aspect of many terrorist, criminal, and gang organizations. For example, news about Osama bin Laden and the Al Qaeda network reveal his, and its, dependence on personal relationships formed over years with "Afghan Arabs" from Egypt and elsewhere. These people are committed to anti-United States terrorism and Islamic fundamentalism.

B-49. To draw an accurate picture of a network, units need to identify ties among its members. Strong bonds formed over time by family, friendship, or organizational association characterize these ties. Units gather information on these ties by analyzing historical documents and records, interviewing individuals, and studying photos and books. It is painstaking work, but there is really no alternative when trying to piece together a network that does not want to be identified. Charts and diagrams lead to understanding the insurgents' means of operations. These same diagrams are also useful for understanding tribal, family, NGO, and transnational terrorist elements. Each diagram and chart may have links to another or several others, but they are not created overnight. It takes time, patience, detailed patrolling, and reporting and recording of efforts.

B-50. As a unit builds its situational awareness, it must create easy-to-understand, adaptable, and accurate diagrams and information sheets. These products feed one another and allow units to maintain and contribute to their understanding of the situation.

B-51. As commanders dispatch patrols to collect information, they can begin to build a graph of the population in the AO. As graphs grow, they may show that traditional, static organizational line charts do not produce viable explanations of insurgent organizational behavior. Individual insurgents may be constantly adapting to the operational environment, their own capabilities, and counterinsurgent tactics. A commander's understanding of the insurgency is only as good as the patrol's last collection.

B-52. Relationships (links) in large data sets are established by similarities between the nodes (people). Figure B-11 shows an example *activities matrix*. People are identified by their participation in independent activities. When graphed, pairs who have engaged in the same activity (columns with dots) are designated with a link.

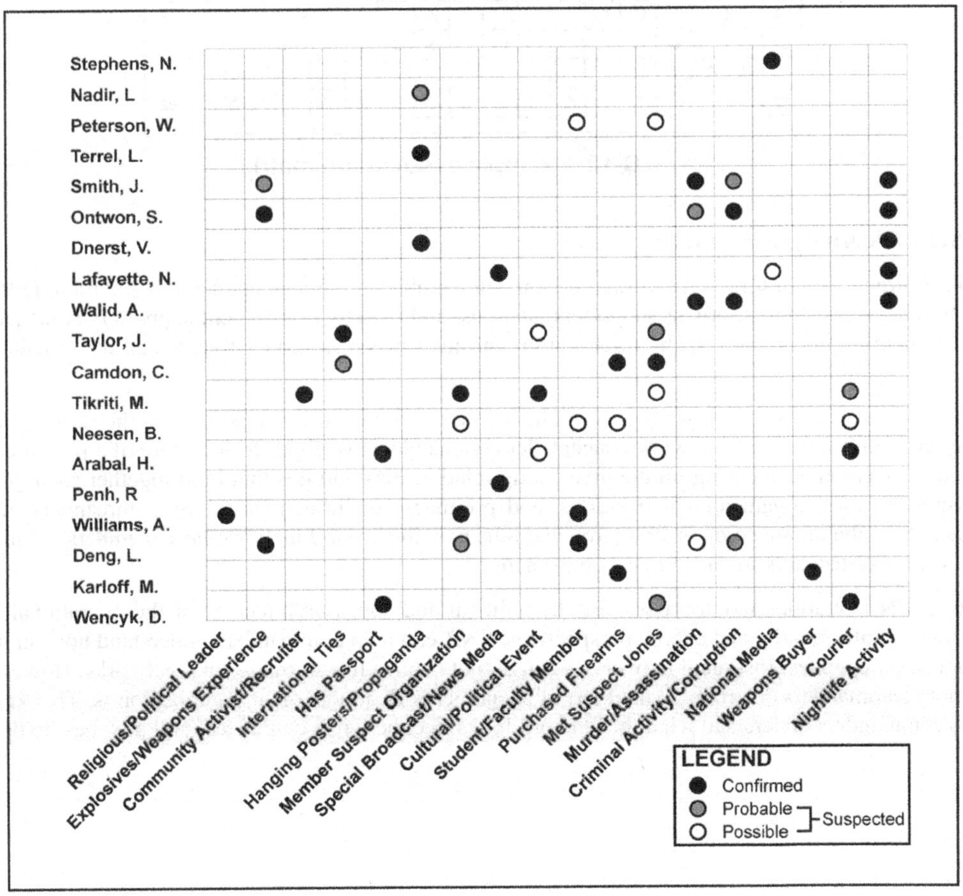

Figure B-11. Example activities matrix

B-53. An *association matrix* portrays the existence of an association, known or suspected, between individuals. (See figure B-12.) Direct connections include such things as face-to-face meetings and confirmed telephonic conversations. Association matrices provide a one-dimensional view of the relationships and tend to focus on the immediate AO. Analysts can use association matrices to identify those personalities and associations needing a more in-depth analysis to determine the degree of relationship, contacts, or knowledge between the individuals. The structure of the insurgent organization is identified as connections between personalities are made.

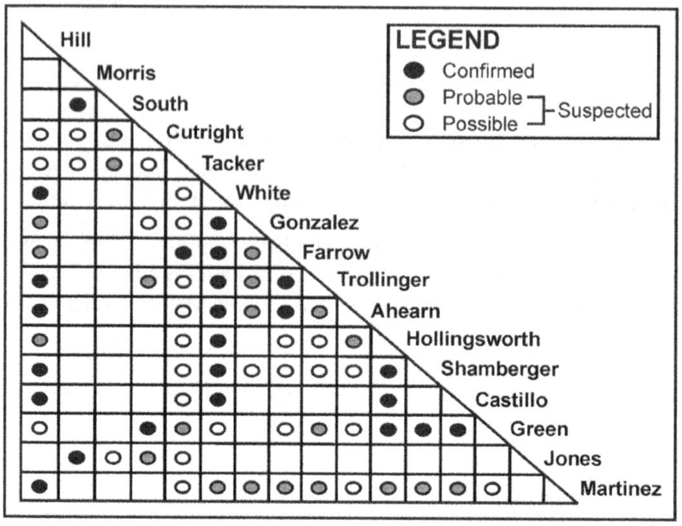

Figure B-12. Example association matrix

Social Network Analysis Summary

B-54. Insurgents often form a networked organization embedded in a sympathetic population. Differentiating between insurgents, insurgent supporters, neutrals, and the HN government supporters is difficult. With every counterinsurgent success, the insurgent organization becomes further fragmented but remains dangerous.

B-55. SNA helps units formalize the informality of insurgent networks by portraying the structure of something not readily observed. Network concepts let commanders highlight the structure of a previously unobserved association by focusing on the preexisting relationships and ties that bind together such groups. By focusing on roles, organizational positions, and prominent or influential actors, commanders may get a sense of how the organization is structured and thus how the group functions, how members are influenced and power exerted, and how resources are exchanged.

B-56. COIN operations require assessing the political and social architecture of the operational environment, from both friendly and enemy perspectives. SNA can help commanders understand how an insurgent organization operates. Insurgent networks often do not behave like normal social networks. However, SNA can help commanders determine what kind of social network an insurgent organization is. That knowledge helps commanders understand what the network looks like, how it is connected, and how best to defeat it.

Historical Time Line

B-57. A *time line* is list of significant dates along with relevant information and analysis. Time lines seek to provide a context to operational conditions. (See figure B-13.) Time lines often contain information related to areas and people as well as events. Some time lines describe population movements (areas) and political shifts (power and authority) that are relevant to the AO. Time lines can also include a brief historical record of the population or area, highlighting the activities (events) of a certain population sector. As analytic tools, time lines might help analysts predict how key population sectors might react to certain circumstances.

B-58. Key local national holidays, historic events, and significant cultural and political events can be extremely important. Soldiers and Marines are often provided with a list of these key dates to identify potential dates of increased or unusual activity. These lists, however, rarely include a description of why these dates are significant and what can be expected to happen on the holiday. In some cases, days of the week are significant.

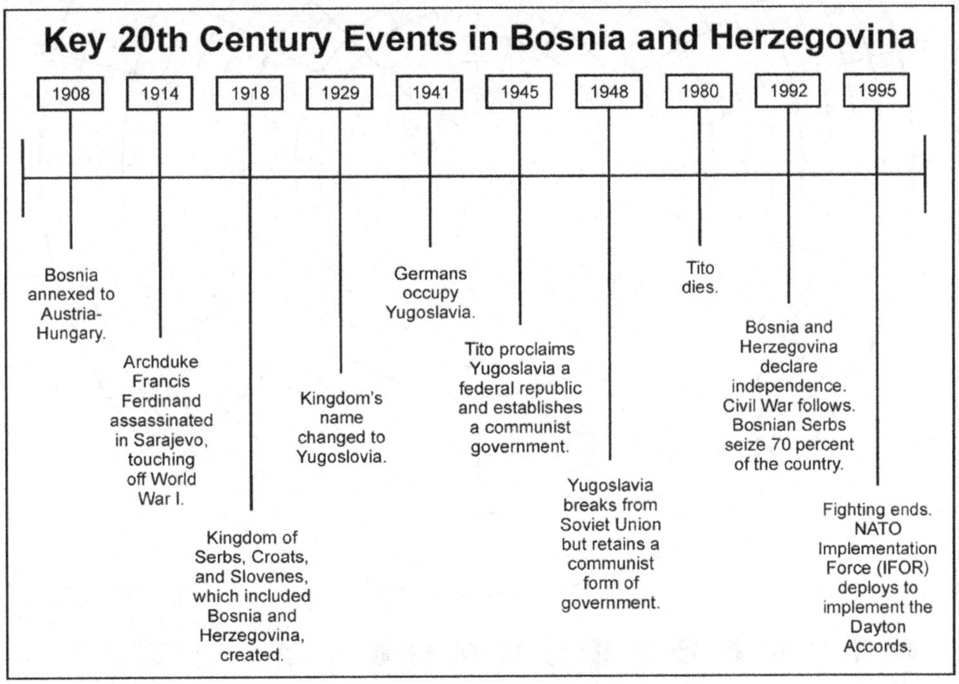

Figure B-13. Example historical time line

Pattern Analysis

B-59. Pattern analysis plot sheets, time-event charts, and coordinates registers are pattern analysis tools used to evaluate a threat and determine threat courses of action. (FM 2-22.3 discusses how use these tools.)

Pattern Analysis Plot Sheet

B-60. Pattern analysis plot sheets focus on the time and date of each serious incident that occurs within the AO. (See figure B-14.) The rings depict days of the month; the segments depict the hours of the day. As shown in the plot sheet's legend, the chart depicts the actual events and identifies each by using an alphanumeric designation that corresponds to the legend used on the coordinates register. (See paragraph B-61.) Another type of pattern analysis plot sheet helps distinguish patterns in activities associated with particular days, dates, or times. When used in conjunction with the coordinates register and doctrinal templates, a pattern analysis plot sheet supplies most of the data needed for an event template. Analysts may choose to modify this product to track shorter periods to avoid clutter and confusion.

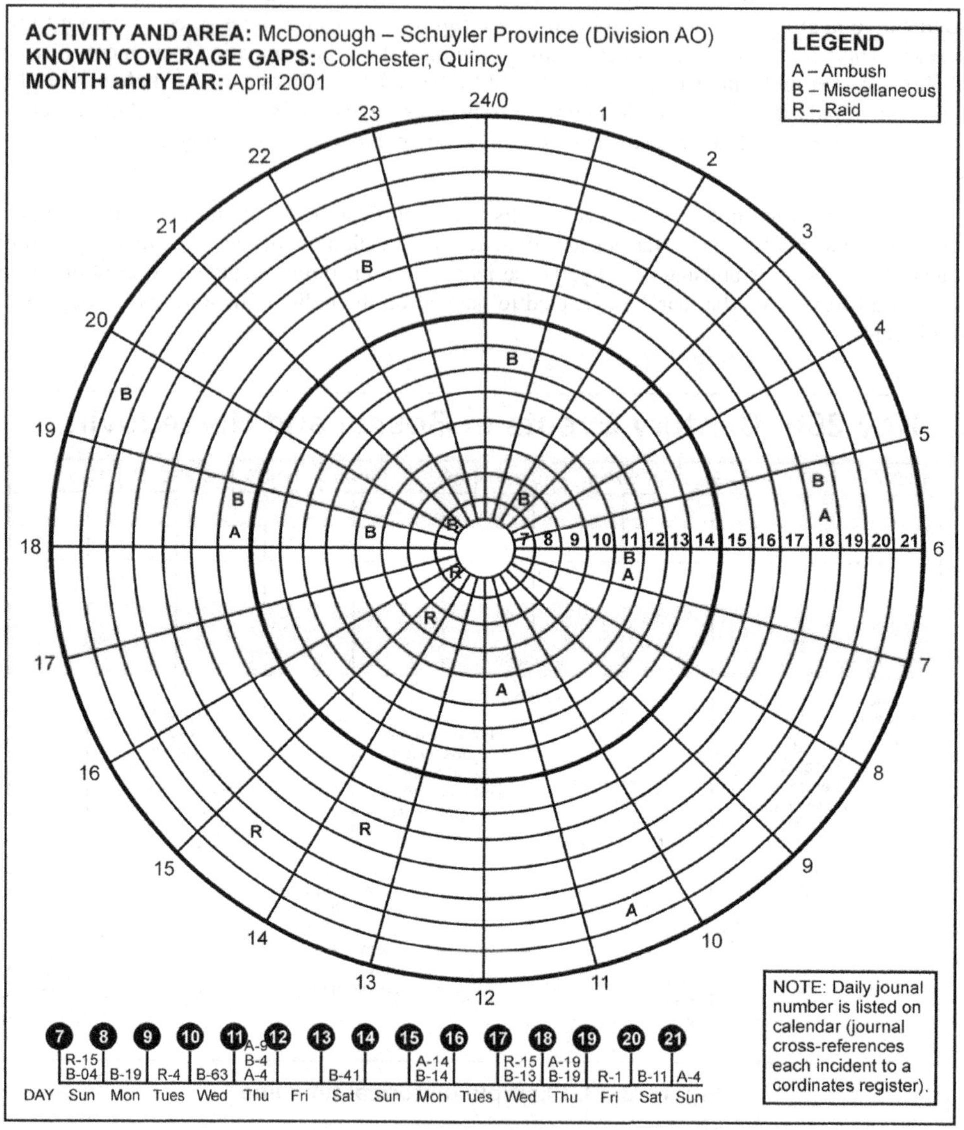

Figure B-14. Example pattern analysis plot sheet

Coordinates Register

B-61. Another pattern analysis tool is the coordinates register, also known as an incident map. (See figure B-15.) A coordinates register illustrates cumulative events that have occurred within an AO. It focuses on the "where" of an event. Analysts may use multiple coordinates registers, each focusing on an individual subject or a blend of subjects. Additionally, a coordinates register includes information like notes or graphics. Analysts should always use the coordinates register in with the pattern analysis plot sheet.

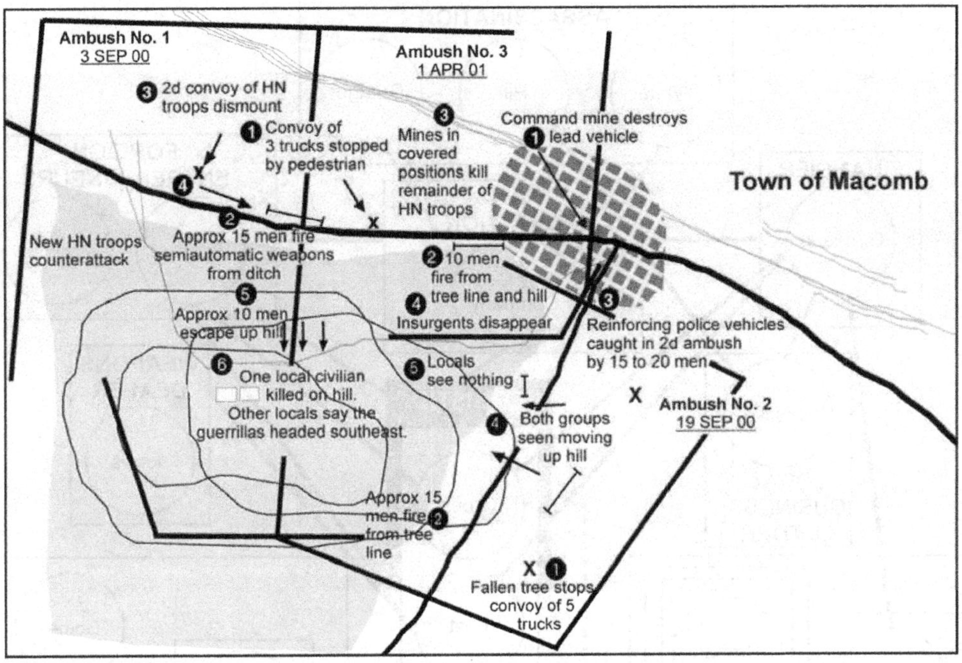

Figure B-15. Example coordinates register

Link Diagrams

B-62. The link diagram graphically depicts relationships between people, events, locations, or other factors deemed significant in any given situation. (See Figure B-16.) Link diagrams help analysts better understand how people and factors are interrelated in order to determine key links. (For more information on link diagrams, see FM 2-22.3.)

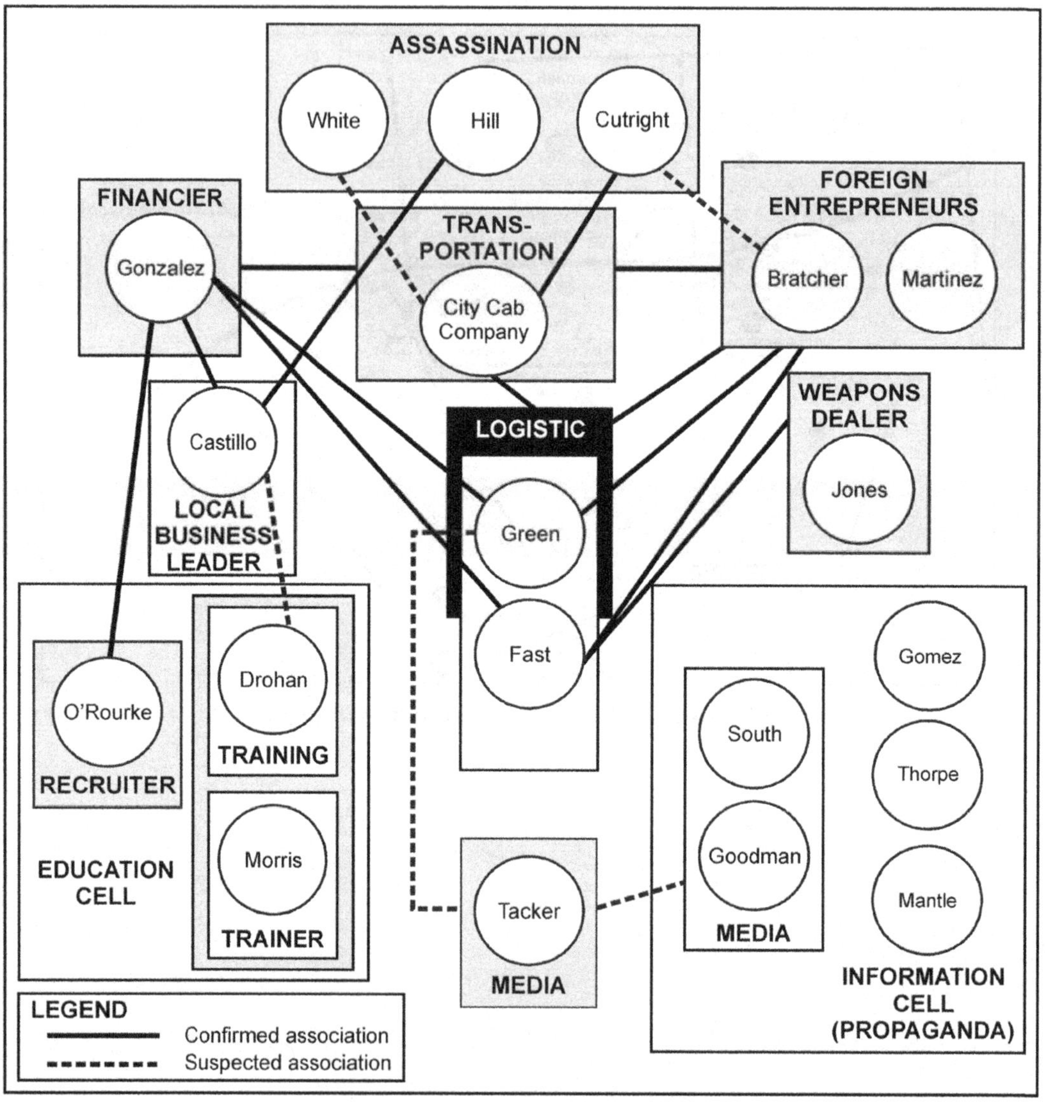

Figure B-16. Example link diagram

Time-Event Charts

B-63. Time-event charts are chronological records of individual or group activities. They are designed to store and display large amounts of information in a small space. Analysts can use time-event charts to help analyze larger scale patterns of such things as activities and relationships. There is great latitude in preparing time-event charts. Some of their common characteristics are as follows:

- The beginning and ends of the chart are shown with triangles.
- Other events are shown with squares.
- Particularly noteworthy events have an X drawn across the square.
- The date is always on the symbol.
- A description is below the symbol.
- The flow is from left to right for each row.

(Figure B-17 is an example showing events surrounding the plot to attack several landmarks in New York City in the early 1990s.)

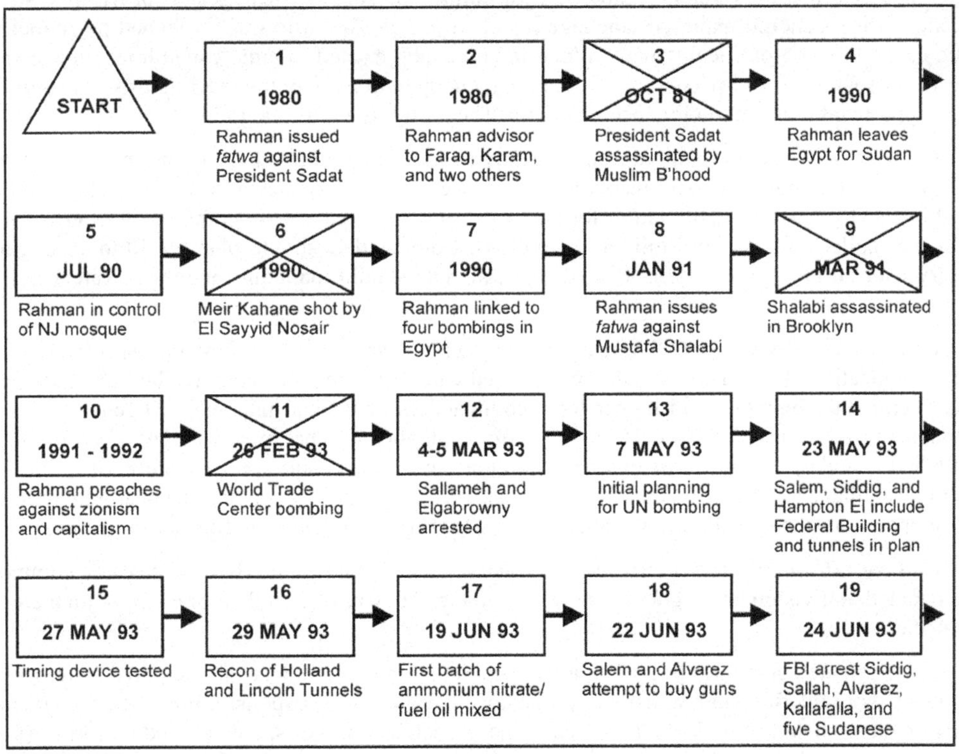

Figure B-17. Example time-event chart

Appendix C
Linguist Support

U.S. forces conducting counterinsurgency operations in foreign nations require linguist support. Military intelligence units assigned to brigade and higher level commands have organic interpreters (linguists) to perform human intelligence and signals intelligence functions. However, the need for interpreters usually exceeds organic capabilities, and commanders should obtain external interpreter support early.

LINGUIST SUPPORT CATEGORIES

C-1. When possible, interpreters should be U.S. military personnel or category II or III linguists. Unit intelligence officers should maintain language rosters at home station to track assigned personnel with linguistic capabilities before deployment. When requirements exceed organic capabilities, unit commanders can hire host-nation (HN) personnel to support their operations. Contracted linguists can provide interpreter support and perform intelligence functions. They fall into three categories.

C-2. *Category I linguists* usually are hired locally and require vetting. They do not have a security clearance. They are the most abundant resource pool; however, their skill level is limited. Category I linguists should be used for basic interpretation for activities such as patrols, base entrance coverage, open-source intelligence collection, and civil-military operations. Commanders should plan for 30 to 40 linguists from category I for an infantry battalion. Brigade headquarters should maintain roughly 15 category I linguists for surge operations.

C-3. *Category II linguists* are U.S. citizens with a secret clearance. Often they possess good oral and written communication skills. They should be managed carefully due to limited availability. Category II linguists interpret for battalion and higher level commanders or tactical human intelligence teams. Brigade commanders should plan for 10 to 15 linguists from category II. That breaks down to one linguist for the brigade commander, one for each infantry battalion commander, and approximately 10 linguists for the supporting military intelligence company. Of those 10, three translate for each tactical human intelligence team or operations management team, and two translate for each signals intelligence collection platform.

C-4. *Category III linguists* are U.S. citizens with a top secret clearance. They are a scarce commodity and often retained at division and higher levels of command. They have excellent oral and written communications skills.

C-5. Some private companies provide linguist support through contracts. The required statement of work or contract should define the linguist's requirements and the unit's responsibilities. Contracted category II and III linguists should provide their own equipment, such as flak vests, Kevlar, and uniforms. (Category I linguists normally do not.) The unit designates a linguist manager to identify language requirements and manage assets. Site managers for the contractor are located at the division level to manage personnel issues such as leave, vacation, pay, and equipment.

C-6. When hiring HN personnel to perform category I linguist requirements as interpreters, units should consider the guidelines under the following categories:

- Selecting interpreters.
- Establishing rapport.
- Orienting interpreters.
- Preparing for presentations.

- Conducting presentations.
- Speaking techniques.

SELECTING INTERPRETERS

C-7. Soldiers and Marines must try to vet interpreters before hiring them. All interpreters must meet a basic set of criteria. They should be native speakers. The target audience should willingly accept their social status. All interpreters should speak English fluently. They should be able to translate correctly. Intelligent translators are mandatory; those with technical knowledge are desired. Interpreters should be reliable, loyal, and compatible with the military personnel. Their gender, age, race, and ethnicity must be compatible with the target audience.

NATIVE SPEAKER

C-8. Interpreters should be native speakers of the socially or geographically determined dialect. Their speech, background, and mannerisms should be completely acceptable to the target audience. The interpreters should not distract the interviewees. The target audience should give no attention to the way interpreters talk, only to what they say. Native speakers can better distinguish dialects of different regions and provinces. This knowledge can help identify interviewees from other countries or from outside the local area.

SOCIAL STATUS AND ETHNO-RELIGIOUS IDENTITY

C-9. If their social standing is considerably lower than that of the audience, interpreters may be limited in their effectiveness. Examples include significant differences in military rank or membership in a shunned ethnic or religious group. Soldiers and Marines must communicate with the local population. They must be tolerant of local prejudices and choose an interpreter who is least likely to cause suspicion or miscommunication. Interpreters should also have a good reputation in the community and not be intimidated when dealing with important audiences.

ENGLISH FLUENCY

C-10. An often overlooked consideration is how well the interpreter speaks English. If the interpreter understands the speaker and the speaker understands the interpreter, then the interpreter's command of English is satisfactory. Soldiers and Marines can check that understanding by speaking something to the interpreter in English and asking the interpreter to paraphrase it.

UNDERSTANDING OF THE AUDIENCE

C-11. Interpreting goes both ways. Interpreters should accurately convey information expressed by interviewees or the target audience. This is especially important when commanders speak with HN civilian leaders and military personnel. Linguists involved in military discussions should understand military terms and doctrine.

INTELLECTUAL CAPABILITIES

C-12. Interpreters should be quick and alert, able to respond to changing conditions and situations. They should be able to grasp complex concepts and discuss them clearly and logically. Although education does not equate to intelligence, it does expose students to diverse and complex topics. As a result, the better educated the interpreters, the better they perform.

TECHNICAL ABILITY

C-13. Sometimes Soldiers and Marines need interpreters with technical training or experience in special subject areas. Such interpreters can translate the meaning as well as the words. For instance, if the subject is nuclear physics, background knowledge is useful.

RELIABILITY

C-14. Soldiers and Marines should avoid a potential interpreter who arrives late for the vetting interview. Throughout the world, the concept of time varies widely. In many countries, timeliness is relatively unimportant. Soldiers and Marines should stress the importance of punctuality with interpreters.

LOYALTY

C-15. If interpreters are local nationals, their first loyalty is probably to the host nation or ethnic group, not to the U.S. military. The security implications are clear. Soldiers should be cautious when they explain concepts. They should limit what information interpreters can overhear. Some interpreters, for political or personal reasons, may have ulterior motives or a hidden agenda. Soldiers and Marines who detect or suspect such motives should tell the commander or security manager.

GENDER, AGE, RACE, AND ETHNICITY

C-16. Gender, age, and race can seriously affect the mission effectiveness of interpreters. In predominantly Muslim countries, cultural prohibitions may cause difficulties with gender. A female interpreter may be ineffective in communicating with males, while a female interpreter may be needed to communicate with females. In regions featuring ethnic strife, such as the Balkans, ethnic divisions may limit the effectiveness of an interpreter from outside the target audience. Since traditions, values, and biases vary from country to country, Soldiers and Marines must thoroughly study the culture to determine the most favorable characteristics for interpreters.

COMPATIBILITY

C-17. The target audience quickly recognizes personality conflicts between Soldiers and Marines and their interpreters. Such friction can undermine the effectiveness of the communication. When selecting interpreters, Soldiers and Marines should look for compatible traits and strive for a harmonious working relationship.

EMPLOYING LINGUISTS

C-18. If several qualified interpreters are available, Soldiers and Marines should select at least two. This is particularly important if the interpreter works during long conferences or courses of instruction. With two interpreters available, they should each work for thirty-minute periods. Due to the mental strain associated with this task, four hours of active interpreting a day is usually the most that interpreters can work before effectiveness declines. During short meetings and conversations with two or more available interpreters, one can provide quality control and assistance for the one translating. This technique is useful when conducting coordination or negotiation meetings, as one interpreter can actively interpret while the other pays attention to the body language and side conversations of the audience. Many times, Soldiers and Marines can learn important auxiliary information from listening to what others are saying among themselves. This information can help in later negotiations.

C-19. Commanders must protect their interpreters. They should emplace security measures to keep interpreters and their families safe. Insurgents know the value of good interpreters and will often try to intimidate or kill interpreters and their family members. Insurgents may also coerce interpreters to gather information on U.S. operations. Soldiers and Marines must actively protect against subversion and espionage, to include using a polygraph.

C-20. Certain tactical situations may require using uncleared HN personnel as interpreters. Commanders should recognize the increased security risk when using such personnel and carefully weigh the risk versus potential gain. If uncleared interpreters are used, Soldiers and Marines must limit discussing sensitive information.

ESTABLISHING RAPPORT

C-21. Interpreters are a vital link among Soldiers, Marines, and the target audience. Without supportive, cooperative interpreters, the mission is jeopardized. Mutual respect and understanding are essential to effective teamwork. Soldiers and Marines should establish rapport early and maintain it throughout the operation. Problems establishing rapport stem mostly from a lack of personal communication skills and misunderstandings regarding culture.

C-22. Before they meet interpreters, Soldiers and Marines study the area of operations and its inhabitants. This process is discussed in chapter 3. Many foreigners have some knowledge about the United States. Unfortunately, much of this comes from commercial movies and television shows. Soldiers and Marines may need to teach the interpreter something realistic about the United States as well.

C-23. Soldiers and Marines working with an interpreter should research and verify the interpreter's background. They should be genuinely interested in the interpreter and the interpreter's family, aspirations, career, and education. Many cultures emphasize family roles differently from the United States, so Soldiers and Marines should first understand the interpreter's home life. Though Soldiers and Marines should gain as much cultural information as possible before deploying, their interpreters can be valuable sources for filling gaps. However, information from interpreters will likely represent the views of the group to which they belong. Members of opposing groups almost certainly see things differently and often view culture and history differently.

C-24. Soldiers and Marines should gain an interpreter's trust and confidence before discussing sensitive issues. These issues include religion, likes, dislikes, and prejudices. Soldiers and Marines should approach these topics carefully. Although deeply held personal beliefs may be revealing and useful in a professional relationship, Soldiers and Marines should draw these out of their interpreters gently and tactfully.

C-25. One way to reinforce the bond between military personnel and their interpreter is to make sure the interpreter has every available comfort. This includes providing personal protection equipment—boots, helmets, and body armor—that the interpreter (especially a category I interpreter) may not already have. Soldiers and Marines must give interpreters the same comforts that military personnel enjoy. Interpreters need the same base comforts—shelter, air conditioning, and heat—as military personnel. If and when an interpreter is assigned to a specific unit, the interpreter ought to live with that organization to develop a bond. If there are several interpreters, it may be more effective for the interpreters to live together on the unit compound.

ORIENTING INTERPRETERS

C-26. Early in the relationship, Soldiers and Marines must explain to interpreters their duties, expected standards of conduct, interview techniques, and any other requirements and expectations. (Table C-1 lists some information to include when orienting interpreters.)

PREPARING FOR PRESENTATIONS

C-27. Sites for interviews, meetings, or classes should be carefully selected and arranged. The physical arrangement can be especially significant for certain groups or cultures.

C-28. Speakers should understand unique cultural practices before interviewing, instructing, or talking with foreign nationals. For example, speakers and interpreters should know when to stand, sit, or cross one's legs. Gestures are a learned behavior and vary from culture to culture. If properly selected, interpreters should be helpful in this regard.

C-29. Interpreters should mirror the speaker's tone and personality. They must not add their own questions or emotions. Speakers should instruct interpreters to inform them discreetly if they notice inconsistencies or peculiarities of speech, dress, and behavior.

Table C-1. Orientation for interpreters

Background for the interpreter includes—
• The current tactical situation. • Information on the source, interviewee, or target audience. • Specific objectives for the interview, meeting, or interrogation. • The conduct of the interview, lesson, or interrogation. • The physical arrangements of interviewing site, if applicable.
Duties of the interpreter include—
• Informing the interviewer about inconsistencies the interviewee uses in language. An example would be someone who claims to be a college professor, yet speaks like an uneducated person. • Possibly assisting in after-action reviews or assessments.
Interview techniques for the interpreter include—
• Simultaneous—when the interpreter listens and translates at the same time (not recommended). • Consecutive—when the interpreter listens to an entire phrase, sentence, or paragraph, and then translates during natural pauses.
Standards of conduct for the interpreter include—
• Being careful not to inject their personality, ideas, or questions. • Mirroring the speaker's tone and personality. • Translating the exact meaning without adding or deleting information.

C-30. Soldiers and Marines must carefully analyze the target audience. This analysis goes beyond the scope of this appendix. Mature judgment, thoughtful consideration of the target audience as individuals, and a genuine concern for their receiving accurate information helps Soldiers and Marines accomplish the mission. Soldiers and Marines should remember, for example, that a farmer from a small village has markedly different expectations and requirements than a city executive.

C-31. Soldiers and Marines who work through an interpreter may take double or triple the time normally required for an event. They may save time if they give the interpreter pertinent information beforehand. This information may include briefing slides, questions to ask, a lesson plan, copies of any handouts, or a glossary of difficult terms.

CONDUCTING PRESENTATIONS

C-32. As part of the initial training for interpreters, Soldiers and Marines emphasize that interpreters follow their speaker's lead. They become a vital communication link between the speaker and target audience. Soldiers and Marines should appeal to the interpreter's professional pride. They clarify how the quality and quantity of the information sent and received directly depends on linguistic skills. Although interpreters perform some editing as a function of the interpreting process, they must transmit the exact meaning without additions or deletions.

C-33. Speakers should avoid simultaneous translations—the speaker and interpreter talking at the same time—when conducting an interview or presenting a lesson. They should talk directly to the individual or audience for a minute or less in a neutral, relaxed manner. The interpreter should watch the speaker carefully. While translating, the interpreter should mimic the speaker's body language as well as interpret verbal meaning. Speakers should observe interpreters closely to detect any inconsistent behaviors. After speakers present one major thought in its entirety, interpreters then reconstruct it in their language. One way to ensure that the interpreter is communicating exactly what the speaker means is to have a senior interpreter observe several conversations. The senior interpreter can provide feedback along with further training.

C-34. Soldiers and Marines should be aware that some interpreters might attempt to save face or to protect themselves by concealing their lack of understanding. They may translate what they believe the speaker or audience said or meant without asking for clarification. This situation can result in misinformation and confusion. It can also impact the speaker's credibility. Interpreters must know that when in doubt they should always ask for clarification.

C-35. During an interview or lesson, if the interviewee asks questions, interpreters should immediately relay them to the speaker for an answer. Interpreters should never attempt to answer questions, even if they know the correct answers. Neither speakers nor interpreters should correct each other in front of an interviewee or class. They should settle all differences away from the subject or audience.

C-36. Establishing rapport with the interpreter is vital; establishing rapport with interviewees or the target audience is equally important. Speakers and their interpreters should concentrate on this task. To establish rapport, interviewees and target audiences should be treated as mature, important people who are worthy and capable.

C-37. Several methods ensure that the speaker communicates directly to the target audience using the interpreter as only a mechanism for that communication. One technique is to have the interpreter stand to the side of and just behind the speaker. This position lets the speaker stand face-to-face with the target audience. The speaker should always look at and talk directly to the target audience, rather than to the interpreter. This method allows the speaker and the target audience to establish a personal relationship.

SPEAKING TECHNIQUES

C-38. An important first step for Soldiers and Marines communicating in a foreign language is to reinforce and polish their English language skills. These skills are important, even when no attempt has been made to learn the HN language. They should use correct words, without idioms or slang. The more clearly Soldiers and Marines speak in English, the easier it is for interpreters to translate exactly. For instance, speakers may want to add words usually left out in colloquial English, such as the "air" in airplane. This ensures they are not misinterpreted as referring to the Great Plains or a carpenter's plane.

C-39. Speakers should not use profanity at all and should avoid slang. Many times, interpreters cannot translate such expressions. Even those they can translate might lose the desired meaning. Terms of surprise or reaction such as "gee whiz" and "golly" are difficult to translate.

C-40. Speakers should avoid using acronyms. While these are part of everyday military language, most interpreters and target audiences do not know them. The interpreter may have to interrupt the interview for clarification. This can disrupt the rhythm of the interview or lesson. If interpreters constantly interrupt the speaker for explanation, they could lose credibility in the eyes of the target audience. Such a reaction could jeopardize the interview or lesson. If speakers use technical terms or expressions, they should be sure interpreters convey the proper meaning. This preparation is best done in advance.

C-41. Before speaking impulsively, Soldiers and Marines should consider what they wish to say. They should break their thoughts into logical bits and articulate them one at a time. Using short, simple words and sentences helps the interpreter to translate quickly and easily. Speakers should never say more in one sentence than they can easily repeat immediately after saying it. Each sentence should contain a complete thought without the extra words.

C-42. Speakers should avoid American "folk" and culture-specific references. Target audiences may have no idea what is being talked about. Even when interpreters understand the reference, they may find it difficult to quickly identify an appropriate equivalent in the target audience's cultural frame of reference.

C-43. Transitional phrases and qualifiers may confuse nonnative speakers and waste valuable time. Examples include "for example," "in most cases," "maybe," and "perhaps."

C-44. Speakers should avoid American humor. Humor is culturally specific and does not translate well. Cultural and language differences can lead to misinterpretations by foreigners.

C-45. Speakers must consider some "dos" and "don'ts" for when working with interpreters. Table C-2 clarifies what speakers should and should not do.

Table C-2. Good and bad practices for speakers

Speakers should—
• Position the interpreter by their side (or a step back). This keeps the subject or target audience from shifting their attention or fixating on the interpreter rather than on the leader. • Always look at and talk directly to the subject or target audience. Guard against the tendency to talk to the interpreter. • Speak slowly and clearly. Repeat as often as necessary. • Speak to the individual or group as if they understand English. Be enthusiastic. Use gestures, movements, voice intonations, and inflections normally used with an English-speaking group. Remember that considerable nonverbal meaning is conveyed through voice and body movements. Encourage interpreters to mimic the same delivery. • Periodically check an interpreter's accuracy, consistency, and clarity. Request a U.S. citizen fluent in the language to check that the interpreter is not distorting the translation, intentionally or unintentionally. Learn some of the language. • Check with the audience whenever a misunderstanding is suspected and clarify immediately. Using the interpreter, ask questions to elicit answers to indicate whether the point is clear. If it is not, rephrase the instruction and illustrate the point differently. Use repetition and examples whenever necessary to facilitate learning. If the interviewees ask few questions, it may mean they have not understood the instruction or the message is unclear. • Ensure interpreters understand they are valuable team members. Recognize them based on the importance of their contributions. Protect interpreters; the insurgents and criminal elements may target them.
Speakers should not—
• Address the subject or audience in the third person through the interpreter. For example, avoid saying, "Tell them I'm glad to be their instructor." Instead, directly address the subject or audience saying, "I am glad to be your instructor." Make continual eye contact with the audience. Watch them, not the interpreter. • Make side comments to the interpreter that are not interpreted. This action is rude, discourteous, and creates the wrong atmosphere. • Distract the audience while the interpreter is translating. Avoid pacing, writing on the blackboard, teetering on the lectern, drinking beverages, or doing any other distracting activity while the interpreter is translating.

This page intentionally left blank.

Appendix D
Legal Considerations

Law and policy govern the actions of the U.S. forces in all military operations, including counterinsurgency. For U.S. forces to conduct operations, a legal basis must exist. This legal basis profoundly influences many aspects of the operation. It affects the rules of engagement, how U.S. forces organize and train foreign forces, the authority to spend funds to benefit the host nation, and the authority of U.S. forces to detain and interrogate. Under the Constitution, the President is commander in chief of the U.S. forces. Therefore, orders issued by the President or the Secretary of Defense to a combatant commander provide the starting point in determining the legal basis. This appendix summarizes some of the laws and policies that bear upon U.S. military operations in support of foreign counterinsurgencies. Laws are legislation passed by Congress and signed into law by the President, as well as treaties to which the United States is party. Policies are executive orders, departmental directives and regulations, and other authoritative statements issued by government officials. No summary provided here can replace a consultation with the unit's supporting staff judge advocate.

AUTHORITY TO ASSIST A FOREIGN GOVERNMENT

D-1. U.S. forces have limited authority to provide assistance to foreign governments. For foreign internal defense, U.S. forces may be authorized to make limited contributions. Assistance to police by U.S. forces is permitted, but not with the Department of Defense (DOD) as the lead governmental department.

AUTHORITY FOR FOREIGN INTERNAL DEFENSE

D-2. The President or Secretary of Defense give the deployment and execution order. Without receiving a deployment or execution order, U.S. forces may be authorized to make only limited contributions in support of a host nation's counterinsurgency (COIN) effort. If the Secretary of State requests and the Secretary of Defense approves, U.S. forces can participate in this action. The request and approval go through standing statutory authorities in Title 22, United States Code. Title 22 contains the Foreign Assistance Act, the Arms Export Control Act, and other laws. It authorizes security assistance, developmental assistance, and other forms of bilateral aid. The request and approval might also occur under various provisions in Title 10, United States Code. Title 10 authorizes certain types of military-to-military contacts, exchanges, exercises, and limited forms of humanitarian and civic assistance in coordination with the U.S. Ambassador for the host nation. In such situations, U.S. military personnel work as administrative and technical personnel. They are part of the U.S. diplomatic mission, pursuant to a status of forces agreement, or pursuant to an exchange of letters. This cooperation and assistance is limited to liaison, contacts, training, equipping, and providing defense articles and services. It does not include direct involvement in operations.

DOD USUALLY NOT LEAD—GENERAL PROHIBITION ON ASSISTANCE TO POLICE

D-3. DOD is usually not the lead governmental department for assisting foreign governments, even for the provision of security assistance—that is, military training, equipment, and defense articles and services—to the host nation's military forces. DOD contribution may be large, but the legal authority is typically one exercised by the Department of State. With regard to provision of training to a foreign government's police or other civil interior forces, the U.S. military typically has no authorized role. The Foreign Assistance Act specifically prohibits assistance to foreign police forces except within carefully circumscribed exceptions, and under a Presidential directive, and the lead role in providing police assistance within those exceptions has been normally delegated to the Department of State's Bureau of International Narcotics and Law Enforcement Affairs. However, the President did sign a decision directive in 2004 granting authority to train and equip Iraqi police to the Commander, U.S. Central Command.

AUTHORIZATION TO USE MILITARY FORCE

D-4. Two types of resolutions authorize involvement of U. S. forces: a Congressional resolution and the 1973 War Powers Resolution. Congressional support is necessary if U.S. forces will be involved in actual operations overseas. The 1973 Resolution lets the President authorize military forces for a limited time.

CONGRESSIONAL RESOLUTION

D-5. Congressional support is needed for any prolonged involvement of U.S. forces in actual operations overseas. Often a Congressional resolution provides the central legal basis for such involvement within domestic law. This is especially likely if U.S. forces are anticipated, at least initially, to be engaged in combat operations against an identified hostile force.

STANDING WAR POWERS RESOLUTION

D-6. The 1973 War Powers Resolution requires the President to consult and report to Congress when introducing U.S. forces into certain situations. There are times though when a specific Congressional authorization for use of force is absent. In the absence of this authorization, the President—without conceding that the 1973 Resolution binds the President's own constitutional authority—makes a report to Congress. The President must make the report within 48 hours of introducing substantial U.S. forces into the host nation. This report details the circumstances necessitating introduction or enlargement of troops. The President bases his or her action on the Constitutional or legislative authority and the estimated scope and duration of the deployment or combat action. The 1973 Resolution states that if Congress does not declare war or specifically authorize the deployment or combat action within 60 days of the report, the President must terminate U.S. military involvement and redeploy U.S. forces.

RULES OF ENGAGEMENT

D-7. *Rules of engagement* (ROE) are directives issued by competent military authority that delineate the circumstances and limitations under which United States forces will initiate and/or continue combat engagement with other forces encountered (JP 1-02). Often these directives are specific to the operation. If there are no operation-specific ROE, U.S. forces apply standing rules of engagement (SROE). When working with a multinational force, commanders must coordinate the ROE thoroughly.

OPERATION-SPECIFIC RULES OF ENGAGEMENT

D-8. In a large-scale deployment, the Secretary of Defense may issue ROE specific to the operation to a combatant commander. The combatant commander and subordinate commanders then issue ROE consistent with the ROE received from the Secretary of Defense. ROE state the circumstances under which Soldiers or Marines may open fire. They may fire when they positively identify a member of a hostile force or they have clear indications of hostile intent. ROE may include rules concerning when civilians may be detained, specify levels of approval authority for using heavy weapons, or identify facilities that may be protected with deadly force. All ROE comply with the law of war. ROE in COIN are dynamic. Commanders must regularly review ROE for their effectiveness in the complex COIN environment. Training counterinsurgents in ROE should be reinforced regularly.

CHAIRMAN OF THE JOINT CHIEFS OF STAFF STANDING RULES OF ENGAGEMENT

D-9. In the absence of operation-specific ROE, U.S. forces apply CJCSI 3121.01B. This instruction refers to SROE for U.S. forces. The SROE establish fundamental policies and procedures governing the actions of U.S. force commanders in certain events. These events include military attacks against the United States and during all military operations, contingencies, terrorist attacks, or prolonged conflicts outside the territorial jurisdiction of the United States. The SROE do not limit a commander's inherent authority and obligation to use all necessary means available. They also do not limit the commander's authority and obligation to take all appropriate action in self-defense of the commander's unit and other U.S. forces in the vicinity. The SROE prescribe how supplemental ROE for specific operations are provided as well as the format by which subordinate commanders may request ROE.

Multinational Rules of Engagement

D-10. When U.S. forces, under U.S. operational or tactical control, operate with a multinational force, reasonable efforts are made to effect common ROE. If such ROE cannot be established, U.S. forces operate under the SROE or operation-specific ROE provided by U.S. authorities. To avoid misunderstanding, commanders thoroughly discuss among multinational forces any differences in ROE or ROE interpretation. They disseminate any differences in ROE to the units involved.

THE LAW OF WAR

D-11. COIN and international armed conflicts often overlap. COIN may take place before, after, or simultaneously with a war occurring between nations. U.S. forces obey the law of war. The law of war is a body of international treaties and customs, recognized by the United States as binding. It regulates the conduct of hostilities and protects noncombatants. The main law of war protections come from the Hague and Geneva Conventions. They apply at the tactical and operational levels and are summarized in ten rules:

- Soldiers and Marines fight only enemy combatants.
- Soldiers and Marines do not harm enemies who surrender. They disarm them and turn them over to their superiors.
- Soldiers and Marines do not kill or torture enemy prisoners of war.
- Soldiers and Marines collect and care for the wounded, whether friend or foe.
- Soldiers and Marines do not attack medical personnel, facilities, or equipment.
- Soldiers and Marines destroy no more than the mission requires.
- Soldiers and Marines treat all civilians humanely.
- Soldiers and Marines do not steal. They respect private property and possessions.
- Soldiers and Marines do their best to prevent violations of the law of war.
- Soldiers and Marines report all violations of the law of war to their superior.

D-12. When insurgency occurs during occupation, the law of war includes rules governing situations in which the military forces of one state occupy the territory of another. Occupation is not a transfer of sovereignty. It does however grant the occupying power the authority and responsibility to restore and maintain public order and safety. The occupying power must respect, as much as possible, the laws in force in the host nation. One of the four Geneva Conventions of 1949—the Geneva Convention Relative to the Protection of Civilian Persons in Time of War—becomes a prominent source of law during occupation.

INTERNAL ARMED CONFLICT

D-13. During COIN operations, commanders must be aware of Common Article 3 of the Geneva Conventions and the status of insurgents under the laws of the host nation.

Geneva Convention, Common Article 3

D-14. Although insurgencies can occur simultaneously with a legal state of war between two nations, they are classically conflicts internal to a single nation, between uniformed government forces and armed elements that do not wear uniforms with fixed distinctive insignia, carry arms openly, or otherwise obey the laws and customs of war. As such, the main body of the law of war does not strictly apply to these conflicts—a legal fact that can be a source of confusion to commanders and Soldiers. It bears emphasis, however, that one article contained in all four of the Geneva Conventions—Common Article 3—is specifically intended to apply to internal armed conflicts:

In the case of armed conflict not of an international character occurring in the territory of one of the High Contracting Parties, each Party to the conflict shall be bound to apply, as a minimum, the following provisions:

(1) Persons taking no active part in the hostilities, including members of armed forces who have laid down their arms and those placed "hors de combat" by sickness, wounds, detention, or any other cause, shall in all circumstances be treated humanely, without any adverse distinction founded on race, colour, religion or faith, sex, birth or wealth, or any other similar criteria.

To this end, the following acts are and shall remain prohibited at any time and in any place whatsoever with respect to the above-mentioned persons:

(a) Violence to life and person, in particular murder of all kinds, mutilation, cruel treatment and torture;

(b) Taking of hostages;

(c) Outrages upon personal dignity, in particular humiliating and degrading treatment;

(d) The passing of sentences and the carrying out of executions without previous judgment pronounced by a regularly constituted court, affording all the judicial guarantees which are recognized as indispensable by civilized peoples.

(2) The wounded and sick shall be collected and cared for.

An impartial humanitarian body, such as the International Committee of the Red Cross, may offer its services to the Parties to the conflict.

The Parties to the conflict should further endeavor to bring into force, by means of special agreements, all or part of the other provisions of the present Convention.

The application of the preceding provisions shall not affect the legal status of the Parties to the conflict.

APPLICATION OF CRIMINAL LAWS OF THE HOST NATION

D-15. The final sentence of Common Article 3 makes clear that insurgents have no special status under international law. They are not, when captured, prisoners of war. Insurgents may be prosecuted legally as criminals for bearing arms against the government and for other offenses, so long as they are accorded the minimum protections described in Common Article 3. U.S. forces conducting COIN should remember that the insurgents are, as a legal matter, criminal suspects within the legal system of the host nation. Counterinsurgents must carefully preserve weapons, witness statements, photographs, and other evidence collected at the scene. This evidence will be used to process the insurgents into the legal system and thus hold them accountable for their crimes while still promoting the rule of law.

D-16. Status of forces agreements establish the legal status of military personnel in foreign countries. Criminal and civil jurisdiction, taxation, and claims for damages and injuries are some of the topics usually covered in a status of forces agreement. In the absence of an agreement or some other arrangement with the host nation, DOD personnel in foreign countries may be subject to its laws.

DETENTION AND INTERROGATION

D-17. Chapters 3, 5, and 7 indicate the need for human intelligence in COIN operations. This need can create great pressure to obtain time-sensitive information from detained individuals. The Detainee Treatment Act of 2005, FM 2-22.3, and other specific standards were created to guide U.S. forces working with detainees.

DETAINEE TREATMENT ACT OF 2005

D-18. U.S. law clearly prohibits U.S. forces, including officials from other government agencies, from using certain methods to obtain information. Instances of detainee abuse, including maltreatment involving interrogation, were documented. In response, Congress passed, and the President signed into law, the Detainee Treatment Act of 2005. (See Table D-1.)

Table D-1. Extract of the Detainee Treatment Act of 2005

Section 1002: Uniform Standards for the Interrogation of Persons Under the Detention of the Department of Defense
(a) In General.—No person in the custody or under the effective control of the Department of Defense or under detention in a Department of Defense facility shall be subject to any treatment or technique of interrogation not authorized by and listed in the United States Army Field Manual on Intelligence Interrogation [FM 2-22.3].
(b) Applicability.—Subsection (a) shall not apply with respect to any person in the custody or under the effective control of the Department of Defense pursuant to a criminal law or immigration law of the United States.
(c) Construction.—Nothing in this section shall be construed to affect the rights under the United States Constitution of any person in the custody or under the physical jurisdiction of the United States.
Section 1003: Prohibition on Cruel, Inhuman, or Degrading Treatment or Punishment of Persons Under Custody or Control of the United States Government
(a) In General.—No individual in the custody or under the physical control of the United States Government, regardless of nationality or physical location, shall be subject to cruel, inhuman, or degrading treatment or punishment.
(b) Construction.—Nothing in this section shall be construed to impose any geographical limitation on the applicability of the prohibition against cruel, inhuman, or degrading treatment or punishment under this section.
(c) Limitation on Supersedure.—The provisions of this section shall not be superseded, except by a provision of law enacted after the date of the enactment of this Act which specifically repeals, modifies, or supersedes the provisions of this section.
(d) Cruel, Inhuman, or Degrading Treatment or Punishment defined.—In this section, the term "cruel, inhuman, or degrading treatment or punishment" means the cruel, unusual, and inhumane treatment or punishment prohibited by the Fifth, Eighth, and Fourteenth Amendments to the Constitution of the United States, as defined in the United States Reservations, Declarations and Understandings to the United Nations Convention Against Torture and Other Forms of Cruel, Inhuman, or Degrading Treatment or Punishment done at New York, December 10, 1984.

Interrogation Field Manual

D-19. The Detainee Treatment Act established FM 2-22.3 as the legal standard. No techniques other than those prescribed by the field manual are authorized by U.S. forces. Commanders must ensure that interrogators receive proper training and supervision.

Standards for Detention and Internment

D-20. Regardless of the precise legal status of those persons captured, detained, or otherwise held in custody by U.S. forces, they must receive humane treatment until properly released. They also must be provided the minimum protections of the Geneva Conventions. Specially trained, organized, and equipped military police units in adequately designed and resourced facilities should accomplish prolonged detention. Such detention must follow the detailed standards contained in AR 190-8/MCO 3461.1. The military police personnel operating such facilities shall not be used to assist in or "set the conditions for" interrogation.

Transfer of Detainees to the Host Nation

D-21. There are certain conditions under which U.S. forces may not transfer the custody of detainees to the host nation or any other foreign government. U.S. forces retain custody if they have substantial grounds to believe that the detainees would be in danger in the custody of others. Such danger could include being subjected to torture or inhumane treatment. (For more information on transferring detainees, see DODD 2310.01E and consult the legal advisor or staff judge advocate.)

ENFORCING DISCIPLINE OF U.S. FORCES

D-22. Despite rigorous selection and training, some personnel require discipline. The Uniform Code of Military Justice is the criminal code of military justice applicable to all military members. Commanders and general officers are responsible for their subordinates and their behavior. Commanders must give clear guidance and ensure compliance. All civilians working for the U.S. Government also must comply with the laws.

UNIFORM CODE OF MILITARY JUSTICE

D-23. Although the vast majority of well-led and well-trained U.S. military personnel perform their duties honorably and lawfully, history records that some commit crimes amidst the decentralized command and control, the strains of opposing a treacherous and hidden enemy, and the often complex ROE that characterize the COIN environment. Uniformed personnel remain subject at all times to the Uniform Code of Military Justice and must be investigated and prosecuted, as appropriate, for violations of orders, maltreatment of detainees, assaults, thefts, sexual offenses, destruction of property, and other crimes, including homicides, that they may commit during COIN.

COMMAND RESPONSIBILITY

D-24. In some cases, military commanders may be deemed responsible for crimes committed by subordinates or others subject to their control. This situation arises when the criminal acts are committed pursuant to the commander's order. Commanders are also responsible if they have actual knowledge, or should have knowledge, through reports received or through other means, that troops or other persons subject to their control are about to commit or have committed a crime, and they fail to take the necessary and reasonable steps to ensure compliance with the law or to punish violators.

GENERAL ORDERS

D-25. Orders issued by general officers in command during COIN likely include provisions, such as a prohibition against drinking alcohol or against entering places of religious worship, important to maintaining discipline of the force, to safeguarding the image of U.S. forces, and to promoting the legitimacy of the host government. These orders are readily enforceable under the Uniform Code of Military Justice.

CIVILIAN PERSONNEL AND CONTRACTORS

D-26. Modern COIN operations involve many DOD civilians as well as civilian personnel employed by government contractors. The means of disciplining such persons for violations differ from the means of disciplining uniformed personnel. These civilians may be made subject to general orders. They are also subject to U.S. laws and to the laws of the host nation. These civilians may be prosecuted or receive adverse administrative action by the United States or contract employers. DOD directives contain further policy and guidance pertaining to U.S. civilians accompanying our forces in COIN.

HUMANITARIAN RELIEF AND RECONSTRUCTION

D-27. In COIN, like all operations, commands require specific authority to expend funds. That authority is normally found in the DOD Appropriations Act, specifically, operation and maintenance funds. In recent COIN operations, Congress appropriated additional funds to commanders for the specific purpose of dealing with COIN. Recent examples include the commander's emergency response program (CERP), the Iraq Relief and Reconstruction Fund, Iraq Freedom Fund, and Commander's Humanitarian Relief and Reconstruction Program funds.

DOD FUNDS GENERALLY NOT EXPENDABLE BY COMMANDERS FOR THIS PURPOSE

D-28. Congress specifically appropriates funds for foreign assistance. The United States Agency for International Development expends such funds under the legal authorities in Title 22, United States Code. Provisions of Title 10 authorize small amounts of money. These funds are appropriated annually for commanders to provide humanitarian relief, disaster relief, or civic assistance in conjunction with military operations. These standing authorities are narrowly defined and generally require significant advance coordination within the DOD and the Department of State. As such, they are of limited value to ongoing COIN operations.

COMMANDERS EMERGENCY RESPONSE PROGRAM

D-29. Beginning in November of 2003, Congress authorized use of a specific amount of operations and maintenance funds for a CERP in Iraq and Afghanistan. The legislation was renewed in successive appropriations and authorization acts. It specified that commanders could spend the funds for urgent humanitarian relief and reconstruction projects. These projects had to immediately assist the Iraqi and Afghan peoples within a commander's area of operations. Congress did not intend the funds to be used as—

- Security assistance such as weapons, ammunition, and supplies for security forces.
- Salaries for Iraqi or Afghan forces or employees.
- Rewards for information.
- Payments in satisfaction of claims made by Iraqis or Afghanis against the United States (specific legislation must authorize such payments).

D-30. The CERP provided tactical commanders a ready source of cash for small-scale projects. They could repair public buildings, clear debris from roadways, provide supplies to hospitals and schools, and meet other local needs. Because Congress had provided special authority for the program, normal federal acquisition laws and regulations did not apply. The reporting requirements were minimal.

D-31. The CERP is not a standing program. Any similar future program should be governed by whatever specific legislative provision Congress chooses to enact. In any program similar to CERP, commanders and staffs must make sound, well-coordinated decisions on how to spend the funds. They must ensure that maximum goodwill is created. Commanders must verify that the extra cash does not create harmful effects in the local economy. One such side effect would be creating unsustainable wages that divert skilled labor from a host-nation (HN) program essential to its legitimacy. Commanders must also ensure that projects can be responsibly administered to achieve the desired objective and that they avoid financing insurgents inadvertently.

TRAINING AND EQUIPPING FOREIGN FORCES

D-32. Effective foreign forces need training and equipment. U.S. laws require Congress to authorize such expenditures. U.S. laws also require the Department of State to verify that the host nation receiving the assistance is not in violation of human rights.

NEED FOR SPECIFIC AUTHORITY

D-33. All training and equipping of foreign security forces must be specifically authorized. Usually, DOD involvement is limited to a precise level of man-hours and materiel requested from the Department of State under the Foreign Assistance Act. The President may authorize deployed U.S. forces to train or advise HN security forces as part of the operational mission. In this case, DOD personnel, operations, and maintenance appropriations provide an incidental benefit to those security forces. All other weapons, training, equipment, logistic support, supplies, and services provided to foreign forces must be paid for with funds appropriated by Congress for that purpose. Examples include the Iraq Security Forces Fund and the Afghan Security Forces Fund of fiscal year 2005. Moreover, the President must give specific authority to the DOD for its role in such "train and equip" efforts. In May of 2004, the President signed a decision directive that made the commander, U.S. Central Command, under policy guidance from the chief of mission, responsible for coordinating all U.S. Government efforts to organize, train, and equip Iraqi Security Forces, including police. Absent such a directive, DOD lacks authority to take the lead in assisting a host nation to train and equip its security forces.

Human Rights Vetting

D-34. Congress typically limits when it will fund training or equipment for foreign security forces. If the Department of State has credible information that the foreign security force unit identified to receive the training or equipment has committed a gross violation of human rights, Congress prohibits funding. Such prohibitions impose a requirement upon Department of State and DOD. These departments must vet the proposed recipient units against a database of credible reports of human rights violations.

CLAIMS AND SOLATIA

D-35. Under certain conditions, the U.S. Government will make payments to HN civilians. The Foreign Claims Act permits claims to be filed against the U.S. Government. In some countries, solatia payments are made.

Foreign Claims Act

D-36. Under the Foreign Claims Act, claims by HN civilians for property losses, injury, or death caused by service members or the civilian component of the U.S. forces may be paid to promote and maintain friendly relations with the host nation. Claims that result from noncombat activities or negligent or wrongful acts or omissions are also payable. Claims that are not payable under the Foreign Claims Act include losses from combat, contractual matters, domestic obligations, and claims which are either not in the best interest of the United States to pay, or which are contrary to public policy. Because payment of claims is specifically governed by law and because many claims prove, upon investigation, to be not payable, U.S. forces must be careful not to raise expectations by promising payment.

Solatia

D-37. If U.S. forces are conducting COIN in a country where payments in sympathy or recognition of loss are common, solatia payments to accident victims may be legally payable. Solatia payments are not claims payments. They are payments in money or in kind to a victim or to a victim's family as an expression of sympathy or condolence. The payments are customarily immediate and generally nominal. The individual or unit involved in the damage has no legal obligation to pay; compensation is simply offered as an expression of remorse in accordance with local custom. Solatia payments should not be made without prior coordination with the combatant command.

ESTABLISHING THE RULE OF LAW

D-38. Establishing the rule of law is a key goal and end state in COIN. Defining that end state requires extensive coordination between the instruments of U.S. power, the host nation, and multinational partners. Additionally, attaining that end state is usually the province of HN authorities, international and intergovernmental organizations, the Department of State, and other U.S. Government agencies, with support from U.S. forces in some cases. Some key aspects of the rule of law include:

- **A government that derives its powers from the governed** and competently manages, coordinates, and sustains collective security, as well as political, social, and economic development. This includes local, regional, and national government.
- **Sustainable security institutions**. These include a civilian-controlled military as well as police, court, and penal institutions. The latter should be perceived by the local populace as fair, just, and transparent.
- **Fundamental human rights**. The United Nations Declaration on Human Rights and the International Convention for Civil and Political Rights provide a guide for applicable human rights. The latter provides for derogation from certain rights, however, during a state of emergency. Respect for the full panoply of human rights should be the goal of the host nation; derogation and

violation of these rights by HN security forces, in particular, often provides an excuse for insurgent activities.

D-39. In periods of extreme unrest and insurgency, HN legal structures—courts, prosecutors, defense assistance, and prisons—may cease to exist or function at any level. Under these conditions, counterinsurgents may need to undertake a significant role in the reconstruction of the HN judicial system in order to establish legal procedures and systems to deal with captured insurgents and common criminals. During judicial reconstruction, counterinsurgents can expect to be involved in providing sustainment and security support. They can also expect to provide legal support and advice to the HN judicial entities. Even when judicial functions are restored, counterinsurgents may still have to provide logistic and security support to judicial activities for a prolonged period. This support continues as long as insurgents continue to disrupt activities that support the legitimate rule of law.

This page intentionally left blank.

Appendix E
Airpower in Counterinsurgency

Counterinsurgency operations are, by their nature, joint operations—and airpower and landpower are interdependent elements of such operations. As this appendix explains, airpower and spacepower are important force multipliers for U.S., multinational, and host-nation forces fighting an insurgency.

OVERVIEW

E-1. Airpower can contribute significant support to land forces conducting counterinsurgency (COIN) operations. Aircraft can, for example, strike insurgents, and that can be enormously important in many situations. However, given the nature of the COIN environment, airpower will most often transport troops, equipment, and supplies and perform intelligence, surveillance, and reconnaissance missions. Rough terrain and poor transportation networks can create serious obstacles for COIN forces while giving advantages to insurgents. Airpower helps counterinsurgents overcome these obstacles. Thus, airpower both serves as a significant force multiplier and enables counterinsurgents to operate more effectively.

E-2. Airpower provides considerable asymmetric advantages to counterinsurgents. If insurgents assemble a conventional force, air assets can respond quickly with precision fires. In a sudden crisis, air mobility can immediately move land forces where they are needed. In numerous COIN operations, airpower has demonstrated a vital supporting role. In Malaya (1948 through 1960) and El Salvador (1980 through 1992), as well as more recently in Colombia and Afghanistan, airpower contributed significantly to successful COIN operations. In these cases, the ability to airlift British and U.S. Army and police units to remote locations proved important in tracking down and eliminating insurgents. Airpower enables counterinsurgents to operate in rough and remote terrain, areas that insurgents traditionally have used as safe havens.

E-3. Effective leaders also use airpower in roles other than delivering ordnance. In Colombia, aerial crop dusters sprayed and eradicated coca fields that provided drug income for insurgents. During the El Salvador insurgency, medical evacuation (MEDEVAC) helicopters provided to the Salvadoran forces played a central role in improving the Salvadorans' fighting capabilities. Salvadoran morale improved noticeably when soldiers knew that, if they were wounded, MEDEVAC helicopters would get them to a hospital in minutes. With this air support, the Salvadoran Army became much more aggressive in tracking down and engaging insurgents.

E-4. Air transport can also quickly deliver humanitarian assistance. In isolated regions, using air transport to airlift or airdrop food and medical supplies to civilians can help win the populace's support. Air transport is also important for COIN logistics. In areas where ground convoys are vulnerable, U.S. forces can airlift supplies, enabling commanders to maintain forces in remote but strategically important locations.

AIRPOWER IN THE STRIKE ROLE

E-5. Precision air attacks can be of enormous value in COIN operations; however, commanders exercise exceptional care when using airpower in the strike role. Bombing, even with the most precise weapons, can cause unintended civilian casualties. Effective leaders weigh the benefits of every air strike against its risks. An air strike can cause collateral damage that turns people against the host-nation (HN) government and provides insurgents with a major propaganda victory. Even when justified under the law of war, bombings that result in civilian casualties can bring media coverage that works to the insurgents' benefit. For example, some Palestinian militants have fired rockets or artillery from near a school or village to draw a retaliatory air strike that kills or wounds civilians. If that occurs, the insurgents display those killed and wounded to the media as victims of aggression.

E-6. Even when destroying an obvious insurgent headquarters or command center, counterinsurgents must take care to minimize civilian casualties. New, precise munitions with smaller blast effects can limit collateral damage. When considering the risk of civilian casualties, commanders must weigh collateral damage against the unintended consequences of taking no action. Avoiding all risk may embolden insurgents while providing them sanctuary. The proper and well-executed use of aerial attack can conserve resources, increase effectiveness, and reduce risk to U.S. forces. Given timely, accurate intelligence, precisely delivered weapons with a demonstrated low failure rate, appropriate yield, and proper fuse can achieve desired effects while mitigating adverse effects. However, inappropriate or indiscriminate use of air strikes can erode popular support and fuel insurgent propaganda. For these reasons, commanders should consider the use of air strikes carefully during COIN operations, neither disregarding them outright nor employing them excessively.

AIRPOWER IN INTELLIGENCE COLLECTION

E-7. Given the challenges faced by human intelligence (HUMINT) assets in finding and penetrating insurgent networks, counterinsurgents must effectively employ all available intelligence collection capabilities. A combination of unmanned aircraft systems, manned aircraft, and space-based platforms can provide counterinsurgents with many collection capabilities.

E-8. When insurgents operate in rural or remote areas, aerial reconnaissance and surveillance proves useful. Working with signals intelligence (SIGINT), aerial reconnaissance and surveillance uses imagery and infrared systems to find hidden base camps and insurgent defensive positions. Persistent aerial surveillance can often identify people, vehicles, and buildings—even when they are hidden under heavy growth. Manned and unmanned aircraft can patrol roads to locate insurgent ambushes and improvised explosive devices. Air-mounted SIGINT collection platforms can detect insurgent communications and locate their points of origin.

E-9. Air assets have proven important in tactical operations and in convoy and route protection. Helicopters have been especially useful in providing overwatch, fire support, alternate communications, and MEDEVAC support. At the tactical level, air support requires a decentralized command and control system that gives supported units immediate access to available combat air assets and to information collected by air reconnaissance and support assets.

E-10. However, intelligence obtained through air and space platforms works best when it is quickly and efficiently routed to a joint intelligence center. This center fuses HUMINT information with that collected by other intelligence disciplines. To provide a complete picture, air and space intelligence must be combined with HUMINT. For example, while SIGINT and aerial surveillance and reconnaissance assets can determine that people are evacuating a village, they cannot explain why the people are leaving.

E-11. HUMINT is also a key enabler of airpower in the strike role. Commanders require the best possible intelligence about a target and its surrounding area when considering an air strike. With proper placement and access to a target, a HUMINT source can often provide the most accurate target data. Details might include optimum strike times, detailed descriptions of the surrounding area, and the presence of sensitive sites like hospitals, churches, and mosques. Target data can include other important factors for collateral damage considerations. Poststrike HUMINT sources equipped with a cell phone, radio, or camera can provide an initial battle damage assessment in near real time. With a thorough debriefing, the HUMINT source can provide an accurate assessment of the functional and psychological effects achieved on the target. Commanders can use this information to assess restrike options.

AIR AND SPACE INFORMATION OPERATIONS

E-12. Air and space forces have information operations (IO) capabilities that include collecting, controlling, exploiting, and protecting information. To make IO most effective, commanders should seamlessly integrate it among all Service components. Air and space forces contribute to the execution of three IO missions:

- Influence operations.
- Electronic warfare.
- Network operations.

E-13. Air and space forces conduct and support many influence operations. These operations include the following:

- Counterpropaganda.
- Psychological operations.
- Military deception.
- Operations security.
- Counterintelligence.
- Public affairs (a related activity of IO).

Commanders must preplan and deconflict these activities to ensure success.

E-14. Airpower and spacepower also contribute to information superiority through electronic warfare operations. Air and space assets are critical in the effort to shape, exploit, and degrade the enemy's electronic devices while protecting and enhancing those of counterinsurgents. The electronic warfare spectrum is not limited to radio frequencies; it includes the optical and infrared regions as well.

E-15. In this context, *network operations* are activities conducted to operate and defend the Global Information Grid (JP 1-02). Commanders enhance these operations by using air and space systems. Such tools help achieve desired effects across the interconnected analog and digital network portions of the Global Information Grid.

HIGH-TECHNOLOGY ASSETS

E-16. Today's high-technology air and space systems have proven their worth in COIN operations. Unmanned aircraft systems, such as the Predator, give counterinsurgents unprecedented capabilities in surveillance and target acquisition. Aerial surveillance platforms with long loiter times can place an entire region under constant surveillance. Tactical air control parties now provide ground commanders beyond-line-of-sight awareness with ROVER (remote operations video enhanced receiver), which links to aircraft targeting pods and unmanned aircraft systems. Predators have been equipped with precision munitions and successfully employed in the strike role against senior terrorist leaders. Air- and space-based SIGINT platforms give U.S. forces and multinational partners important information collection capabilities. Modern munitions, such as the joint direct attack munition, can guide accurately through clouds and bad weather to destroy insurgent targets under adverse conditions.

LOW-TECHNOLOGY ASSETS

E-17. Today's low-technology aspects of airpower have also proven effective in COIN operations. Light, slow, inexpensive civilian aircraft often have successfully patrolled border areas. In the 1980s, Guatemala mobilized its civilian light aircraft, formed them into an air force reserve, and used them to patrol main roads to report suspected ambushes. This successfully deterred insurgent attacks along Guatemala's major routes. In Africa in the 1980s, South African forces used light aircraft to locate small groups of insurgents trying to infiltrate Namibia from Angola. In Iraq, light aircraft fly patrols to spot insurgents crossing the border. Israel and the United States have even used stationary balloons equipped with video cameras and infrared sensors to watch for border incursions. These unmanned balloons are a simple, inexpensive, and effective means to monitor activity in remote areas.

E-18. The United States and many small nations have effectively used aerial gunships as close air support weapons in COIN operations. A gunship is a transport aircraft modified to carry and fire heavy guns and light artillery from fixed mounts. Many gunship models exist. They range from the Air Force's AC-130 to smaller transports modified to carry weapons ranging from .50-caliber machine guns to 40-millimeter rapid-fire cannons. The gunship's major limitation is its vulnerability to antiaircraft weapons and missiles. Gunships require a relatively benign environment to operate.

AIRLIFT

E-19. Airlift provides a significant asymmetric advantage to COIN forces, enabling commanders to rapidly deploy, reposition, sustain, and redeploy land forces. While land forces can execute these basic missions alone, airlift bypasses weaknesses insurgents have traditionally exploited. For example, airlift enables land forces to operate in rough and remote terrain and to avoid lines of communications (LOCs) targeted by insurgents. During Operation Iraqi Freedom, airlift has provided protected LOCs through convoy mitigation flights. These flights rerouted typical convoy supplies and vehicles. Since insurgents frequently attacked ground convoys, convoy mitigation flights saved lives.

E-20. Sources of airlift include multinational and HN rotary- and fixed-wing assets. Special operations forces provide specialized airlift capabilities for inserting and extracting troops. Strategic intertheater airlift platforms can provide a logistic pipeline. This pipeline moves large quantities of time-critical equipment, supplies, and personnel into and out of a theater. Modern strategic airlift can often provide a direct delivery capability, landing at relatively short, austere fields formerly serviced only by intratheater airlift.

E-21. Modes of airlift include airland and airdrop. Each mode provides advantages and disadvantages, depending on the environment. Airland missions carry greater payloads, resulting in less potential for damage. They also provide backhaul capability (critical for MEDEVAC), troop rotation, equipment repair, and repositioning and redeployment of COIN forces. Fixed-wing assets on airland missions require longer and better prepared landing surfaces. Vertical-lift assets on airland missions can operate from much smaller, more austere fields; however, they fly at slower speeds and often have smaller payloads and shorter ranges. Airdrop missions require the least amount of infrastructure at the receiving end and allow for rapid buildup of forces—up to brigade size. Equally important, airdrop can provide precision insertion and sustainment of numerous small units. Advances in precision-guided, steerable parachutes increase the capability of high-value airdrop missions.

E-22. Airlift is more costly than surface transportation. It is usually a small percentage of the overall transportation network during major combat operations; however, in particularly challenging situations, airlift may become the primary transportation mode for sustainment and repositioning.

E-23. Airlift supports every logical line of operations. For example, it supports IO when COIN forces provide humanitarian airlift to a battered populace. It clearly supports combat operations. Likewise, airlift supports the essential services, governance, and economic development logical lines of operations. HN security forces thus should include airlift development as the host nation's first component of airpower and spacepower.

THE AIRPOWER COMMAND STRUCTURE

E-24. COIN operations require a joint, multinational command and control architecture for air and space that is effective and responsive. The joint structure applies to more than just U.S. forces; it involves coordinating air assets of multinational partners and the host nation. COIN planning must thus establish a joint and multinational airpower command and control system and policies on the rules and conditions for employing airpower in the theater.

E-25. During COIN operations, most planning occurs at lower echelons. Ideally, components at the operational level fully coordinate these plans. Air and space planners require visibility of actions planned at all echelons to provide the most effective air and space support. Furthermore, COIN planning is often fluid and develops along short planning and execution timelines, necessitating informal and formal coordination and integration for safety and efficiency.

E-26. U.S. and multinational air units, along with HN forces, will likely use expeditionary airfields. COIN planners must consider where to locate airfields, including those intended for use as aerial ports of debarkation and other air operations. Factors to consider include—

- Projected near-, mid- and long-term uses of the airfield.
- Types and ranges of aircraft to be operated.
- Shoulder-launched, surface-to-air-missile threats to aircraft.

- Stand-off threats to airfields.
- Proximity to other threats.
- Proximity to land LOCs.
- Availability of fuels.

Airpower operating from remote or dispersed airfields may present a smaller signature than large numbers of land forces, possibly lessening HN sensitivities to foreign military presence. Commanders must properly protect their bases and coordinate their defense among all counterinsurgents.

BUILDING HOST-NATION AIRPOWER CAPABILITY

E-27. U.S. and multinational operations strive to enable the host nation to provide its own internal and external defense. Planners therefore need to establish a long-term program to develop a HN airpower capability. The HN air force should be appropriate for that nation's requirements. For conducting effective COIN operations, a HN air force requires the following basic capabilities:

- Aerial reconnaissance and surveillance.
- Air transport.
- Close air support for land forces.
- Helicopter troop lift.
- MEDEVAC.
- Counterair.
- Interdiction.

E-28. The first step in developing HN airpower is developing the right organizational model for a HN air force. Planning should identify gaps in the host nation's ability to command, control, and employ airpower in COIN operations.

E-29. The next step is to help the host nation develop its aviation infrastructure under a long-term plan. Most developing nations need considerable assistance to develop an appropriate organization, a suitable force structure, and basing plans. As airpower assets represent a large cost for a small nation, an effective airfield security program is also necessary.

E-30. An important training asset is the U.S. Air Force Special Operations Command. This command has teams qualified to operate the most common equipment used in developing nations. These teams also have the language and cultural training to effectively support aircrew and personnel training. The Air Force can also train HN pilots and aircrews through the International Military Education and Training Program.

E-31. Planners should consider HN economic and technological resources when selecting equipment. In most cases, the host nation acquires, or the U.S. and multinational partners provide, a small air force. Although this air force often has limited resources, the host nation still should effectively operate and maintain its aircraft and supporting systems. Multinational support in training and equipping the HN air force can be very important. U.S. aircraft have tremendous capabilities, but they can be too expensive and too complex for some developing nations to operate and maintain. Multinational partners with capable, but less expensive and less sophisticated, aircraft can often help equip the host nation.

E-32. Training and developing a capable HN air force takes considerable time due to the requirements to qualify aircrews, maintenance personnel, and other specialists. Working effectively in joint operations and coordinating support to land forces requires a high skill level. Even when the HN army and police are trained, U.S. personnel will likely stay with HN forces to perform liaison for supporting U.S. air assets and to advise HN forces in the use of their own airpower.

E-33. Developing capable air forces usually takes longer than developing land forces. As a result, Air Force units, advisors, and trainers will likely remain after land force trainers and advisors have completed their mission. Effective air and land operations are complex and require many resources. Often host nations continue to rely on U.S. air liaison personnel, land controllers, and aircraft for an extended period. Thus, COIN planners must consider the long-term U.S. air support requirements in comprehensive COIN planning.

This page intentionally left blank.

Source Notes

These are the sources used, quoted, or paraphrased in this publication. They are listed by page number. Where material appears in a paragraph, both page and paragraph number are listed. Boldface indicates the titles of historical vignettes. Web sites were accessed during December 2006.

ix "This is a game of wits and will…": Peter J. Schoomaker, quoted in "Serving a Nation at War: A Campaign Quality Army with Joint and Expeditionary Capabilities," (Statement before the House Armed Services Committee, 21 Jul 2004), 17. United States House of Representatives Web site < http://www.house.gov/hasc/openingstatementsandpressreleases/108thcongress/ >

1-1 "Counterinsurgency is not just thinking man's warfare…": This sentence appeared in an electronic mail message from a special forces officer serving in Iraq to the military assistant to the deputy secretary of defense in 2005.

1-4 para 1-19. "The printing press is…": T.E. Lawrence, "Evolution of a Revolt," *The Army Quarterly* (Devon, United Kingdom) 1, 1 (Oct 1920; reprint, Fort Leavenworth, KS: U.S. Army Command and General Staff College, Combat Studies Institute, 1990): 11. Combined Arms Research Library Web site < http://cgsc.leavenworth.army.mil/carl/download/csipubs >.

1-12 1-67. …planners assumed that combatants required a 10 or 15 to 1 advantage…: Andrew F. Krepinevich, *The Army and Vietnam* (Baltimore: Johns Hopkins University Press, 1986), 157; A better force requirement gauge…" John J. McGrath, *Boots on the Ground: Troop Density in Contingency Operations*, Global War on Terrorism Occasional Paper 16 (Fort Leavenworth, KS: Combat Studies Institute Press, 2006), 1, 6n1.

1-22 para 1-123. …revolutionary war was 80 percent political action and only 20 percent military: cited in David Galula, *Counterinsurgency Warfare: Theory and Practice* (1964; reprint New York: Praeger, 2005), 89 (hereafter cited as Galula); …he was involved with establishing special schools…: Walter Sullivan, "China's Communists Train Political Corps to Aid Army," *The New York Times*, 4 Jul 1949.

1-27 1-154. "There's very clear evidence,…": Creighton W. Abrams Jr., Commander's Weekly Intelligence Estimate (Headquarters, Military Assistance Command, Vietnam, 21 Nov 1970), audiotape, U.S. Army Military History Institute, Carlisle Barracks, PA; Lewis Sorley, editor, *Vietnam Chronicles: The Abrams Tapes 1968–1972* (Lubbock: Texas Tech University Press, 2004), 633 (hereafter cited as Sorley); "Do not try and do too much…": T.E. Lawrence, "Twenty-seven Articles," *Arab Bulletin* (20 August 1917), in T.E. Lawrence, *Secret Despatches* [sic] *from Arabia*, A. W. Lawrence, ed. (London, Golden Cockerel Press, 1939), 126–133. T.E. Lawrence.net Web site < http://telawrence.net/telawrencenet/works/ >.

2-1 "Essential though it is,…": Galula, 89.

2-3 **"Hand Shake Con" in Operation Provide Comfort**: Anthony C. Zinni, "Non-Traditional Military Missions: Their Nature, and the Need for Cultural Awareness and Flexible Thinking," chapter 4 in *Capital "W" War: A Case for Strategic Principles of War (Because Wars Are Conflicts of Societies, Not Tactical Exercises Writ Large)*, by Joe Strange, Perspectives on Warfighting 6 (Quantico, VA: Marine Corps University, 1998), 265–266. (Operation Provide Comfort began on April 6, 1991 and ended July 24, 1991. Operation Provide Comfort II was a show of force to deter new Iraqi attacks on the Kurds and had only limited humanitarian aspects. Provide Comfort II began 24 July 1991 and ended 31 December 1996.)

2-9 para 2-42. "To confine soldiers…": Galula, 88.

2-10 para 2-48. The Foreign Service Act assigns…: Foreign Service Act of 1980, Public Law 96-465, 96th Congress, 2d session (17 Oct 1980), sections 102(3) and 207a(1).

2-12 **Provincial Reconstruction Teams in Afghanistan**: Prepared based on the experiences of several participants in operations in Afghanistan between 2003 and 2005.

2-12 **Civil Operations and Rural Development Support and Accelerated Pacification in Vietnam**: Richard A. Hunt, *Pacification: the American Struggle for Vietnam's Hearts and Minds* (Boulder, CO: Westview Press, 1995), 86–279; Thomas W. Scoville, *Reorganizing for Pacification Support* (Washington, DC: Center of Military History, 1991), 43–83.

2-14 para 2-57. "You [military professionals] must know…": John F. Kennedy, *Public Papers of the Presidents of the United States: John F. Kennedy, Containing the Public Messages, Speeches, and Statements of the President, January 20 to December 31,1961* (Washington, DC: Government Printing Office, 1962), 448. (The remarks were made at Annapolis, MD, to the graduating class of the U.S. Naval Academy on 7 June 1961.)

3-1 "Everything *good* that happens…": Sorley, 506.

3-18 **Asymmetric Tactics in Ireland**: James Fintan Lalor to d'Arcy McGee, 30 March 1847, *Collected Writings of James Fintan Lalor*, edited by L. Fogarty (Dublin, 1918), 83. Quoted in *Handbook for Volunteers of the Irish Republican Army: Notes on Guerrilla Warfare* (Boulder, CO: Paladin Press, 1985), 3; Charles Townshend, *Political Violence in Ireland: Government and Resistance Since 1848* (New York: Oxford University Press, 1983), 32.

4-1 "The first, the supreme,…" Carl von Clausewitz, *On War*, ed. and trans. Michael Howard and Peter Paret (Princeton, NJ: Princeton University Press, 1976; reprint 1984), 88–89.

4-1 **Campaign Assessment and Reassessment**: George W. Smith, Jr., "Avoiding a Napoleonic Ulcer: Bridging the Gap of Cultural Intelligence (Or, Have We Focused on the Wrong Transformation?)" (Quantico, VA: Marine Corps War College, 2004). < http://www.mcu.usmc.mil/MCWAR/IRP/default.htm >

4-7 **Iterative Design During Operation Iraqi Freedom II**: Lieutenant General James N. Mattis, USMC, interviewed by Colonel Douglas M. King, USMC, Marine Corps Combat Development Command, Feb-Mar, 2006.

5-1 "It is a persistently methodical approach…": Sir Robert Thompson, *Defeating Communist Insurgency: The Lessons of Malaya and Vietnam* ([New York: Praeger, 1966]; reprint with a foreword by Robert Bowie, St. Petersburg, FL: Hailer Publishing, 2005), 171 (hereafter cited as Thompson).

5-6 **The Importance of Multiple Lines of Operations in COIN**: Thomas A Marks, *Counter-revolution in China: Wang Sheng and the Kuomintang* (London: Frank Cass, 1998), 77–117.

5-21 **Clear-Hold-Build in Tal Afar, 2005–2006**: Office of the Assistant Secretary of Defense (Public Affairs), News Transcript, News Briefing with COL H.R. McMaster, 27 Jan 2006, DOD News: News Briefing with COL H.R. McMaster Web site. < http://www.defenselink.mil /transcripts/2006/tr20060127-12385.html >

5-24 **Combined Action Program**: Matthew Danner, "The Combined Action Platoon: Seeds of Success in Iraq," United States Marine Corps Combined Action Platoon (USMC CAP) Web site, < http://capmarine.com/personal/iraq/danner.htm >.

5-25 "The two best guides,…": Thompson, 170.

6-1 "[H]elping others to help themselves…": "Quadrennial Defense Review Report," (Washington, DC: Office of the Secretary of Defense, 6 Feb 2006), 11.

6-18 **Multinational Security Transition Command–Iraq**: John R. Martin, "Training Indigenous Security Forces at the Upper End of the Counterinsurgency Spectrum," *Military Review* 86, 6 (Nov-Dec 2006): 58–64.

6-21 **Developing a Police Force in Malaya**: James S. Corum, *Training Indigenous Forces in Counterinsurgency: A Tale of Two Insurgencies* (Carlisle, PA: Strategic Studies Institute, U.S. Army War College, Mar, 2006), 4–24 < http://www.StrategicStudiesInstitute.army.mil >.

7-1 "Leaders must have a strong sense…" MCDP 1, *Warfighting* (20 June 1997), 57.

7-3 **Defusing a Confrontation**: Excerpted from Dan Baum, "Battle Lessons, What the Generals Don't Know," *The New Yorker*, Jan 17, 2005. The New Yorker : fact: content Web site < http://www.newyorker.com/fact/content/?050117fa_fact >.

7-4 **Patience, Presence, and Courage**: Adapted from Nelson Hernandez, "Attacks Rock Foundation That Marines Built in Anbar," *Washington Post*, Feb. 7, 2006: 1.

7-7 para 7-39. "No person in the custody…": Detainee Treatment Act of 2005, Public Law 109-148, 109th Congress, 1st Session, (30 Dec 2005), sections 1002(a), 1003(a).

7-8 **Lose Moral Legitimacy, Lose the War**: Lou DiMarco, "Losing the Moral Compass: Torture and *Guerre Revolutionnaire* in the Algerian War," *Parameters* 36, 2 (Summer, 2006): 63–76.

8-1 "In my experience in previous wars,…: Senior Officer Debriefing Report: MG J.M. Heiser, Jr., Commanding General, 1st Logistical Command, Period 2 August 1968 to 23 August 1969 (Department of the Army: Office of the Adjutant General, 23 Sep 1969), 343.

8-3 **What is Different: Insurgent Perceptions of Military Logistics**: "We have a claim…" Mao Tse-Tung, "Problems of Strategy in China's Revolutionary War," *Selected Military Writings of Mao Tse-Tung* (Peking: Foreign Language Press, 1966), 147; "Weapons are not difficult to obtain.…": Ming Fan, "A Textbook on Guerrilla Warfare," translated by and reprinted in Gene Z. Hanrahan, ed. *Chinese Communist Guerrilla Warfare Tactics* (Boulder, CO: Paladin Press, 1974), 76–77.

8-3 "In one moment in time,…": Charles C. Krulak, "The Three Block War: Fighting in Urban Areas," *Vital Speeches of the Day* 64, 5 (New York: December 15, 1997): 139–142. (Speech given before the National Press Club, October 10, 1997. Quote appears on page 139.)

8-6 **Vietnam: Meeting the Enemy and Convoy Security**: John H. Hay, *Vietnam Studies: Tactical and Materiel Innovations* (Washington, DC: Department of the Army, 1989), 154–155.

8-8 **Air Delivery in Iraq: Maximizing Counterinsurgency Potential**: Marine Corps Center for Lessons Learned, "Initial Observations Report–Operation Iraqi Freedom II-1, Subject: Air Delivery Operations," (March, 2005).

8-10 **Building a Military: Sustainment Failure**: Julian Thompson, *The Lifeblood of War: Logistics in Armed Conflict* (New York: Brassey's, 1991), 217–218; Dong Van Khuyen, *RVNAF Logistics* (Washington, DC: U.S. Army Center of Military History, 1980).

8-17 **Host-Nation Contracting: A Potential Double-Edged Sword**: Interview between Major Sean Davis, USA, and Lieutenant Colonel Marian Vlasak, USA (Fort Leavenworth, KS, 1 May 2006).

A-1 Appendix A is based on David Kilcullen, "Countering Global Insurgency," *Journal of Strategic Studies* 28, 4 (August 2005): 597–617.

B-14 **The Capture of Saddam Hussein**: Brian J. Reed, "Formalizing the Informal: A Network Analysis of an Insurgency" (Ph.D. dissertation, Department of Sociology, University of Maryland, College Park, MD, 2006), 81–88.

D-3 para D-14. "In the case of armed conflict not…": Geneva Convention Relative to the Treatment of Prisoners of War, Article 3, August 12, 1949, 6 United States Treaties 3114, 75 United Nations Treaty Series 31.

D-5 table D-1. Detainee Treatment Act of 2005, Public Law 109-148, 109th Congress, 1st Session, (30 Dec 2005), sections 1002, 1003.

This page intentionally left blank.

Glossary

The glossary lists acronyms and terms with Army, multi-Service, or joint definitions, and other selected terms. Where Army and joint definitions are different, *(Army)* follows the term. The proponent manual for other terms is listed in parentheses after the definition. Terms for which the Army and Marine Corps have agreed on a common definition are followed by *(Army-Marine Corps)*.

SECTION I – ACRONYMS AND ABBREVIATIONS

ACR armored cavalry regiment

AO area of operations

AR Army regulation

ASCOPE A memory aid for the characteristics of civil considerations: area, structures, capabilities, organizations, people, and events. *See also* METT-TC.

CERP commander's emergency response program

CJCSI Chairman of the Joint Chiefs of Staff instruction

CMO civil-military operations

CMOC civil-military operations center

COIN counterinsurgency

CORDS civil operations and revolutionary (rural) development support

COTS commercial off-the-shelf

DA Department of the Army

DOCEX document exploitation

DOD Department of Defense

DODD Department of Defense Directive

DOTMLPF Memory aid for the force development domains: doctrine, organization, training, materiel, leadership and education, personnel, and facilities.

FARC Fuerzas Armadas Revolucionarias de Colombia (Revolutionary Armed Forces of Columbia)

FBI Federal Bureau of Investigation

FID foreign internal defense

FM field manual

FMFRP Fleet Marine Force reference publication

FMI field manual interim

G-4 assistant chief of staff, logistics

GEOINT geospatial intelligence

HMMWV high-mobility, multipurpose, wheeled vehicle

HN host-nation

HUMINT	human intelligence
IGO	intergovernmental organization
IMET	international military education and training
IMINT	imagery intelligence
IO	information operations
IPB	intelligence preparation of the battelfield
ISR	intelligence, surveillance, and reconnaissance
JIACG	joint interagency coordination group
JP	joint publication
LLO	logical line of operations
LOC	line of communications
MACV	Military Assistance Command, Vietnam
MASINT	measurement and signature intelligence
MCDP	Marine Corps doctrinal publication
MCIP	Marine Corps interim publication
MCO	Marine Corps order
MCRP	Marine Corps reference publication
MCWP	Marine Corps warfighting publication
MEDEVAC	medical evacuation
METT-TC	A memory aid for mission, enemy, terrain and weather, troops and support available, time available, civil considerations used in two contexts: (1) In the context of information management, the major subject categories into which relevant information is grouped for military operations (FM 6-0); (2) In the context of tactics, the major factors considered during mission analysis (FM 3-90). [Note: The Marine Corps uses METT-T: mission, enemy, terrain and weather, troops and support available, time available.]
MNSTC-I	Multinational Security Transition Command–Iraq
MOE	measure of effectiveness
MOP	measure of performance
NATO	North Atlantic Treaty Organization
NCO	noncommissioned officer
NGO	nongovernmental organization
NSC	National Security Council
OCS	officer candidate school
OSINT	open-source intelligence
PIR	priority intelligence requirement
PRC	purchase request and committal
PRT	provincial reconstruction team
ROE	rules of engagement
S-2	intelligence staff officer
S-4	logistics staff officer
SIGINT	signals intelligence

SNA	social network analysis
SOF	special operations forces
SROE	standing rules of engagement
TAREX	target exploitation
UN	United Nations
U.S.	United States
USA	United States Army
USAID	United States Agency for International Development
USMC	United States Marine Corps

SECTION II – TERMS AND DEFINITIONS

all-source intelligence

(joint) Intelligence products and/or organizations and activities that incorporate all sources of information, most frequently including human resources intelligence, imagery intelligence, measurement and signature intelligence, signals intelligence, and open-source data in the production of finished intelligence. (JP 1-02)

area of interest

(joint) That area of concern to the commander, including the area of influence, areas adjacent thereto, and extending into enemy territory to the objectives of current or planned operations. This area also includes areas occupied by enemy forces who could jeopardize the accomplishment of the mission. (JP 1-02)

area of operations

(joint) An operational area defined by the joint force commander for land and maritime forces. Areas of operations do not typically encompass the entire operational area of the joint force commander, but should be large enough for component commanders to accomplish their missions and protect their forces. (JP 1-02)

area security

A form of security operations conducted to protect friendly forces, installation routes, and actions within a specific area. (FM 3-90)

assessment

(Army) The continuous monitoring and evaluation of the current situation and progress of an operation. (FMI 5-0.1)

board

A temporary grouping of selected staff representatives delegated decision authority for a particular purpose or function. (FMI 5-0.1)

center of gravity

(joint) The source of power that provides moral or physical strength, freedom of action, or will to act. (JP 1-02)

civil considerations

How the manmade infrastructure, civilian institutions, and attitudes and activities of the civilian leaders, populations, and organizations within an area of operations influence the conduct of military operations. (FM 6-0) *See also* METT-TC.

clear

(Army) A tactical mission task that requires the commander to remove all enemy forces and eliminate organized resistance in an assigned area. (FM 3-90)

coalition

(joint) An ad hoc arrangement between two or more nations for common action. (JP 1-02)

combatant commander

(joint) A commander of one of the unified or specified combatant commands established by the President. (JP 1-02)

command and control system

(joint) The facilities, equipment, communications, procedures, and personnel essential to a commander for planning, directing, and controlling operations of assigned forces pursuant to the missions assigned. (JP 1-02) (Army) The arrangement of personnel, information management, procedures, and equipment and facilities essential for the commander to conduct operations. (FM 6-0)

commander's intent

(Army) A clear, concise statement of what the force must do and the conditions the force must meet to succeed with respect to the enemy, terrain, and civil considerations that represent the operation's desired end state. (FMI 5-0.1) (Marine Corps) A commander's clear, concise articulation of the purpose(s) behind one or more tasks assigned to a subordinate. It is one of two parts of every mission statement which guides the exercise of initiative in the absence of instructions. (MCRP 5-12A)

commander's visualization

The mental process of developing situational understanding, determining a desired end state, and envisioning how the force will achieve that end state. (FMI 5-0.1)

command post cell

A grouping of personnel and equipment by warfighting function or purpose to facilitate command and control during operations. (FMI 5-0.1)

common operational picture

(joint) A single identical display of relevant information shared by more than one command. A common operational picture facilitates collaborative planning and assists all echelons to achieve situational awareness. (JP 1-02) (Army) An operational picture tailored to the user's requirements, based on common data and information shared by more than one command. (FM 3-0)

counterinsurgency

(joint) Those military, paramilitary, political, economic, psychological, and civic actions taken by a government to defeat insurgency. (JP 1-02)

counterintelligence

(joint) Information gathered and activities conducted to protect against espionage, other intelligence activities, sabotage, or assassinations conducted by or on behalf of foreign governments or elements thereof, foreign organizations, or foreign persons, or international terrorist activities. (JP 1-02) (Army) Counterintelligence counters or neutralizes intelligence collection efforts through collection, counterintelligence investigations, operations, analysis and production, and functional and technical services. Counterintelligence includes all actions taken to detect, identify, exploit, and neutralize the multidiscipline intelligence activities of friends, competitors, opponents, adversaries, and enemies; and is the key intelligence community contributor to protect United States interests and equities. (FM 2-0)

counterterrorism

(joint) Operations that include the offensive measures taken to prevent, deter, preempt, and respond to terrorism. (JP 1-02)

decisive point

(joint) A geographic place, specific key event, critical system or function that, when acted upon, allows commanders to gain a marked advantage over an enemy or contribute materially to achieving success. (JP 1-02)

dislocated civilian

(joint) A broad term that includes a displaced person, an evacuee, an expellee, an internally displaced person, a migrant, a refugee, or a stateless person. (JP 1-02)

end state

(joint) The set of required conditions that defines achievement of the commander's objectives. (JP 1-02)

execute

To put a plan into action by applying combat power to accomplish the mission and using situational understanding to assess progress and make execution and adjustment decisions. (FM 6-0)

foreign internal defense

(joint) Participation by civilian and military agencies of a government in any of the action programs taken by another government or other designated organization to free and protect its society from subversion, lawlessness, and insurgency. (JP 1-02)

forward operations base

(joint) In special operations, a base usually located in friendly territory or afloat that is established to extend command and control or communications or to provide support for training and tactical operations. Facilities may be established for temporary or longer duration operations and may include an airfield or an unimproved airstrip, an anchorage, or a pier. A forward operations base may be the location of special operations component headquarters or a smaller unit that is controlled and/or supported by a main operations base. (JP 1-02) [Note: Army special operations forces term is "forward operational base."]

full spectrum operations

The range of operations Army forces conduct in war and military operations other than war. (FM 3-0) [Note: A new definition for this term is being staffed for the revision of FM 3-0. Upon publication of FM 3-0, the definition it contains will replace this definition.]

host nation

(joint) A nation that receives the forces and/or supplies of allied nations, coalition partners, and/or NATO organizations to be located on, to operate in, or to transit through its territory. (JP 1-02)

human intelligence

(Army) The collection of information by a trained human intelligence collector from people and their associated documents and media sources to identify elements, intentions, composition, strength, dispositions, tactics, equipment, personnel, and capabilities (FM 2-22.3). [Note: Trained HUMINT collectors are Soldiers holding military occupational specialties 97E, 351Y {formerly 351C}, 351M {formerly 351E}, 35E, and 35F, and Marines holding the specialty 0251.]

information environment

(joint) The aggregate of individuals, organizations or systems that collect, process, or disseminate or act on information. (JP 1-02)

information operations

(joint) The integrated employment of the core capabilities of eletronic warfare, computer network operations, psychological operations, military deception, and operations security, in concert with specified supporting and relted capabilities, to influence, disrupt, corrupt, or usurp adversarial human and automated decision making while protecting our own. (JP 1-02) (Army) The employment of the core capabilities of electronic warfare, computer network operations, psychological operations, military deception, and operations security, in concert with specified supporting and related capabilities, to affect and defend information and information systems and to influence decisionmaking. (FM 3-13)

insurgency

(joint) An organized movement aimed at the overthrow of a constituted government through the use of subversion and armed conflict. (JP 1-02)

intelligence discipline

(joint) A well-defined area of intelligence collection, processing, exploitation, and reporting using a specific category of technical or human resources. There are seven major disciplines: human intelligence, imagery intelligence, measurement and signature intelligence, signals intelligence, open-source intelligence, technical intelligence, and counterintelligence. [Note: The Army definition replaces "all-source analysis and production" with "open-source intelligence."] (JP 1-02)

intelligence preparation of the battlefield

The systematic, continuous process of analyzing the threat and environment in a specific geographic area. Intelligence preparation of the battlefield (IPB) is designed to support the staff estimate and military decision-making process. Most intelligence requirements are generated as a result of the IPB process and its interrelation with the decision-making process. (FM 34-130)

interagency coordination

(joint) Within the context of Department of Defense involvement, the coordination that occurs between elements of Department of Defense and engaged U.S. Government agencies for the purpose of achieving an objective. (JP 1-02)

intergovernmental organization

(joint) An organization created by a formal agreement (e.g. a treaty) between two or more governments. It may be established on a global, regional, or functional basis for wide-ranging or narrowly defined purposes. Formed to protect and promote national interests shared by member states. Examples include the United Nations, North Atlantic Treaty Organization, and the African Union. (JP 1-02)

intuitive decisionmaking

(Army-Marine Corps) The act of reaching a conclusion which emphasizes pattern recognition based on knowledge, judgment, experience, education, intelligence, boldness, perception, and character. This approach focuses on assessment of the situation vice comparison of multiple options. (FM 6-0; MCRP 5-12A)

line of communications

(joint) A route, either land, water, and/or air, that connects an operating military force with a base of operations and along which supplies and military forces move. (JP 1-02)

line of operations

(joint) 1. A logical line that connects actions on nodes and/or decisive points related in time and purpose with an objective(s). 2. A physical line that defines the interior or exterior orientation of the force in relation to the enemy or that connects actions on nodes and/or decisive points related in time and space to an objective(s). (JP 1-02)

measure of effectiveness

(joint) A criterion used to assess changes in system behavior, capability, or operational environment that is tied to measuring the attainment of an end state, achievement of an objective, or creation of an effect. (JP 1-02)

measure of performance

(joint) A criterion to assess friendly actions that is tied to measuring task accomplishment. (JP 1-02)

mission command

The conduct of military operations through decentralized execution based upon mission orders for effective mission accomplishment. Successful mission command results from subordinate leaders at all echelons exercising disciplined initiative within the commander's intent to accomplish missions. It requires an environment of trust and mutual understanding. (FM 6-0)

narrative

The central mechanism, expressed in story form, through which ideologies are expressed and absorbed.

nongovernmental organization

(joint) A private, self-governing, not-for-profit organization dedicated to alleviating human suffering; and/or promoting education, health care, economic development, environmental protection, human rights, and conflict resolution; and/or encouraging the establishment of democratic institutions and civil society. (JP 1-02)

open-source intelligence

(joint) Information of potential intelligence value that is available to the general public. (JP 1-02)

operating tempo

The annual operating miles or hours for the major equipment system in a battalion-level or equivalent organization. Commanders use operating tempo to forecast and allocate funds for fuel and repair parts for training events and programs. (FM 7-0) [Usually OPTEMPO.]

operational environment

(joint) A composite of the conditions, circumstances, and influences that affect the employment of capabilities and bear on the decisions of the commander. (JP 1-02)

operational picture

A single display of relevant information within a commander's area of interest. (FM 3-0)

personnel tempo

The time a service member is deployed. [Usually PERSTEMPO.]

planning

The process by which commanders (and staffs, if available) translate the commander's visualization into a specific course of action for preparation and execution, focusing on the expected results. (FMI 5-0.1)

preparation

Activities by the unit before execution to improve its ability to conduct the operation, including, but not limited to, the following: plan refinement, rehearsals, reconnaissance, coordination, inspection, and movement. (FM 3-0)

reachback

(joint) The process of obtaining products, services, and applications, or forces, or equipment, or material from organizations that are not forward deployed. (JP 1-02)

riverine area

(joint) An inland or coastal area comprising both land and water, characterized by limited land lines of communications, with extensive water surface and/or inland waterways that provide natural routes for surface transportation and communications. (JP 1-02)

rules of engagement

(joint) Directives issued by competent military authority that delineate the circumstances and limitations under which United States forces will initiate and/or continue combat engagement with other forces encountered. (JP 1-02)

running estimate

A staff section's continuous assessment of current and future operations to determine if the current operation is proceeding according to the commander's intent and if future operations are supportable. (FMI 5-0.1)

security

(joint) 1. Measures taken by a military unit, an activity or installation to protect itself against all acts designed to, or which may, impair its effectiveness. 2. A condition that results from the establishment and maintenance of protective measures that ensure a state of inviolability from hostile acts or influences. (JP 1-02)

situational awareness

Knowledge of the immediate present environment, including knowledge of the factors of METT-TC. (FMI 5-0.1)

situational understanding

(Army) The product of applying analysis and judgment to the common operational picture to determine the relationship among the factors of METT-TC. (FM 3-0) (Marine Corps) Knowledge and understanding of the current situation which promotes timely, relevant, and accurate assessment of friendly, enemy, and other operations within the battlespace in order to facilitate decisionmaking. An informational perspective and skill that foster an ability to determine quickly the context and relevance of events that are unfolding. (MCRP 5-12A)

stability operations

(joint) An overarching term encompassing various military missions, tasks, and activities conducted outside the United States in coordination with other instruments of national power to maintain or reestablish a safe and secure environment, provide essential governmental services, emergency infrastructure reconstruction, and humanitarian relief. (JP 1-02)

staff estimate

See running estimate.

status of forces agreement

(joint) An agreement that defines the legal position of a visiting military force deployed in the territory of a friendly state. Agreements delineating the status of visiting military forces may be bilateral or multilateral. Provisions pertaining to the status of visiting forces may be set forth in a separate agreement, or they may form a part of a more comprehensive agreement. These provisions describe how the authorities of a visiting force may control members of that force and the amenability of the force or its members to the local law or to the authority of local officials. To the extent that agreements delineate matters affecting the relations between a military force and civilian authorities and population, they may be considered as civil affairs agreements. (JP 1-02)

strike

(joint) An attack to damage or destroy an objective or capability. (JP 1-02)

subordinates' initiative

The assumption of responsibility for deciding and initiating independent actions when the concept of operations or order no longer applies or when an unanticipated opportunity leading to the accomplishment of the commander's intent presents itself. (FM 6-0)

tempo

(Army) The rate of military action. (FM 3-0) (Marine Corps) The relative speed and rhythm of military operations over time with respect to the enemy. (MCRP 5-12A)

theater of war

(joint) Defined by the Secretary of Defense or the geographic combatant commander, the area of air, land, and water that is, or may become, directly involved in the conduct of the war. A theater of war does not normally encompass the geographic combatant commander's entire area of responsibility and may contain more than one theater of operations. (JP 1-02)

warfighting function

A group of tasks and systems (people, organizations, information, and processes) united by a common purpose that commanders use to accomplish missions and training objectives. (FMI 5-0.1)

working group

A temporary grouping of predetermined staff representatives who meet to coordinate and provide recommendations for a particular purpose or function. (FMI 5-0.1)

Annotated Bibliography

This bibliography is a tool for Army and Marine Corps leaders to help them increase their knowledge of insurgency and counterinsurgency. Reading what others have written provides a foundation that leaders can use to assess counterinsurgency situations and make appropriate decisions. The books and articles that follow are not the only good ones on this subject. The field is vast and rich. They are, however, some of the more useful for Soldiers and Marines. (Web sites were accessed during December 2006.)

THE CLASSICS

Calwell, Charles E. *Small Wars: Their Principles and Practice*. Lincoln, NE: University of Nebraska Press, 1996. (Reprint of *Small Wars: A Tactical Textbook for Imperial Soldiers* [London: Greenhill Books, 1890]. A British major general who fought in small wars in Afghanistan and the Boer War provides lessons learned that remain applicable today.)

Galula, David. *Counterinsurgency Warfare: Theory and Practice.* London: Praeger, 1964. (Lessons derived from the author's observation of insurgency and counterinsurgency in Greece, China, and Algeria.)

Gurr, Ted Robert. *Why Men Rebel*. Princeton, NJ: Princeton University Press, 1971. (Describes the relative deprivation theory, which states that unmet expectations motivate those who join rebel movements.)

Hoffer, Eric. *The True Believer: Thoughts on the Nature of Mass Movements*. New York: Harper Perennial Modern Classics, 2002. (This book, originally published in 1951, explains why people become members of cults and similar groups.)

Horne, Alistair. *A Savage War of Peace*. New York: Viking, 1977. (One of the best analyses of the approaches and problems on both sides during the war in Algeria. For more on this conflict, see *The Battle of Algiers*, a troubling and instructive 1966 movie.)

Jeapes, Tony. *SAS Secret War*. London: Greenhill Books, 2005. (How the British Special Air Service raised and employed irregular tribal forces to counter a communist insurgency in Oman during the 1960s and 1970s.)

Kitson, Frank. *Low Intensity Operations: Subversion, Insurgency and Peacekeeping*. London: Faber and Faber, 1971. (Explanation of the British school of counterinsurgency from one of its best practitioners.)

Komer, Robert. *Bureaucracy Does Its Thing: Institutional Constraints on U.S.-GVN Performance in Vietnam*. Washington, D.C.: RAND, 1972. Rand Corporation Web site < http://www.rand.org/pubs/reports/R967/ > (Bureaucracies do what they do—even if they lose the war.)

Larteguy, Jean. *The Centurions*. New York: Dutton, 1962. (A fact-based novel about the French experience in Vietnam and Algeria that depicts the leadership and ethical dilemmas involved in counterinsurgency. The sequel *The Praetorians* is also a classic depiction of the impact of ethical erosion on a military organization.)

Lawrence, T.E. *Seven Pillars of Wisdom: A Triumph*. New York: Anchor, 1991. (Reprint of 1917 book published in London by George Doran. Autobiographical account of Lawrence of Arabia's attempts to organize Arab nationalism during World War I.)

———. "The 27 Articles of T.E. Lawrence." *The Arab Bulletin* (20 Aug 1917). Defense and the National Interest Web site < http://www.d-n-i.net/fcs/lawrence_27_articles.htm > (Much of the best of *Seven Pillars of Wisdom* in easily digestible bullet points.)

Linn, Brian McAllister. *The Philippine War, 1899–1902*. Lawrence, KS: University Press of Kansas, 2002. (The definitive treatment of successful U.S. counterinsurgency operations in the Philippines.)

Mao Zedong. *On Guerrilla Warfare*. London: Cassell, 1965. (Mao describes the principles which he used so well in seizing power in China and which have inspired many imitators.)

McCuen, John J. *The Art of Counter-Revolutionary War*. St. Petersburg, FL: Hailer Publishing, 2005. (Originally published by Harrisburg, PA: Stackpole Books, 1966. Discusses theory, practice, and historical keys to victory.)

Race, Jeffrey. *War Comes to Long An: Revolutionary Conflict in a Vietnamese Province.* Berkeley, CA: University of California Press, 1972. (Counterinsurgency is scalable. Depicts the evolution of insurgency in one province in Vietnam.)

Thompson, Robert. *Defeating Communist Insurgency*. St. Petersburg, FL: Hailer Publishing, 2005. (Written in 1966. Provides lessons from the author's counterinsurgency experience in Malaya and Vietnam.)

Trinquier, Roger. *Modern Warfare: A French View of Counterinsurgency*. New York: Praeger, 1964. (The French school of counterinsurgency with a focus on "whatever means necessary.")

United States Marine Corps. *Small Wars Manual.* Washington, D.C.: Government Printing Office, 1987. Air War College Gateway to the Internet Web site < http://www.au.af.mil/au/ > (This book, originally published in 1940, covers lessons learned from the Corps' experience in the interwar years.)

West, Bing. *The Village*. New York: Pocket Books, 1972. (A first-person account of military advisors embedded with Vietnamese units.)

OVERVIEWS AND SPECIAL SUBJECTS IN COUNTERINSURGENCY

Asprey, Robert. *War in the Shadows: The Guerrilla in History*. 2 vols. New York: William Morrow, 1994. (First published in 1975. Presents the history of guerrilla war from ancient Persia to modern Afghanistan.)

Baker, Ralph O. "The Decisive Weapon: A Brigade Combat Team Commander's Perspective on Information Operations." *Military Review* 86, 3 (May-Jun 2006), 13–32. (A brigade combat team commander in Iraq in 2003–2004 gives his perspective on information operations.)

Corum, James and Wray Johnson. *Airpower in Small Wars: Fighting Insurgents and Terrorists*. Lawrence, KS: University Press of Kansas, 2003. (Depicts uses and limits of airpower and technology in counterinsurgency.)

Davidson, Phillip. *Secrets of the Vietnam War*. Novato, CA: Presidio Press, 1990. (MACV commander General Westmoreland's intelligence officer provides an insightful analysis of the intricacies of the North Vietnamese strategy of dau tranh ["the struggle"].)

Ellis, John. *From the Barrel of a Gun: A History of Guerrilla, Revolutionary, and Counter-insurgency Warfare from the Romans to the Present*. London: Greenhill, 1995. (A comprehensive short overview of counterinsurgency.)

Hammes, T.X. *The Sling and the Stone: On War in the 21st Century*. Osceola, WI: Zenith Press, 2004. (The future of warfare for the West is insurgency and terror according to a Marine with Operation Iraqi Freedom experience.)

Krepinevich, Andrew Jr. *The Army and Vietnam*. Baltimore: Johns Hopkins University Press, 1986. (Argues that the Army never adapted to the insurgency in Vietnam, preferring to fight the war as a conventional conflict with an emphasis on firepower.)

Merom, Gil. *How Democracies Lose Small Wars: State, Society, and the Failures of France in Algeria, Israel in Lebanon, and the United States in Vietnam*. New York: Cambridge University Press, 2003. (Examines the cases of Algeria, Lebanon, and Vietnam. Determines that great powers lose small wars when they lose public support at home.)

Nagl, John A. *Learning to Eat Soup with a Knife: Counterinsurgency Lessons from Malaya and Vietnam.* Chicago: University of Chicago Press, 2005. (How to learn to defeat an insurgency. Foreword by Peter J. Schoomaker.)

O'Neill, Bard E. *Insurgency and Terrorism: From Revolution to Apocalypse*. Dulles, VA: Potomac Books, 2005. (A framework for analyzing insurgency operations and a good first book in insurgency studies.)

Sepp, Kalev I. "Best Practices in Counterinsurgency." *Military Review* 85, 3 (May-Jun 2005), 8–12. (Historical best practices for success in counterinsurgency.)

Shy, John and Thomas W. Collier. "Revolutionary War" in Peter Paret, ed. *Makers of Modern Strategy: From Machiavelli to the Nuclear Age*. Princeton, NJ: Princeton Univ. Press, 1986. (One of the best overview of the various counterinsurgency schools, discussing both the writings and the contexts in which they were developed.)

Sorley, Lewis. *A Better War: The Unexamined Victories and Final Tragedy of America's Last Years in Vietnam*. New York: Harvest/HBJ, 2000. (Describes the impact of General Creighton Abrams on the conduct of the war in South Vietnam. While he improved unity of effort in counterinsurgency, the North Vietnamese were successfully focusing on facilitating American withdrawal by targeting will in the United States.)

Taber, Robert. *War of the Flea: The Classic Study of Guerrilla Warfare*. Dulles, VA: Potomac Books, 2002. (Explains the advantages of the insurgent and how to overcome them.)

CONTEMPORARY EXPERIENCES AND THE WAR ON TERRORISM

Alwin-Foster, Nigel R.F. "Changing the Army for Counterinsurgency Operations." *Military Review* 85, 6 (Nov-Dec 2005), 2–15. (A provocative look at U.S. counterinsurgency operations in Iraq in 2003-2004 from a British practitioner.)

Barno, David W. "Challenges in Fighting a Global Insurgency." *Parameters* 36, 2 (Summer 2006), 15–29. (Observations from a three-star commander in Afghanistan.)

Chiarelli, Peter W. and Patrick R. Michaelis. "Winning the Peace: The Requirement for Full-Spectrum Operations," *Military Review* 85, 4 (Jul-Aug 2005), 4–17. (The commander of Task Force Baghdad in 2004 describes his lessons learned.)

Collins, Joseph J. "Afghanistan: Winning a Three Block War." *The Journal of Conflict Studies* 24, 2 (Winter 2004), 61–77. (The former deputy assistant secretary of defense for stability operations provides his views on achieving success in Afghanistan.)

Crane, Conrad and W. Andrew Terrill. *Reconstructing Iraq: Insights, Challenges, and Missions for Military Forces in a Post-conflict Scenario*. Carlisle Barracks, PA: U.S. Army War College, 2003. < http://www.strategicstudiesinstitute.army.mil/pubs > (Prescient look at the demands of rebuilding a state after changing a regime.)

Filkins, Dexter. "What the War Did to Colonel Sassaman." *The New York Times Magazine* (23 Oct 2005), 92. (Case study of a talented 4th Infantry Division battalion commander in Iraq in 2003-2004 who made some questionable ethical decisions that ended his career.)

Gunaratna, Rohan. *Inside Al Qaeda: Global Network of Terror*. Berkeley, CA: University of Berkeley Press, 2003. (The story behind the rise of the transnational insurgency.)

Hoffman, Bruce. *Insurgency and Counterinsurgency in Iraq*. Santa Monica, CA: RAND, 2004. Rand Corporation Web site < http://www.rand.org/pubs/occasional_papers/OP127/ > (Analysis of America's efforts in Iraq in 2003 informed by good history and theory.)

Kepel, Gilles. *The War for Muslim Minds: Islam and the West*. Cambridge, MA: Belknap Press, 2004. (A French explanation for the rise of Islamic extremism with suggestions for defeating it.)

Kilcullen, David. "Countering Global Insurgency: A Strategy for the War on Terrorism." *Journal of Strategic Studies* 28, 4 (Aug 2005), 597–617. (Describes the war on terrorism as a counterinsurgency campaign.)

———. "'Twenty-Eight Articles': Fundamentals of Company-level Counterinsurgency." *Military Review* 86, 3 (May-Jun 2006), 103–108. (Australian counterinsurgent prescribes actions for captains in counterinsurgency campaigns.)

———. "Counterinsurgency *Redux*," *Survival* 48, 4 (Winter 2006-2007), 111–130. (Discusses insurgency's evolution from the classic Maoist form to the modern transnational, shifting coalitions that challenge the United States today.)

Lewis, Bernard. *The Crisis of Islam: Holy War and Unholy Terror*. New York: Modern Library, 2003. (A controversial but important analysis of the philosophical origins of transnational insurgency.)

McFate, Montgomery. "Iraq: The Social Context of IEDs." *Military Review* 85, 3 (May-Jun 2005), 37–40. (The insurgents' best weapon doesn't grow next to roads—it's constructed and planted there. Understanding who does that, and why, helps defeat improvised explosive devices.)

Metz, Steven and Raymond Millen, *Insurgency and Counterinsurgency in the 21st Century: Reconceptualizing Threat and Response*. Carlisle Barracks, PA: U.S. Army War College, 2004. (Longtime scholars of counterinsurgency put the war on terrorism in historical context.)

Multi-national Force–Iraq. "Counterinsurgency Handbook," 1st ed. Camp Taji, Iraq: Counterinsurgency Center for Excellence, May, 2006. (Designed to help leaders at all levels conduct counterinsurgency operations but focused at the company, platoon, and squad levels. Contains a variety of principles, considerations, and checklists.)

Packer, George. *The Assassins' Gate: America in Iraq*. New York: Farrar, Straus and Giroux, 2005. (A journalist for *The New Yorker* talks to Iraqis and Americans about Operation Iraqi Freedom.)

———. "The Lesson of Tal Afar: Is It Too Late for the Administration to Correct Its Course in Iraq?" *The New Yorker* (10 Apr 2006), 48–65. (The 2005 success of the 3d Armored Cavalry Regiment with the clear-hold-build tactic in Tal Afar.)

Petraeus, David. "Learning Counterinsurgency: Observations from Soldiering in Iraq." *Military Review* 86, 1 (Jan-Feb 2006), 2–12. (Commander of the 101st and Multinational Security Transition Command–Iraq passes on his personal lessons learned from two years in Iraq.)

Sageman, Marc. *Understanding Terror Networks*. Philadelphia, PA: University of Pennsylvania Press, 2004. (A former foreign service officer with Afghanistan experience explains the motivation of terrorists—not deprivation, but the need to belong.)

Military References

Army publications that are assigned a Marine Corps number are indicated with an asterisk.

REQUIRED PUBLICATIONS

These documents must be available to intended users of this publication.

*FM 1-02/MCRP 5-12A. *Operational Terms and Graphics.* 21 Sep 2004.

JP 1-02. *Department of Defense Dictionary of Military and Associated Terms.* 4 Dec 2001. (DOD Dictionary of Military Terms Web site < http://www.dtic.mil/doctrine/jel/doddict/ >)

RELATED PUBLICATIONS

These sources contain relevant supplemental information.

JOINT AND DEPARTMENT OF DEFENSE PUBLICATIONS

CJCSI 3121.01B. *Standing Rules of Engagement for U.S. Forces.* 15 Jan 2000.

DODD 2310.01E. *The Department of Defense Detainee Program.* 5 Sep 2006.

DODD 5105.38M. *Security Assistance Management Manual.* 3 Oct 2003. (Published by the Defense Security Cooperation Agency. Chapter 8 addresses end-use monitoring. AR 12-1 implements for the Army.)

JP 1. *Joint Warfare of the Armed Forces of the United States.* 14 Nov 2000.

JP 3-0. *Joint Operations.* 17 Sep 2006.

JP 3-07.1. *Joint Tactics, Techniques, and Procedures for Foreign Internal Defense.* 30 Apr 2004.

JP 3-08. *Interagency, Intergovernmental Organization, and Nongovernmental Organization Coordination During Joint Operations.* 2 vols. 17 Mar 2006.

JP 3-60. *Joint Doctrine for Targeting.* 17 Jan 2002.

JP 3-13. *Information Operations.* 13 Feb 2006.

JP 3-61. *Public Affairs.* 9 May 2005.

SERVICE PUBLICATIONS

AR 12-1. *Security Assistance, International Logistics, Training, and Technical Assistance Support Policy and Responsibilities.* 24 Jan 2000. (Marine Corps follows DODD 5105.38M.)

*AR 190-8/MCO 3461.1. *Enemy Prisoners of War, Retained Personnel, Civilian Internees and Other Detainees.* 1 Oct 1997.

FM 2-0 (34-1). *Intelligence.* 17 May 2004.

FM 2-22.3 (34-52). *Human Intelligence Collector Operations.* 6 Sep 2006. (See MarAdmin 458.06 for Marine Corps policy and guidance on intelligence interrogations.)

FM 3-0. *Operations.* 14 Jun 2001. (Under revision. Projected for republication during fiscal year 2007.)

FM 3-05 (100-25), *Army Special Operations Forces.* 20 Sep 2006.

FM 3-05.40 (41-10). *Civil Affairs Operations.* 29 Sep 2006. (MCRP 3-33.1 contains Marine Corps civil affairs doctrine.)

*FM 3-05.301/MCRP 3-40.6A. *Psychological Operations Tactics, Techniques, and Procedures.* 31 Dec 2003. (Distribution limited to government agencies only.)

*FM 3-05.401/MCRP 3-33.1A. *Civil Affairs Tactics, Techniques, and Procedures.* 23 Sep 2003.

*FM 3-09.31 (6-71)/MCRP 3-16C. *Tactics, Techniques, and Procedures for Fire Support for the Combined Arms Commander.* 1 Oct 2002.

FM 3-13 (100-6). *Information Operations: Doctrine, Tactics, Techniques, and Procedures.* 28 Nov 2003. (Appendix E addresses information operations targeting.)

FM 3-61.1. *Public Affairs Tactics, Techniques, and Procedures.* 1 Oct 2000.

FM 3-90. *Tactics.* 4 Jul 2001.

FM 4-0 (100-10). *Combat Service Support.* 29 Aug 2003.

FM 4-02 (8-10). *Force Health Protection in a Global Environment.* 13 Feb 2003. (NAVMED P-117, chapter 19, contains corresponding Marine Corps doctrine.)

FM 5-0 (101-5). *Army Planning and Orders Production.* 20 Jan 2005. (MCDP 5 contains Marine Corps planning doctrine.)

FM 5-104. *General Engineering.* 12 Nov 1986. (Will be republished as FM 3-34.300.)

FM 5-250. *Explosives and Demolitions. 30 Jul 1998.* (Will be republished as FM 3-34.214.)

FM 6-0. *Mission Command: Command and Control of Army Forces.* 11 Aug 2003.

FM 6-20-10. *Tactics, Techniques, and Procedures for the Targeting Process.* 8 May 1996.

FM 6-22 (22-100). *Army Leadership.*12 Oct 2006.

*FM 6-22.5 (22-9)/MCRP 6-11C. *Combat Stress.* 23 Jun 2000.

FM 7-98. *Operations in a Low-Intensity Conflict.* 19 Oct 1992. (Contains tactical-level guidance for brigade and battalion operations in an irregular warfare and peace operations environment.)

FM 20-32. *Mine/Countermine Operations.* 29 May 1998. (Will be republished as FM 3-34.210, *Explosive Hazards Operations*.)

FM 27-10. *The Law of Land Warfare.* 18 Jul 1956.

FM 31-20-3. *Foreign Internal Defense: Tactics, Techniques, and Procedures for Special Forces.* 20 Sep 1994. (Will be republished as FM 3-05.202.)

*FM 34-130/FMFRP 3-23-2. *Intelligence Preparation of the Battlefield.* 8 Jul 1994. (Will be republished as FM 2-01.3/MRCP 2-3A.)

FM 46-1. *Public Affairs Operations.* 30 May 1997.

FM 90-8. *Counterguerrilla Operations.* 29 Aug 1986.

FMI 2-91.4. *Intelligence Support to Operations in the Urban Environment.* 30 Jun 2005. (Expires 30 Jun 2007. Distribution limited to government agencies only. Available in electronic media only. Army Doctrine and Training Digital Library Web site < www.adtdl.army.mil >).

*FMI 3-34.119/MCIP 3-17.01. *Improved Explosive Device Defeat.* 21 Sep 2005. (Expires 21 Sep 2007. Distribution limited to government agencies only. Available in electronic media only. Army Doctrine and Training Digital Library Web site < www.adtdl.army.mil >.)

FMI 5-0.1. *The Operations Process.* 31 Mar 2006. (Expires 31 Mar 2008. When FM 3-0 is republished, it will address the material in FMI 5-0.1 that is relevant to this publication.)

MarAdmin (Marine Administrative Message) 458/06. "USMC Interim Policy and Guidance for Intelligence Interrogations." 22 Sep 2006. (States that FM 2-22.3 provides DOD-wide doctrine on intelligence interrogations. Lists sections of FM 2-22.33 that apply. Marine Corps Publications Web site < http://www.usmc.mil/maradmins/ >.)

MCDP 1. *Warfighting.* 20 Jun 1997.

MCDP 4. *Logistics.* 21 Feb 1997.

MCDP 5. *Planning.* 21 July 1997.

MCDP 6. *Command and Control.* 4 Oct 1996.

MCRP 3-33.1A. *Civil Affairs Operations.* 14 Feb 2000. (FM 3-05.40 contains Army civil affairs doctrine.)

MCWP 4-12. *Operational-Level Logistics.* 30 Jan 2002.

NAVMED P-117. *Manual of the Medical Department, U.S. Navy.* Chapter 19, "Fleet Marine Force," change 117. 21 Jun 2001. (Article 19-24 discusses levels of care. FM 4-02 contains the corresponding Army doctrine. When published, MCRP 4-11.1G will supersede this publication.)

PRESCRIBED FORMS

None

REFERENCED FORMS

None

This page intentionally left blank.

Index

Entries are by paragraph number unless a page is specified.

Entries are by paragraph number unless a page is specified.

Entries are by paragraph number unless a page is specified.

Entries are by paragraph number unless a page is specified.

Entries are by paragraph number unless a page is specified.

Entries are by paragraph number unless a page is specified.

Entries are by paragraph number unless a page is specified.

Entries are by paragraph number unless a page is specified.

Entries are by paragraph number unless a page is specified.

Entries are by paragraph number unless a page is specified.

Entries are by paragraph number unless a page is specified.

Entries are by paragraph number unless a page is specified.

Entries are by paragraph number unless a page is specified.

Entries are by paragraph number unless a page is specified.

Entries are by paragraph number unless a page is specified.

Entries are by paragraph number unless a page is specified.

Entries are by paragraph number unless a page is specified.

This page intentionally left blank.

FM 3-24
MCWP 3-33.5
15 December 2006

By Order of the Secretary of the Army:

PETER J. SCHOOMAKER
General, United States Army
Chief of Staff

Official:

JOYCE E. MORROW
Administrative Assistant to the
Secretary of the Army
0634603

By Direction of the Commandant of the Marine Corps:

JAMES F. AMOS
Lieutenant General, U.S. Marine Corps
Deputy Commandant
Combat Development and Integration

DISTRIBUTION:

Active Army, Army National Guard, and U.S. Army Reserve: To be distributed in accordance with the initial distribution number 121724, requirements for FM 3-24.

Made in the USA
Monee, IL
08 November 2026

67384884R00155